NanoScience and Technology

NanoScience and Technology

Series Editors:
P. Avouris B. Bhushan K. von Klitzing H. Sakaki R. Wiesendanger

The series NanoScience and Technology is focused on the fascinating nano-world, mesoscopic physics, analysis with atomic resolution, nano and quantum-effect devices, nanomechanics and atomic-scale processes. All the basic aspects and technology-oriented developments in this emerging discipline are covered by comprehensive and timely books. The series constitutes a survey of the relevant special topics, which are presented by leading experts in the field. These books will appeal to researchers, engineers, and advanced students.

Nanoelectrodynamics
Electrons and Electromagnetic Fields in Nanometer-Scale Structures
Editor: H. Nejo

Single Organic Nanoparticles
Editors: H. Masuhara, H. Nakanishi, K. Sasaki

Epitaxy of Nanostructures
By V.A. Shchukin, N.N. Ledentsov, D. Bimberg

Applied Scanning Probe Methods I
Editors: B. Bhushan, H. Fuchs, S. Hosaka

Nanostructures
Theory and Modeling
By C. Delerue, M. Lannoo

Nanoscale Characterisation of Ferroelectric Materials
Scanning Probe Microscopy Approach
Editors: M. Alexe, A. Gruverman

Magnetic Microscopy of Nanostructures
Editors: H. Hopster, H.P. Oepen

Silicon Quantum Integrated Circuits
Silicon-Germanium Heterostructure Devices: Basics and Realisations
By E. Kasper, D.J. Paul

The Physics of Nanotubes
Fundamentals of Theory, Optics and Transport Devices
Editors: S.V. Rotkin, S. Subramoney

Single Molecule Chemistry and Physics
An Introduction
By C. Wang, C. Bai

Atomic Force Microscopy, Scanning Nearfield Optical Microscopy and Nanoscratching
Application to Rough and Natural Surfaces
By G. Kaupp

Applied Scanning Probe Methods II
Scanning Probe Microscopy Techniques
Editors: B. Bhushan, H. Fuchs

Applied Scanning Probe Methods III
Characterization
Editors: B. Bhushan, H. Fuchs

Applied Scanning Probe Methods IV
Industrial Application
Editors: B. Bhushan, H. Fuchs

Nanocatalysis
Editors: U. Heiz, U. Landman

Roadmap 2005 of Scanning Probe Microscopy
Editor: S. Morita

Scanning Probe Microscopy
Atomic Scale Engineering by Forces and Currents
By A. Foster, W. Hofer

A. Foster W. Hofer

Scanning Probe Microscopy

Atomic Scale Engineering by Forces and Currents

With 116 Figures

Springer

Adam Foster
Laboratory of Physics
Helsinki University of Technology
Helsinki, Finland
asf@fyslab.hut.fi

Werner Hofer
Surface Science Research Centre
The University of Liverpool
Liverpool L69 3BX
Britain
whofer@liverpool.ac.uk

ISSN 1434-4904

ISBN-10 0-387-40090-7
ISBN-13 978-0387-40090-7

Library of Congres Control Number: 2005936713

Printed in the United States of America.

9 8 7 6 5 4 3 2 1

springer.com

Preface

This monograph on scanning probe microscopes (SPM) has three aims: to present, in a coherent way, the theoretical methods necessary to interpret experiments; to demonstrate how experimental results are in fact enhanced by theoretical analysis; and to describe the physical processes in solids that can be analyzed by this experimental method. In all these aims we focus on high-resolution experiments as the cutting edge in SPM, offering access to physical phenomena at the atomic scale.

The presentation is directed at an audience of practitioners in the field and newcomers alike. For one group, it presents an overview of methods, which are found in a widely disparate range of publications. Moreover, the immediate relevance for the physics of scanning probe microscopes is not usually obvious. For these practitioners, we aim at providing them with a toolbox that can be used in conjunction with existing numerical methods in solid state physics. For the other group, we seek to define the range of phenomena in solid state physics where scanning probe microscopes provide the best analytical tool at present. We also aim at demonstrating, in a step-by-step fashion, how physical problems in this field can be treated experimentally, and clarified with the help of state-of-the-art theoretical methods.

The monograph has four distinct parts: Part I, which includes Chapters 1 and 2, covers the basic physical principles and the experimental implementation of the instrument. Part II, Chapters 3–5, contains the core of the theoretical framework. Part III, Chapters 6–9, explains how the theoretical results can be used to analyze experimental data. We conclude the presentation with an outlook on the field, as it presents itself today, and try to estimate its potential development in the near future.

A systematic study of the present state in scanning probe microscopy is impossible without help from a large number of experimenters and theorists. In this respect the authors are grateful to their collaborators over the years in the field, and for the insights gained in many discussions. In particular we would like to thank the following individuals:

Wolf Allers, Andres Arnau, Clemens Barth, Alexis Baratoff, Roland Bennewitz, Richard Berndt, Flemming Besenbacher, Matthias Bode, Harald Brune, Giovanni Comelli, Pedro Echenique, Sam Fain, Roman Fasel, Andrew Fisher, Fernando Flores, Andrey Gal, Aran Garcia-Lekue, Franz Giessibl, Sebastian Gritschneder, Peter Grutter, Claude Henry, Regina Hoffmann, Lev Kantorovich, Josef Kirschner, Jeppe Lauritsen, Petri Lehtinen, Alexander Livshits, Christian Loppacher, Nicolas Lorente, Edvin Lundgren, Ernst Meyer, Rodolfo Miranda, Herve Ness, Risto Nieminen, Georg Olesen, Riku Oja, Olli Pakarinen, Krisztian Palotas, Ruben Pérez, John Pethica, John Polanyi, Josef Redinger, Michael Reichling, Jeff Reimers, Neville Richardson, Federico Rosei, Alexander Shluger, Alexander Schwarz, Udo Schwarz, Peter Sushko, Peter Varga, Matt Watkins, Roland Wiesendanger, and Robert Wolkow.
A first draft of the book was sent out to several colleagues for their comments, criticism, and suggestions for possible improvements. Their feedback was invaluable for improving and clarifying the presentation, both from a theoretical angle, and from the viewpoint of experiments. We would like to thank them particularly for the time and effort they devoted to this careful reading.

Contents

Mathematical Symbols

Symbol	Name	Unit	Chapter
V	Bias potential	volt (V)	4
B	Magnetic field	tesla (T) = V s/m^2	3
μ	Magnetic moment	$\mu_B = e\hbar/2mc$	3
	Chemical potential	eV	4
H	Hamiltonian	eV	3
ψ_μ,χ_ν	Eigenvector	$(1/\text{Å})^{3/2}$	3
$\Gamma\mu\nu$	Transition rate	1/s	3
Γ	Contact	eV	4
I, $I_{\mu\nu}$	Current	ampere (A)	3
E_μ, E_ν	Eigenvalues	eV	3
E_F	Fermi energy	eV	4
σ	Broadening	eV	3
$\rho(\mathbf{r})$, $n(\mathbf{r})$	Electron density	$(1/\text{Å})^3$	3
k	Electron wavevector, mode	1/Å	4
k_F	Fermi wavevector	1/Å	4
$f(E)$	Fermi distribution	unity	4
v_k	Electron velocity	m/s	4
R_C	Contact resistance	ohm(Ω)	4
G, σ	Conductance	Ω^{-1}	4
Σ	Self energy	eV	4
T	Transmission	unity	4
S	Scattering matrix	unity	4
t	Transmission coefficient	unity	4
r	Reflection coefficient	unity	4
$\bar{T}$	Transmission function	unity	4

Symbol	Name	Unit	Chapter
G_{in}, G_{out}	Incoming and outgoing Green's function	$(\mathrm{eV})^{-1}$	4
$G^R(=G_{out})$	Retarded Green's function (GF)	$(\mathrm{eV})^{-1}$	4
$G^A(=G_{in})$	Advanced Green's function (GF)	$(\mathrm{eV})^{-1}$	4
ϵ_i	Eigenvalue	eV	4
U	Potential	eV	4
Σ^R	Retarded self-energy (SE)	eV	4
Σ^A	Advanced self-energy (SE)	eV	4
Γ_R	Retarded contact	eV	4
Γ_A	Advanced contact	eV	4
A	Spectral function	$(\mathrm{eV})^{-1}$	4
$\Sigma^<$	Nonequilibrium SE (less)	eV	4
$\Sigma^>$	Nonequilibrium SE (more)	eV	4
$G^<$	Nonequilibrium GF (less)	$(\mathrm{eV})^{-1}$	4
$G^>$	Nonequilibrium GF (more)	$(\mathrm{eV})^{-1}$	4
D	Phonon correlation function	eV	4
$\mathbf{J}$	Current density	$\mathrm{A/m^2}$	4
f	Force	newton (N)	3
V	Potential	electron volt (eV)	3
E	Energy	eV $= 1.6\times10^{-19}$ joule	3
C	Capacitance	farad (F)	3
k	Cantilever spring constant	(N/m)	6
ω_0,f_0	Cantilever equilibrium frequency	(s^{-1})	6
A_0, A	Cantilever amplitude	(m)	6
Q	Quality factor	unity	6
U_{bias}	Compensating bias in SFM	(V)	6
H	Hamaker constant	(joule)	6
R	Tip radius	(m)	6
h	Equilibrium height of cantilever	(m)	6
Δf	Frequency shift	(Hz)	6
γ_0	Normalized frequency shift	(fN $\sqrt{m}$)	6

1

The Physics of Scanning Probe Microscopes

The objects of most scientific disciplines cover a relatively small length scale. While physicists quite frequently revel in the extension of their subject, ranging from the Planck length (10^{-33} m) to the diameter of the universe (10^{26} m or 10^{10} light years), other sciences have to make do with more humble ranges. Chemistry (10^{-10} m to 10^{-3} m, the size of macromolecules), biology (10^{-10} m to 10^{2} m, the size of the largest organisms), and geology (10^{-10} m to 10^{7} m, the size of a planet) all cover only a tiny fraction of this range. Based on this comparison, physicists sometimes imply that theirs is the most universal science. On closer scrutiny this claim loses some of its initial appeal, because events on the subnuclear as well as the galactic scale usually do not have much impact on human conditions. The actual scale of physical research that is important in an everyday context then encompasses roughly the range from 10^{-12} to 10^{7} m. This range, incidentally, is the range of materials science. Today, at the beginning of the twenty-first century, the basic natural sciences-physics, chemistry, and biology-are gradually merging into a single discipline, which aims at understanding processes at the very elementary level of atoms. This reflects a trend in current technology, which tries to mimic nature's elegant and subtle methods rather than employing brute force. For this reason, physics is confronted today by an unprecedented challenge on the precision and accuracy of its theoretical descriptions.

Materials science deals with the structure, the properties, and the interactions in systems composed of atoms and molecules. In principle, there is no limit to the size of a system. This limit is usually defined by the required precision with which small changes on the atomic scale need to be described. Finite element methods, for example, which have been used by engineers for decades, predict the properties of large chunks of material employed in the construction of buildings, ships, or airplanes. Detailed state-of-the-art calculations covering only a few dozen atoms are at the other end of the precision range, dealing with the minute interactions between single atoms. But amazingly, these methods are able to predict the property of, for example, the earth's core: a large chunk of material indeed. Why does this work, one might ask? The answer, if any single answer can be given, is the ubiquity of electrons. Electrons are the glue, which holds molecules and crystals together. Their density in solids is roughly constant (about one electron every 4 Å^3), they interact with their environment via charge ($-$e) and spin ($\hbar/2$). The fundamental interactions in materials science are thus interactions via electric and magnetic fields. This has a profound impact on theoretical descriptions, since all that needs to be known are the position of the nuclei and the charge and velocity distribution of electrons in order to describe material properties. From this fact derives much of the simplicity of current theoretical models.

Progress in physics depends on an intricate balance between experimental and theoretical methods. In the 1920s and 1930s of the last century progress was due to the rise of quantum mechanics and the interest it created in atomic research. In the 1950s and 1960s, the development of solid state technology boosted extensive research into material properties. Finally, in the 1980s and 1990s, the eventual availability of computer technology and precise theoretical models allowed one to contemplate subtle material changes, chemical reactions, and even biological processes. The scanning probe microscope (SPM) was invented and perfected in this period. More than any other instrument it reflects the close ties between physics, chemistry, and biology. It is the only instrument that can be found in the labs of all three disciplines around the world. The theoretical description of its operation is the topic of this monograph. But even in a mainly theoretical exposition, it is useful to regard its experimental merits and shortcomings in the context of other methods. Also, and even primarily so, it is important to understand the physical principles and processes involved on a rather basic level.

1.1 Experimental methods

The wealth of experimental methods in materials science lies in the details of their application, because fundamentally, all standard methods to probe into material properties utilize only five basic physical phenomena:

- **Adsorption:** A probe particle is adsorbed by a material; the adsorption is detected by a characteristic lack of intensity or through secondary emissions.
- **Emission:** The spatial distribution of particles emitted from a material is used to gain information about the material's structural properties.
- **Transmission:** Particles are transmitted through a material; the spatial distribution of collected particles allows an analysis of its structural properties.
- **Diffraction:** The wave properties of particles are used to gain information about the spatial distribution of diffracting structures like ion cores.
- **Scattering:** A probe particle is scattered by the material; this allows an analysis of the spatial distribution of scatterers.

The particles can be ions (H^+, He^+) [1, 2], neutrons [3, 4], electrons [5], or photons. Of the twenty to thirty experimental methods most common in surface science, more than two-thirds are based on electrons and photons. The experimental preference, for instance over ions, has two reasons: (i) Unlike ions they interact with a material without substantial impact and therefore more or less nondestructively. (ii) Their wave properties can be tuned over a wide energy range. For photons, this range covers all wavelengths from the infrared (energy meV) to x-rays (energy keV). At one end, this range is sufficient to probe the scale of core-level electrons, at the other, phonon excitations characteristic of chemical bonds. For electrons, the range is from eV to hundreds of keV. Their small wavelength makes them suitable, in one regime, for delivering transmission images of thin films with a resolution well above that of x-ray methods. This is the principle of transmission electron microscopes. In the other regime their wavelength is comparable to the length scale of crystal lattice parameters; in this case diffraction images allow a precise determination of structural properties. This is the mode of operation of low-energy electron diffraction (LEED) methods, the de facto standard of structural surface analysis until the 1980s. An introduction to experimental methods is given in a number of excellent textbooks. See, for example, the books by Ashcroft and Mermin [6], and Zangwill [7].

1.2 Theoretical methods

Perhaps the most obvious change in the work of materials scientists over the last few decades involves the interpretation of experimental results. Compare, for example, the figures in the groundbreaking paper of Davisson and Germer [8], in which they announced the discovery of wave properties of electrons, to the intricate I/V curves in modern LEED experiments on Cu(100) [9]. In one case, the interpretation is straightforward: electrons are diffracted by a crystal in the same way as x-rays; *therefore* electrons possess wave properties. In the other case the interpretation has to pass a complicated theoretical evaluation procedure: in the simulations electrons are scattered by a geometrical distribution of ions in the same way as in the experiments, *therefore* the geometrical arrangement of nuclei is the same as in the simulations. While in one case there is no question about the significance of the result, nor its uniqueness (after all, it is the definition of waves to be subject to diffraction and interference), for LEED, once the theoretical model embarks on a rather complicated parameter space, this is not always the case. Indeed, the theoretical models for the analysis of LEED data can fail spectacularly, as the same calculations for the structural properties of the Si(111) surface show [10, 11]. Here, two completely different structural models lead to the same theoretical predictions. The result advocates caution: theoretical methods are usually unsuitable for an unambiguous analysis of experiments, unless these experiments-and ideally the theoretical models-cover a number of different methods. This is a crucial point, equally valid in SPM, which we will revisit throughout the text.

The ambiguity in the theoretical methods is one of the main reasons for traditional methods of experimental analysis having become less fashionable in recent years. Since in these methods particles interact with systems on a large scale, the preparation of homogeneous systems in the experiments becomes an important condition for theoretical analysis. Theoretical methods rely on computationally expensive quantum-mechanical models, which can be used only for a limited number of atoms and a limited parameter space. Complicated reconstructions on surfaces and moderately disordered systems surpass the ability of most theoretical methods with any sufficient degree of precision. Experiments on such systems, even though they might be potentially very interesting, frequently lack theoretical backing for their unambiguous interpretation. At present, this cannot be helped. There exists a tradeoff between precision and system size, which can be changed only by the advance of more efficient theoretical methods and increased computing power. The most efficient methods today can treat a few thousand atoms; the space covered by such a system is still only a cube of less than 5 nm size. This is too small to treat the system size necessary to cover all interactions of the probe particles, since the resolution of these methods is typically less than 100 nm.

A second reason that standard methods have become less popular falls outside the scope of natural sciences and may actually have a cultural background. If by anything, today's culture is defined by the dominance of images over words. Standard methods deliver either complicated graphs, which have to be interpreted to describe the actual processes, or images of abstract (e.g., reciprocal) space, not real space. In a culture in which events are frequently tied to images of these events, this is seen as a deficiency.

1.3 Local probes

From a physical perspective the common denominator of all standard methods is the large distance between the actual measuring device, e.g., the fluorescent screen of an LEED, the photo diodes of a detector, or an energy analyzer, and the sample. The distance from the particle source to the sample is typically of the same length scale. This is also the reason for the poor resolution. In patterning methods with ions or electrons, used for the production of silicon chips, the obtainable resolution today is in the range of 50 nm. Increasing this precision seems technically infeasible. This entails that methods aiming at a higher resolution have to be based on a different physical principle. Fortunately, such a principle was detected in the 1970s [12, 13, 14], and its feasibility proved in a series of groundbreaking experiments in the 1980s [15, 16, 17]. This principle is the local probe.

To apprehend the novelty of the concept imagine that an observer could reduce his size to that of an atom and position himself (or herself) *inside* a material. His environment then consists of singular massive structures, the ion cores of atoms, in imperceptibly slow motion due to thermal conditions. In between ion cores, fluctuating electrons, which connect the separate ions via their oscillations in chemical bonds, react to any change of electrostatic conditions by readjusting their local distribution, and create complicated patterns due to their correlated motion. The motion of electrons defines a natural time scale of events in condensed systems; if the typical electron process is thought to last about one day, then the motion of ion cores must be measured in years. The typical energy scale for electron processes in this environment is in the range of a few meV (magnetic properties) to about 80 meV (ambient thermal conditions). Electron phonon interactions and electron hole creation occur within the same energy scale.

Most standard experimental techniques cause mayhem in such an environment. The energy range of the probe particles, typically orders of magnitude above bond energies, is sufficient for excitations on a massive scale. The intricate balance that characterizes material structures on the atomic level is in effect destroyed. The reason that these methods still allow one to detect material properties is the limited duration of their interaction and the long time between single events. However, it is inconceivable that these methods could be used to analyze the subtle processes occurring during the formation of chemical bonds, the migration of atoms between different sites, or the excitation of single phonon modes. The only standard methods comparatively free of this problem are infrared adsorption spectroscopy (IRAS)[18] and electron energy loss spectroscopy (EELS)[19]. There, electrons or photons incident on a surface possess energies comparable to or less than bond energies (for EELS, typically around 5 eV), and their energy losses detected at specific angles after the scattering event can be referred to inelastic processes due to phonon excitations. Characteristically, these methods are limited to surface analysis due to the small energies of the probe particles.

Now consider that instead of probing material properties by targeting a sample with particles of comparatively high energy, you could do so by taking hold of a single atom and changing its position relative to atoms of a sample in a continuous way. Obviously, this is feasible only for the surface atoms of a sample. But this limitation is more than balanced by the ability to probe into the properties of surfaces while keeping the interaction between the atom and the surface at the lowest level of detection. The degree of interaction depends on the details of the experimental implementation and the actual measurement. Historically, it was determined only by a combination of experimental and theoretical methods. However, it will be seen in the course of this presentation that a large class of experiments are in fact done without substantial interaction between the surface and the probe tip. In this case, experiments can be directly related to structural and electronic properties of the sample. This, in fact, is the principle of the scanning probe microscope. At the limit of its remarkable precision lies the detection of slight changes in the electronic environment of single atoms, sufficient, for example, to detect the changes in the valance band structure due to a different chemical environment, and even the minute interactions between electrons and phonons in a molecular bond.

1.3.1 Principles of local probes

Considering the potential of local probes, the physical conditions for their operation are remarkably simple. H. Rohrer, in his article for the first volume of the three-volume survey on scanning probe microscopes [20], defines four technical requirements for such an instrument:

1. strong distance dependency of the interaction
2. close proximity of probe and object
3. very sharp probe tip
4. stable positioning device

Concerning the first point, one might ask, what is meant by a "strong distance dependency." We shall investigate this question in detail in Chapter 3, by comparing the distance-dependency of different interactions like electrostatic or van der Waals forces and their obtainable resolution in the context of the interactions that local probe instruments utilize at present. Atomic structures can be resolved only if the interaction changes by a measurable amount when the distance is changed by about one atomic diameter. The only interactions that fulfill this condition are chemical forces (changing from about 0.2 to 3.0 nN within a distance of 0.2 nm), and tunneling currents (changing by one order of magnitude within 0.1 nm). Both interactions are limited to a very close proximity of sample and probe (less than one nm), so that in fact, for high resolution experiments the first requirement already implies the second. From a historical perspective, it was the experimental proof of the feasibility of vacuum tunneling [12, 13] that triggered the development of the SPM. The first SPM was therefore a scanning tunneling microscope (STM) [15, 16, 17]. Only after the technical problems in its development were solved could the same principle be applied to the scanning force microscope (SFM). The SFM was consequently realized only a few years after the STM [21].
A strong distance dependency is obviously not enough if single structures on a surface with an extension of less than a few nm are to be resolved. A flat probe, in this case, would be insensitive to the structure, since the interaction would be independent of its lateral position. Therefore only very sharp probes are suitable for attaining high resolution images. Methods of manufacturing probes vary for different groups and experiments; the only common denominator seems to be that the tip of the probe has a diameter of less than 100 nm, and that the pinnacle of the tip presents an atomic-scale apex. The actual structure of the tip will be discussed throughout this monograph, since it is one of the main features determining the image in simulations. Experimentally, however, this is still fairly uncharted territory, because only very few experimental results have been published where the tip was known in any detail.

The final point of the technical requirements proved to be the most difficult to realize experimentally. As will be discussed in Chapter 2, the main obstacle for stable positioning of the probe turned out to be vibrational coupling to the environment. However, this problem was finally solved, and the ingenuity of the solution can be measured by the obtainable resolution with today's best instruments. This resolution can be as high as 0.5–1.0 pm, or about 10^{-12} m. Local probe instruments thus resolve the structure of surfaces down to the lowest level required in materials science (so far).

1.3.2 Surface preparation

In most experimental SPM publications the focus is usually on the presentation of the images and their interpretation in terms of surface properties. The actual preparation of the surface or the probe figures less prominently. It is confined to the more technical aspects, taking up considerably less than half of an average paper in this field. Judging from the actual experimental procedures, this seems somewhat unbalanced.
Surface preparation is probably the single most important ingredient in successful experiments, some experimenters spend weeks or even months to condition a surface for the actual measurement. It is thus due only to the accumulation of a vast body of techniques to this end that the instrument could become so successful. The conditions necessary to obtain high quality, high-resolution SPM images on a surface are the following:

1. flat surface with terraces wider than a few hundred Å
2. low surface contamination
3. high degree of surface ordering
4. low mobility of surface atoms

Large terraces are obtained by crystal cleavage (e.g., for polar surfaces of insulators or semiconductors) or by removing the surface layers with high-energy ions and subsequent annealing to near-melting temperatures (most metals and alloys).
Surface contamination is a serious problem on most metals, which usually contain a high concentration of carbon and oxygen. In this case the usual procedure is to study the segregation of contaminants and to perform repeated temperature programmed heating cycles with intermediate removal of surface layers by ion bombardment, until the immediate vicinity of the surface is clean (less than 1% contamination). On semiconductors, contaminants are removed by chemical methods. Information about methods of surface preparation can be found in the review papers [22, 23, 24] and references therein.
Most surfaces reconstruct spontaneously in order to minimize the free surface energy, e.g., by dimerization of bonds (semiconductors) or reconstruction of the atomic arrangement (metals). Catalytic reactions, that is, the chemisorption of gas molecules on a surface, their dissociation and recombination, are usually connected to massive reconstructions. The surface order

can be changed with a different surface coverage quite substantially. There exists, indeed, a vast body of experimental work from the 1990s, when all these effects were recorded in great detail.

Dynamic effects on surfaces proved an obstacle only as long as the SPM was operated at ambient temperatures. Today, low-temperature instruments can reach temperatures of less than 2 K. In this environment the migration of atoms is virtually frozen. The surface electronic structure under these conditions is stable enough to detect effects on the energy scale of a few meV. The most spectacular effects in this energy range are surface-charge waves and the magnetic effects of single electrons.

1.4 Summary

Scanning probe microscopes are based on two strongly distance-dependent processes: electron tunneling and chemical bonding. They are generally limited to the analysis of surface properties. The obtainable resolution in an SPM is below the range of atomic dimensions; they are sensitive to electronic and chemical surface structures on the atomic scale. Their field of application covers physical, chemical, and biological research.

The undisputable success of scanning probe methods has been attributed to many individual features, most of them related to the technical details of the method. There can be no doubt that these advantages contribute to its wide range of applications. However, from a more general perspective, it seems that its importance can also be seen in a different context.

Contrary to many other analytical tools, the SPM is a *soft* technique. It allows us to analyze events on the atomic level with minimum disturbance of the system under scrutiny. More than other methods, it allows us therefore to study the events and processes occurring in a close to *natural* environment. With this property it is well in line with investigative methods in other sciences, becoming more and more important as scientists aim at a deeper understanding of how natural systems really work.

Further reading

Introduction

C. Julian Chen, *Introduction to Scanning Tunneling Microscopy*, Oxford University Press, Oxford (1993).

Roland Wiesendanger, *Scanning Probe Microscopy and Spectroscopy: Methods and Applications*, Cambridge University Press, Cambridge (1994).

Intermediate

R. J. Behm, N. Garcia, and H. Rohrer, *Scanning Tunneling Microscopy and Related Methods*, Kluwer, Dordrecht (1990).
D. A. Bonnell, *Probe Microscopy and Spectroscopy: Theory, Techniques, and Applications*, Wiley and Sons, New York (2000).
Ernst Meyer, *Atomic Force Microscopy: Fundamentals to Most Advanced Applications*, Springer-Verlag, New York (2002).

In depth

H. J. Güntherodt and R. Wiesendanger (editors), *Scanning Tunneling Microscopy, Volumes I–III*, 2nd edition. Springer-Verlag, Berlin (1996).
H. J. Güntherodt, D. Anselmetti, and E. Meyer (editors), *Forces in Scanning Probe Methods*, Kluwer, Dordrecht (1995).
Roland Wiesendanger (editor), *Scanning Probe Microscopy: Analytical Methods*, Springer-Verlag, Berlin (1998)
R. Wiesendanger, S. Morita, and E. Meyer (editors), *Noncontact Atomic Force Microscopy*, Springer-Verlag, Berlin (2002).
Gewirth, R. J. Colton, J. E. Frommer, A. Engel, and H. E. Gaub (editors), *Procedures in Scanning Probe Microscopies* Wiley and Sons, New York (1998).
V. J. Morris, A. P. Gunning, A. R. Kirby, *Atomic Force Microscopy for Biologists*, Imperial College Press, London (1999).

References

1. T. M. Buck. *Methods of Surface Analysis.* Elsevier, Amsterdam, 1975.
2. W. M. Gibson. *Chemistry and Physics of Solids*, volume 5. Springer-Verlag, Berlin, 1984.
3. G. E. Bacon. *Neutron Diffraction.* 3rd edition, Adam Hilger, 1987.
4. S. Lovesey. *Theory of Neutron Scattering from Condensed Matter.* Clarendon Press, 1987.
5. B. Fultz and J. M. Hove. *Transmission Electron Microscopy and Diffractometry of Materials.* Springer-Verlag, Berlin, 2001.
6. N. A. Ashcroft and N. D. Mermin. *Solid State Physics.* Saunders, Philadelphia, 1976.
7. A. Zangwill. *Physics at Surfaces.* Cambridge University Press, Cambridge, 1988.
8. C. J. Davisson and L. H. Germer. *Phys. Rev.*, 30:705, 1927.
9. H. L. Davis and J. R. Noolan. *J. Vac. Sci. Tech.*, 20:842, 1982.
10. D. M. Zehner, J. R. Noolan, H. L. Davis, and C. W. White. *J. Vac. Sci. Tech.*, 18:852, 1981.
11. G. J. R. Jones and B. W. Holland. *Solid State Commun.*, 53:45, 1985.
12. R. Young, J. Ward, and F. Scire. *Phys. Rev. Lett.*, 27:922, 1971.
13. R. Young, J. Ward, and F. Scire. *Rev. Sci. Instrum.*, 43:999, 1972.
14. E. C. Teague. *Room Temperature Gold-Vacuum-Gold Tunneling Experiments.* PhD thesis, North Texas State University, 1978.

15. G. Binnig and H. Rohrer. *Helv. Phys. Acta*, 55:726, 1982.
16. G. Binnig, H. Rohrer, Ch. Gerber, and E. Weibel. *Appl. Phys. Lett.*, 40:178, 1982.
17. G. Binnig, H. Rohrer, Ch. Gerber, and E. Weibel. *Phys. Rev. Lett.*, 49:57, 1982.
18. J. C. Tully, Y. J. Chabal, K. Raghavachari, J. M. Bowman, and R. R. Lucchese. *Phys. Rev. B*, 31:1184, 1985.
19. M. R. Barnes and R. F. Willis. *Phys. Rev. Lett.*, 41:1729, 1978.
20. H. J. Guntherodt and R. Wiesendanger (eds.). *Scanning Tunneling Microscopy I–III, 2nd edition.* Springer-Verlag, Berlin, 1996.
21. G. Binnig, C. F. Quate, and C. Gerber. *Phys. Rev. Lett.*, 56:930, 1986.
22. F. J. Himpsel, J. E. Ortega, G. J. Mankey, and F. F. Willis. *Adv. Phys.*, 47:511, 1998.
23. R. A. Wolkow. *Annu. Rev.Phys. Chem.*, 50:413, 1999.
24. J. Shen and J. Kirschner. *Surf. Sci.*, 500:300, 2002.

2

SPM: The Instrument

The main obstacle local probe instruments faced in their development was the vibration of surfaces in an everyday environment. Usually, this vibration does not affect standard experimental methods because of the different time scales. Surfaces oscillate due to mechanical coupling with their environment. Compared to the time scale for electron processes in solids (time scale typically 10^{-14} to 10^{-15} s) or even the substantially slower phonon processes (time scale 10^{-12} s) they are very slow indeed (time scale 10^{-1} to 10^{-2} s). However, local probes scan across a surface of 100 to 1000 Å in about 1 ms (the typical duration for a scanline). Under these conditions the amplitudes of a few nm due to surface oscillations make scans in principle impossible if the tip of the local probe is less than one nm from the surface. The first successful tunneling experiments were consequently performed in a metal–oxide–metal junction rather than metal–vacuum–metal [1, 2]. As Giaever explained in his Nobel Prize lecture of 1973, "To be able to measure a tunneling current the two metals must be spaced no more than 100 Å apart, and we decided early in the game not to attempt to use air or vacuum between the two metals because of problems with vibration."

2.1 SPM Setups

The experimental setup of scanning probes such as STM and SFM [3, 4, 5, 6, 7, 8, 9] is determined mainly by the desired thermal and chemical environment. For traditional applications in surface science such as the research of surface reconstructions, surface growth, surface dynamics, and surface chemistry, the instrument is suspended in a soft damping system and in ultrahigh vacuum (UHV) chambers of less than 10^{-9} torr. The UHV chamber and the analytical instruments themselves are mounted on a rack, which is either mounted on specially damped concrete blocks, or suspended from the laboratory ceiling by elastic coils. The purpose of this elaborate scheme is to eliminate all vibrations from the environment, which would make the periodic motion of an SPM tip of less than 1 Å invisible due to background noise. The best instruments today, which are mostly home-built, are capable of a vertical resolution better than 1 pm, or one two-hundredth of an atomic diameter.

For biological applications, e.g., the research of DNA and single cells, as well as for electrochemical purposes, the SPM is operating under liquid conditions (see, e.g., [7, 10, 11, 12, 13, 14, 15, 16]). From an experimental point of view these conditions substantially limit the obtainable information and spatial resolution at a given surface structure. It is, however, an important step toward a realistic environment. In biological applications a liquid is the environment of all living organisms, and is therefore in a sense indispensable. However, theoretically this condition is poorly researched. Therefore we shall not consider it, but assume in the following that the STM or SFM operates in UHV.

The only experimental limitation for an STM is the requirement of conducting surfaces. Insulator interfaces for STM analysis are therefore grown to a few monolayers on a metal base (e.g., NaCl [17] or MgO [18]). Provided the tunneling current is still detectable, the insulator can be scanned in the same way as conducting crystal interfaces. An SFM is generally free from these limitations and could be used to study any surface. However, for achieving atomic resolution it seems crucial that surfaces be smooth enough and that there be no strong long-range tip surface forces, e.g., due to charging. In recent years, the emphasis in both STM and SFM studies is gradually shifting from the research of surface topography and surface reconstructions [19, 20] to surface chemistry [21, 22, 23, 24] and surface dynamics [25, 26, 27, 28, 29, 30].

2.1.1 STM setup

Most STM experiments on semiconductors are done at room temperature, while high-resolution scans on metals rely, with but a few exceptions [31], on a low-temperature environment of 4–16 K. Low-temperature SFM is still a less common practice. However, several home-built instruments have already demonstrated great improvement in resolution with respect to room-temperature instruments [32, 33] and there are commercial low-temperature SFMs on the market. In this case the sample and the whole SPM system are cooled by liquid helium. Thermal motion in this temperature range is greatly reduced, and high-resolution images of close-packed atomic structures can then be obtained much more routinely. Figure 2.1 shows the setup of an STM. In most cases the STM is built into a UHV chamber. Its main components are a sample holder, on which the surface under study is mounted; a piezotube, which holds the STM tip; an electronic feedback loop; and a computer to monitor and record the operation.

2.1.2 SFM setup

Measuring very small forces and force variations over a surface places more emphasis on cantilever and tip. Most observations are made by monitoring

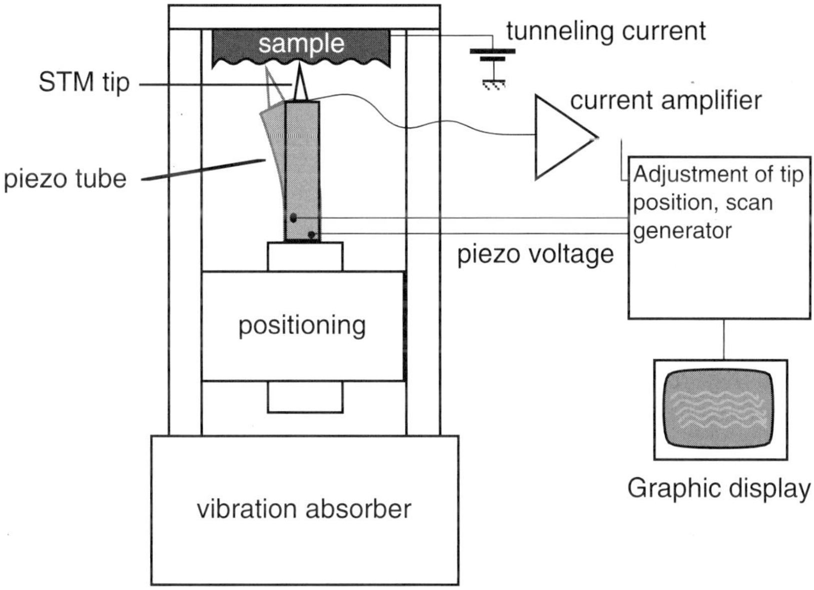

Fig. 2.1. Setup of a scanning tunneling microscope. The tip is mounted on a piezo-tube, which is deformed by applied electric fields. This deformation translates into lateral and vertical manipulation of the tip. Via an electronic feedback loop the position of the tip is adjusted according to the tunneling current (constant current mode) and a two-dimensional current contour recorded. This contour encodes all the information about the measurement. *Courtesy of M. Schmid* [34].

normal and torsional cantilever deflections induced by the tip–surface interaction using various optical methods [7, 8, 35]. In initial SFM designs the tip was pressed to a surface either by the van der Waals force or by external elastic force of the cantilever, and imaging was performed in the so-called contact mode. Although providing interesting insights into nanotribology and adhesion physics, this technique proved unreliable for imaging in atomic resolution. In contact the tip and surface were constantly exchanging material during scanning, changing the nature of the interactions [7, 9, 36, 37]. Attempts to avoid "hard" contact were thwarted by the tip's propensity to jump-to-contact even at large tip–surface distances, the generally attractive van der Waals force overcoming the stiffness of the cantilever within a certain distance. However, relatively recently it has been demonstrated that one can obtain much better sensitivity in measuring force variations on the atomic scale by employing dynamic force microscopy (DFM). In this case the cantilever is vibrated at a certain frequency above the surface, greatly reducing (but not eliminating) the problems of jump-to-contact and tip crashes. Stable operation is now possible if the following two conditions are met [38]:

$$\left|\frac{d^2\phi}{dz^2}\right|_{\max} < k, \tag{2.1}$$

$$\left|-\frac{d\phi}{dz}\right|_{\max} < kA_0, \tag{2.2}$$

where z is the tip–surface distance, ϕ is the tip–surface interaction potential, k is the spring constant of the cantilever, and A_0 is the amplitude of the oscillations. Since in this case the tip is thought not to be in direct hard contact with the surface, this technique is often also called noncontact SFM (NC-SFM).

NC-SFM was originally based on the amplitude modulation (AM) mode [39], where the cantilever is driven by a fixed amplitude at a fixed frequency. Upon approach to the surface, the tip–surface interaction causes a change in the amplitude and phase of the cantilever oscillations, providing a measurable signal. In practice, the response of the cantilever in this mode was found to be rather slow [40], and it was replaced by the frequency modulation (FM) mode [41] in atomic resolution studies. However, the AM mode has proved rather successful in "tapping mode" studies in air and liquids [40]. In general, the best mode of operation is determined by the resolution required and the system itself [42]. True atomic resolution in SFM has been achieved only in FM mode, which will be the focus of this book.

In FM mode NC-SFM, a cantilever with an eigenfrequency of f_0 and spring constant k is maintained in oscillations at a constant amplitude A_0 via a feedback loop (see Figure 2.2). The cantilever can be considered as a self-driven oscillator. The actual frequency of oscillations depends on f_0, the quality factor Q of the cantilever, and the phase shift θ between the driving excitation and the deflection of the cantilever. For $\theta = \pi/2$ the system oscillates at $f = f_0$. Generally, during experiments the tip–surface distance is varied in order to achieve a constant frequency change Δf, and the resulting topography map provides the image of the surface. It is also possible to image at constant height, where now the change in Δf provides the imaging signal.

For reliable imaging, there is one further aspect of the experimental setup that is important: as in STM, a bias U is normally applied between tip and sample in SFM experiments. Undoped semiconductor and insulating surfaces will usually contain significant localized charges after preparation, especially cleaved ionic surfaces. These produce significant long-range electrostatic forces (see Chapter 3), as well as sudden changes in the tip-surface force during scanning. For conducting surfaces, the work-function difference between tip and surface will contribute a long-range capacitive force. These additional forces make scanning more difficult by reducing the relative contribution of short-range forces and increasing the possibilities of tip crashes. Reducing the effect of electrostatic forces can be achieved by minimizing Δf as a function of applied bias at a certain point on the surface. An example of this process can be seen in Figure 2.3.

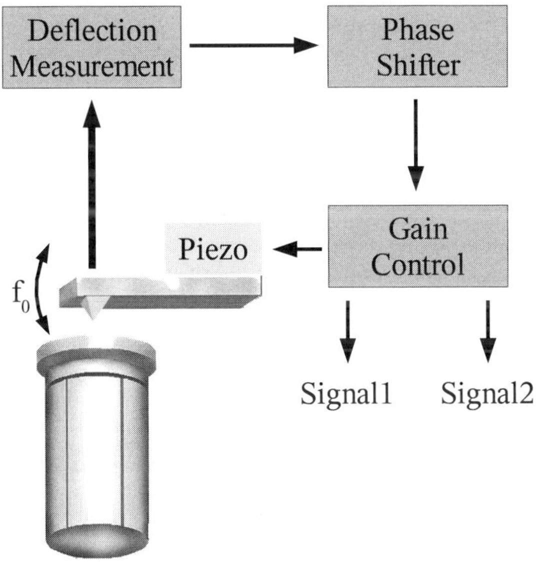

Fig. 2.2. Schematic diagram showing the feedback loop in standard SFM operation. Adapted from ref. [43].

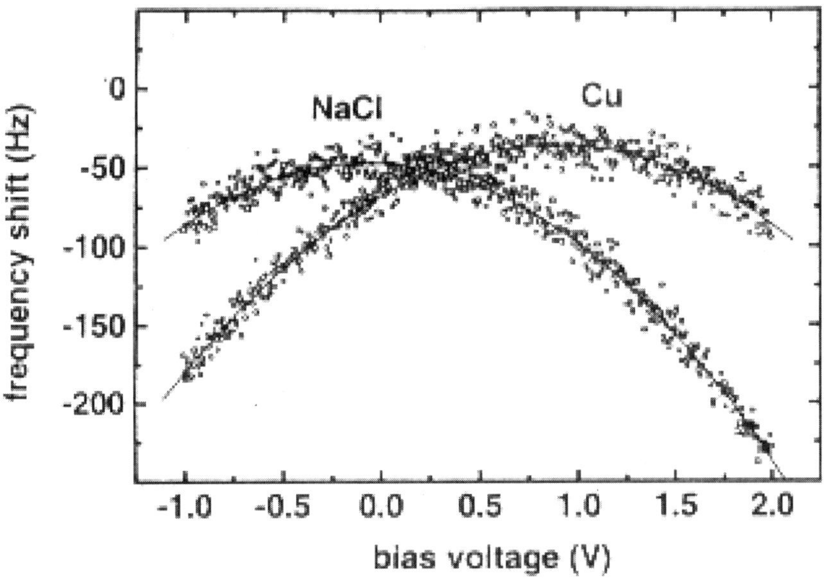

Fig. 2.3. Frequency shift vs bias voltage curves recorded at constant height over a Cu(111) and over an NaCl thin film on Cu(111). R. Bennewitz and M. Bammerlin and M. Guggisberg and C. Loppacher and A. Baratoff and E. Meyer and H.-J. Güntherodt, Surface and Interface Analysis 27, 462 (1999), reprinted with permission.

2.1.3 Tip and surface preparation

Not every surface can be imaged in STM or SFM with high resolution. To achieve atomic resolution, the surface in most cases needs extensive preparation. Sputtering (bombardment with ions, mostly Ar^+), and annealing (heating to the point where the surface defects are smoothed out) over weeks and even months in controlled cycles is not uncommon, e.g. on metal surfaces [44]. Surface preparation in itself is a sophisticated art, and one of the keys to successful imaging [20, 45, 46]. Contrary to k-space methods such as ion scattering and electron diffraction, a surface need not be ordered to be imaged by SPM. In fact, single impurities and step edges on a surface are often used by experimenters to check the quality of their images. Such an impurity is imaged only as a single structure, assuming no distorting effects like double tips are present.

The tip is the crucial part in imaging in all SPM methods. STM tips are often made from a pure metal (tungsten, iridium [20]), a metal alloy (PtIr [47]), or a metal base coated with 10–20 layers of a different material (e.g., Gd or Fe on polycrystalline tungsten [48]), often produced in the lab from metal wire. In some cases heavily doped Si tips are also used for STM imaging. Although similar tips could also be used for SFM measurements, this is very rare. This is due to the fact that the cantilever holding of a tip plays a very important role in monitoring force changes in SFM: (i) in many SFM realizations cantilever deflections are measured by detecting light reflected from the back of the cantilever; (ii) cantilever spring constant, tip shape, and tip sharpness all play crucial roles in image formation. Therefore standard cantilevers are required. In most cases these are produced from silicon by microfabrication in very much the same way as semiconductor chips.

In some cases the tip is modified by controlled adsorption of molecules [49, 50, 51, 52]. In STM it has been shown that this affects the apparent height of molecules on a surface [49, 50]. The exact geometry of the tip is commonly unknown except for some outstanding STM measurements, where the tip structure was determined before and after a scan by field-ion microscopy [53]. To complicate matters further, the tip geometry is decisive for reproducible scanning tunneling spectroscopy (STS) measurements [54]; unfortunately, the tip most suitable for STS has been shown to be unsuited for topographic measurements, because it does not yield a high enough resolution [55]. In SFM, some attempts have been made to produce clean silicon tips [56], even with specific orbital configurations at the apex [57], but images have yet to be produced on anything other than silicon surfaces: hence evidence of real control is lacking. Currently the most widely held opinion is that SPM tips consist of a base with rather low curvature [58] and an atomic tip cluster of a few layers with a single atom at the foremost position.

In STM, all the current in the tunneling junction is transported via this "apex" atom; the area of conductance is consequently rather small and in the range of a few Å^2 [20] (see Figure 2.4). This is the origin of STM precision, because it makes the current very sensitive to the electronic environment of a very small area of the surface. Variations in the interaction of the last several atoms of the sharp tip apex with the surface atoms also determine the image contrast in SFM images. However, this sensitivity to atomic structure is also the origin of the features that make the interpretation of STM and SFM images so difficult because the actual geometry and chemistry of the tip apex which influences the conductance in the vacuum barrier between surface and tip and also determines the tip surface forces, cannot usually be determined. Even for simple metal surfaces like Cu(100) and NaCl(100) this leads to different experimental results for different scans [30, 59].

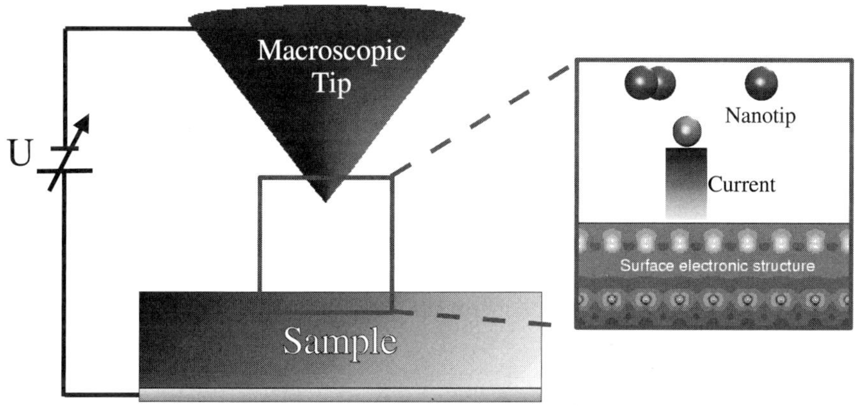

Fig. 2.4. Tunneling current in a scanning tunneling microscope. The surface of the tip is generally not smooth. A microtip of a few atoms will bear the bulk of the tunneling current; due to this spatial limitation of current flow the electronic properties of a scanned surface can be extremely well resolved (resolution laterally better than 1 Å).

Since the determining factors in SPM experiments are not fully known, their relevance needs to be inferred from simulations. Simulations need to be done in a systematic manner, e.g., by studying the effect of adsorbates on the electronic structure of model tips [60, 61], and by modeling the effect of these adsorbates on STM scans [62]. Experimentally, the difficulty is circumvented, at least in careful measurements, by recording a series of scans and presenting decisive measures such as the surface corrugation as a statistical average.

2.2 Experimental development

Since its invention in the early 1980s, SPM experiments have come a long way. While initially the emphasis in experiments was mainly the resolution of atomic positions, today experimental results can provide information on such heterogeneous topics as the chemical composition of surfaces, collective effects mediated by intramolecular interactions in molecular overlayers, activation barriers for chemical reactions or molecular diffusion, long range electron interactions and lifetimes of different states, and noncollinear and anisotropic effects due to magnetic confinement.

Ideally, experimental data are self-evident. A set of data admits one and only one interpretation. However, this is generally not the case, as already emphasized in Chapter 1. Two different sets of model calculations, with widely varying atomic positions, lead to the same predicted LEED images on Si(111) [63, 64]. It was partly the ability to "see atoms" that made SPM instruments such a success. But does one really "see" atoms, e.g., in STM scans on flat surfaces? Clearly, if the interpretation of a given SPM experiment is highly nontrivial, then the more subtle effects increasingly probed and manipulated today require extensive analysis and a high level of understanding about a system to be correctly interpreted. This, in turn, requires that experimenters as well as theorists be aware of the possible shortcomings in a given method and that they be able to address issues excluded by one method by other means. Here we consider several example systems from STM and SFM that demonstrate both the capabilities and interpretational problems of the techniques:

- Measurements with atomic resolution on flat metal surfaces like Au(110) and Au(111) were among the first to be undertaken in experiments. The interval from the first STM experiments on the missing row reconstruction of Au(110) and the close packed Au(111) surface was less than five years. During this period the STM was developed from a tool to image monoatomic steps on a surface to a tool capable of resolving the position of single atoms.
- The development of tunneling spectroscopy experiments with high local resolution on magnetic surfaces marks the change of focus from the analysis of surface topography to a detailed analysis of surface electronic structures.
- The silicon (111) 7×7 surface was the first imaged in atomic resolution by SFM and it remains a benchmark surface for experiments. This is in part due to tradition, but also due to its distinctive and complex surface structure, which provides a clear test for atomic resolution. The story of SFM is also the story of imaging Si(111) 7×7, and so it is a good example of experimental development.
- Of course, we cannot really discuss the development of SFM without considering an insulating surface. In fact, a class of insulating materials provides probably the best cross-section of experiments; simple cubic crystals such as NaCl, MgO, and NiO have offered some of the greatest challenges to both experiment and theory.

2.2.1 STM Case 1: Au(110) and Au(111)

The very first demonstration of the STM's ability was published by Binnig and coworkers in a paper in 1982, where they showed the exponential decay of the tunneling current on a platinum plate (see Figure 2.5). Subsequently, they included a two-dimensional scan mechanism, and scanned the surface of Au(110) in 1982 and again in 1983 [4, 5, 65]. Comparing the quality of the images of the two separate publications, which appeared less than one year apart, one already notes a substantial improvement in the resolution. While the first image of the Au(110) surface (frame (b)) allows only a rather vague identification of the underlying atomic structure, the second image (frame(c)) already allows the authors to resolve two different reconstructions: the 1×2 reconstruction arises from the two-row facets along the $[1\bar{1}1]$ direction of the surface, while three-row facets lead to a 1×3 reconstruction. At the same time, the STM was successful in imaging the 7×7 reconstruction of the Si(111) surface. However, at this stage the instrument was still far from its ability today. If one takes a typical area of today's high-resolution scans (about 2.5 nm × 2.5 nm; see frame (c)), then it becomes clear that the lateral resolution was at best 0.5 nm, enough for a semiconductor surface like Si(111), where the Si surface atoms are quite far apart, but not quite sufficient for a close-packed metal surface, where distances between two atoms are on the order of 0.2–0.3 nm.

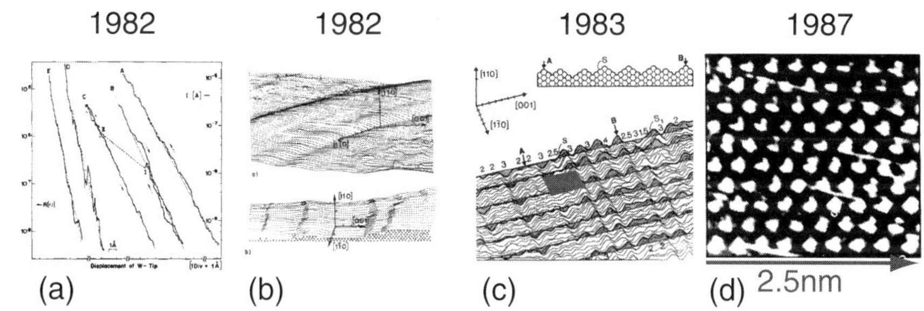

Fig. 2.5. Development of an STM's ability to image single atoms on metal surfaces. From the first demonstration of an exponential decay in the tunneling gap, frame (a), to the first demonstration of a two-dimensional scan on Au(110), frame (b), to the resolution of two different reconstructions on the same surface, frame (c), it took less than one year. However, the ability to resolve single atoms on a close packed metal surface took five years to develop and was demonstrated only in 1987 (frame (d). To appreciate the gain in resolution we have sketched the whole area of frame (d) in image (c). Reprinted with permission from [4, 5, 65, 66]. Copyright by the American Physical Society.

The ability to image single atoms was widely exploited in the late 1980s and early 1990s. In principle, it was now possible to resolve any structure of a metal surface with atomic resolution, with the possible exeption of some particularly difficult materials like some $3d$ transition metals. This is, with hindsight, not a problem of the instrument's resolution, but a problem of the corrugation height of single atoms. It will be shown in the applications of STM theory, presented in later chapters, that some of these surfaces possess a surface corrugation that is less than 2 pm under normal tunneling conditions. This makes imaging these surfaces with atomic resolution not so much a problem of lateral resolution as a problem of the instrument's stability and vibration damping. The next development, influencing the focus of research, was the advent of low-temperature STMs. Low temperatures remove several of the key obstacles to accurate images: The first is the mobility of adatoms and adsorbates, in particular on metals. The second is the statistical nature of many physical properties under ambient conditions. A room-temperature STM will provide only an average of these properties, e.g., magnetic characteristics, and is thus not suitable for studying the local correlation of these properties.

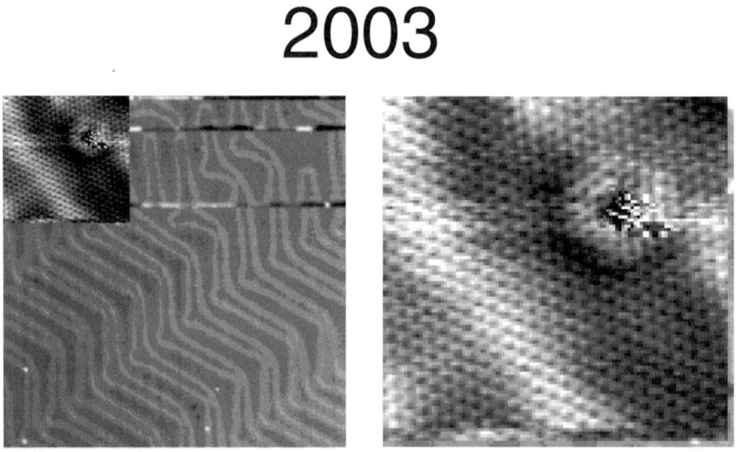

Fig. 2.6. Low-temperature images of the reconstruction on Au(111) (left), and an atomically resolved detail of the original image. With the advent of low temperature STM, imaging close packed metal surfaces became fairly routine. P. Han and E. C. H. Sykes and T. P. Pearl and P. S. Weiss, J. Phys. Chem. A 107, 8124 (2003). Copyright (2003) American Chemical Society, reprinted with permission.

Low-temperature STM also has a slightly improved resolution, as seen in images of the Au(111) surface, as shown in Figure 2.6, which shows a recent experiment [67]. Today, atomic resolution on close-packed metal surfaces is routinely achieved in many labs around the world. These studies even provide, in single cases, a clear picture of single electronic states. However, from a theoretical point of view, the development caught up with this experimental ability with a delay of about fifteen years. The reason in this case was the need for powerful modeling tools for the surface electronic structure as well as the physical processes under tunneling conditions. The history of these development and the state of the art today are essentially the topic of this book.

2.2.2 STM Case 2: Resolution of Spin States

Once experiments were able to resolve surface structures with a lateral resolution of 0.1 nm (the best instruments today probably have a lateral resolution of about 0.05 nm), it became possible to analyze in detail not only the atomic configuration, but also the local extent of single electron states. This gave rise to a wealth of new experimental data. In particular the question whether an STM would also be able to resolve the spin state of an electron occupied the imagination of experimenters and theorists alike from the late 1980s, since it had been shown by Pierce [68] that a scanning electron microscope could resolve magnetic domains with a resolution of about 100 nm. This led to the search for a suitable combination of surface material and STM tip that would maximize the effect.

It is well known that Cr(001) possesses a surface state in the minority band, and also that it orders antiferromagnetically. In this case one expects that a step edge will possess a surface state of spin-up electrons at the upper terrace, and of spin-down electrons at the lower terrace. The surface atoms themselves cannot usually be resolved on Cr(001). The reason is that the charge density contour of this surface is very flat. We shall present an analysis of topographies on this surface in later chapters. Here, we wish only to make the point that even if one could not resolve the atoms of the surface directly, one could still resolve step edges. And if, as one could expect, the lower terrace contributes mainly spin-up electrons to the tunneling current while the upper terrace yields primarily spin-down electrons, then it could be expected that one should notice a difference in a topographic image of two adjacent step edges. However, this assumption is justified only if the STM tip itself is spin-polarized. The apex atom of the STM tip in this case has to possess a different density of states at the Fermi level for spin-up and spin-down electrons. This immediately raises the question how one might fabricate such a tip.

In the first publication, which claimed to have resolved the two different terrace types, by Roland Wiesendanger in 1990 [69], the experimentalists used a CrO_2 tip made by vacuum deposition on Si(111). The experimental result is shown in Figure 2.7(a). The experimental result consisted of only four scanlines, where a characteristic variation of the step height from 0.12 to 0.16 nm was found on Cr(001), supporting the theoretical model of chromium antiferromagnetism from one layer to the next. It should be noted that the experiments were performed at room temperature, variable- or low-temperature STMs not being available at this time.

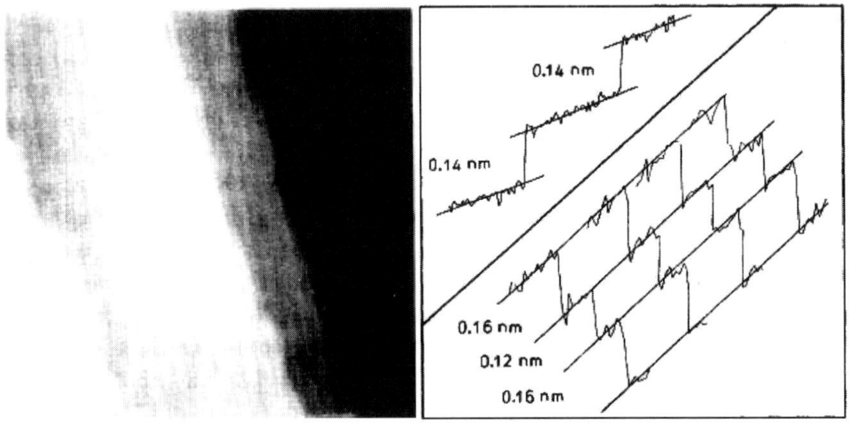

Fig. 2.7. (Left) Step edges on a Cr(001) surface measured by STM. (Right) One linescan with a tungsten tip, and four linescans with CrO_2 tip across three steps. The step height varies by 0.01 to 0.02 nm compared to the step height measured with a tungsten tip; the variation, moreover, seems oscillating. R. Wiesendanger and H.-J. Güntherodt and G. Güntherodt and R. J. Gambino and R. Ruf, Phys. Rev. Lett. 65, 247 (1990). Copyright (1990) by the American Physical Society, reprinted with permission.

Given the rather sketchy evidence and the problem of spin fluctuations at room temperature, it seemed not too surprising that quite a few experimentalists remained sceptical. In their view the experiment did not amount to proof that the technique really was working. Initially, their skepticism seemed justified. From 1990 to about 1998, no new paper on the ability to detect the spin of tunneling electrons was published. However, during this period the whole field underwent quite a change. The variable temperature STM made experiments, in particular on metal surfaces, much more controllable and improved the resolution of the obtained images quite generally. In addition, new vacuum deposition techniques were making it gradually possible to study any material compound, since the surface-science community experimented with a large variety of ultrathin films in a search for new effects. These two improvements facilitated the development and operation of new instruments, where the sample usually consisted of a magnetic array with reduced dimensions, and the STM tip was tailored to maximize the difference between the conductance properties of spin-up and spin-down electrons.
Initially, STM tips were made by vacuum deposition of magnetic metals or metal oxides on surfaces such as silicon (111). In this case the deposed layer had to be physically removed from the substrate and attached to a metal tip. This proved to be too complicated, given that an STM tip can be easily destroyed in an experiment. Early work with CrO_2 tips [70] was therefore soon given up, and the right tip material for STM and STS experiments on

magnetic structures became the subject of a thorough analysis. Given that the tip should act as a spin-valve in the experiments, without influencing or even changing the magnetic properties of the surface atoms, experimentalists tried to fabricate suitable tips that would comply with the following conditions:

- The apex atom of the tip has a high spin polarization.
- The bulk material of the tip is nonmagnetic, to reduce stray fields, which could influence the surface magnetic structure.
- The tip is clean of adsorbates and chemically inert.
- The tip can be magnetized and the magnetization axis changed periodically.
- The magnetization axis can be changed from in-plane to out-of-plane.

In principle, polycrystalline wires made of manganese or chromium would be suitable, since the invidual layers of the tip material couple antiferromagnetically. However, to date no successful experiments using these wires have been reported. The required STM tip properties can be obtained using two different methods: (i) A very sharp magnetic tip is posed in an oscillating magnetic field, the spin signal is determined by lock-in techniques and subtracting the signal intensities at the endpoints of a magnetization cycle; or (ii) by coating a nonmagnetic tip with a few (up to 20) layers of a magnetic material and performing high-resolution tunneling spectroscopies. Both of these techniques have been developed in different labs. Magnetic domains of Co(0001) were measured with a very sharp tip made of amorphous FeCoSiB. The scanning electron microscope image of such a tip, produced by slow etching, is shown in Figure 2.8(a).

If such a tip is periodically magnetized by an external magnetic field, the magnetic axis at the apex atom will change its orientation. For perpendicular magnetization, it points to the surface, and the spin-up and spin-down components projected onto this direction will be periodically reversed. The effect itself is rather small, typically only a few percent of the nonmagnetic background signal [71]. However, as shown in Figure 2.8, this is sufficient to separate the magnetic components of the surface electronic structure. In Figure 2.8(b) a topographic image of the Co(0001) surface has been taken without any external magnetic tip field. In this case the surface appears flat with a few impurities. As the magnetic field is switched on and the spin-up and spin-down components of the surface charge are measured at the endpoints of the magnetic cycle, their difference reveals a domain wall crossing the previously flat surface.

The STM tip in this case was carefully chosen for its low coercivity and saturation magnetization. Due to the shape anisotropy of the material, the tip is magnetized along its axis. These two features, joined with the low diameter of the tip, make it possible to reduce the external magnetic field and the number of coils around the tip shaft, which in turn lead to a minimization of the tip's stray fields. This method was mainly developed at the Max-Planck-Institut in Halle, Germany.

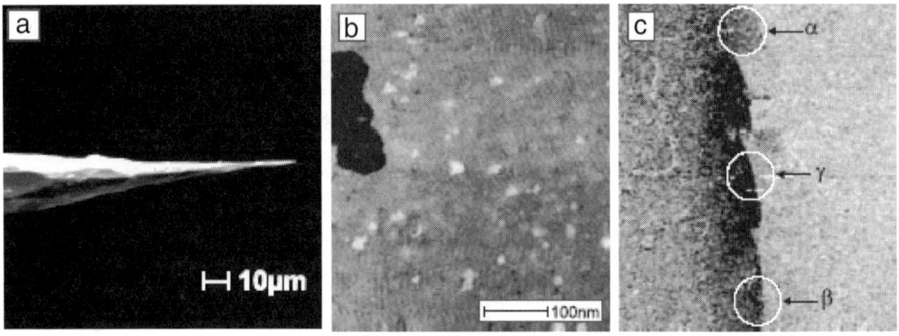

Fig. 2.8. (a) Magnetic tip made of FeCoSiB by slow etching. (b) Topography of a Co(0001) surface. (c) Topographic image obtained by subtracting the two signals at the endpoint of a magnetization cycle with a periodic magnetic field of $H = 70$ μT and a magnetization axis perpendicular to the surface. The image clearly shows a domain wall, the height variation is typically a few percent of the nonmagnetic background signal. W. Wulfhekel and H. F. Ding and W. Lutzke and G. Steierl and M. Vazquez and P. Marin and A. Hernando and J. Kirschner, J. Appl. Phys. 72, 463 (2001), reprinted with permission.

The second method, using nonmagnetic tips coated by a few layers of magnetic material, was pioneered by a different group at the University of Hamburg, also in Germany. A detailed account of the development is given in a recent review by Bode [72]. A detailed account of the fabrication of coated tips on a tungsten base is given in [73]. In Figure 2.9 we show scanning electron images of a coated tungsten tip (a), a high-resolution image of the tip apex (b), and a sketch of the apex curvature with the magnetic coating (c). The magnetic coating of about 2 nm is very thin compared to the apex radius of the tip (in the range of 1000 nm). This particular geometry has different effects on the magnetization, depending on the chemical nature of the material. For 3–10 monolayers of Fe, the tip is usually sensitive to in-plane magnetization. For 7–9 monolayers of Gd, or 25–45 monolayers of Cr, it is sensitive to out-of-plane magnetization. The change is essentially due to a competition between shape anisotropy of the tip, which favors an orientation of the magnetic field parallel to the tip axis, and surface anisotropy, which frequently favors an orientation parallel to the surface.

However, given the large apex radius in the SEM images, it does not seem very clear how atomic resolution could be obtained, unless the apex contains at some point a protrusion made of one or only a few atoms of the magnetic material. In this case the large magnetic field of the surface layer should force the electrons of the protruding atoms to adjust to the magnetization direction in the coating. The effect is the same as for magnetic material in an external magnetic field: the symmetry of spin states is broken and the spin-up and spin-down charge contributions will be aligned along the magnetic axis.

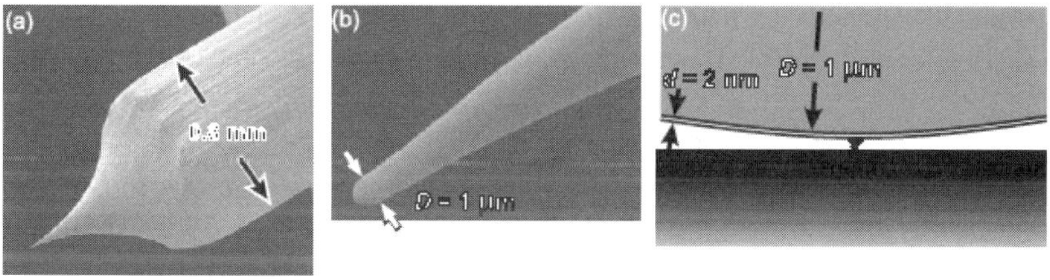

Fig. 2.9. (a) SEM image of a coated tungsten tip. (b) Tip apex; the diameter of the apex is about 1000 nm. (c) Sketch of the apex, coated with a thin film (2 nm) of magnetic material. M. Bode, Rep. Progr. Phys. 66, 523 (2003), reprinted with permission.

Once the fabrication of STM tips with defined magnetic properties was accomplished, the measurements of spin-states in low temperature experiments became feasible. However, the magnetic properties of thin films themselves are not too exciting, apart from the effect of anisotropy on the direction of the magnetic field. Therefore experiments focused initially on systems, where theory predicted large anisotropy effects as in low-dimensional structures. In this field, the group in Hamburg has accomplished some pioneering research. As an example, state-of-the-art experiments on Cr(001) that resolve the antiferromagnetic coupling of chromium layers by spin polarized STS show how the technique has developed within the last fifteen years.

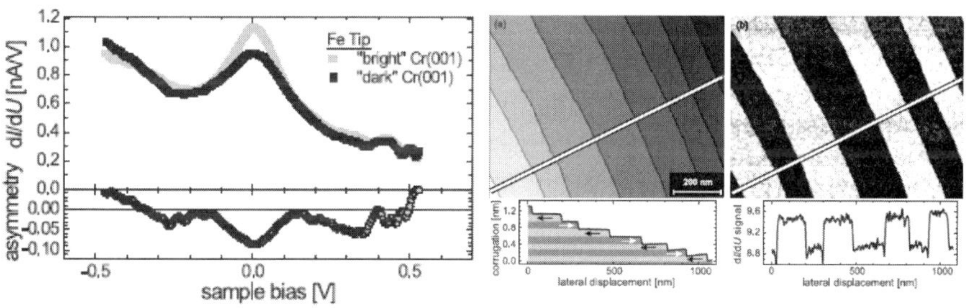

Fig. 2.10. (Left) Tunneling spectrum on Cr(001) measured with an iron-coated tip. The peak of the Cr(001) surface state is detected near the Fermi level; the height of the peak depends on the terrace on which it is measured. (Center) Topographic image of Cr(001) terraces. (Right) Height of the dI/dV value at successive terraces at $-$ 290mV: the oscillation of the feature clearly demonstrates the influence of magnetic properties on the tunneling spectrum and can be seen as a proof of antiferromagnetic ordering. Reprinted with permission from [72].

Experimental results are shown in Figure 2.10. The Cr(001) surface possesses a surface state, which shows up as a distinct peak of the dI/dV spectrum near the Fermi level. The spectrum to the left, taken with an iron-coated tungsten tip, reveals this feature, but the height of the surface state changes from one terrace to the next, clear indication that the magnetic properties of adjacent terraces are different. A topographic image shows that the step height is actually fairly constant, even though adjacent terraces will have a reversed magnetic moment. This indicates that the magnetic contrast is usually too low to show up in topographic images. However, if the height of the conductance is measured at adjacent terraces, it changes by about 10%: this change can be made visible in a local map of the conductance (shown on the right), and clearly demonstrates the periodic changes of the magnetic orientation.

2.2.3 SFM Case 1: silicon (111) 7 × 7

The need for almost flawless samples of silicon in the microelectronics industry has driven the refinement of techniques for preparing clean, smooth silicon surfaces efficiently. Consequently, the availability of good silicon samples has encouraged its use as a benchmark surface for adsorption, growth, manipulation, and, of course, in SPM techniques. As mentioned previously, the first atomically resolved images in both STM [65] and SFM [74] were achieved on the silicon (111) surface. Even now, atomic resolution on silicon (111) remains the first goal of any novice SPM adventurer. As such, it provides an excellent example of how the quality of images and the information that can be extracted from an experiment have developed.

The stable surface of silicon at room temperature is the rather complex (111) 7×7 reconstruction, first observed by transmission electron microscopy (TEM) and diffraction (TED) [75], and shown in Figure 2.11. The initial atomically resolved SFM image of this surface (shown in Figure 2.12) demonstrates many of the features that remain important discussion topics to this day. Although the bright spots in the image have the periodicity of the atomic lattice of Si(111), this is not the end of the story. The contrast pattern itself is not consistent across the image, with the resolution poor at the bottom of the image, disappearing at some points, but also becoming very vivid at other points. This behavior is characteristic of tip instabilities and changes [74]. Contrast in SFM is very sensitive to the microscopic nature of the end of the tip, and if this changes, the image contrast will change. The topic of tip instabilities will be revisited several times during the course of this book.

A second equally important aspect of SFM imaging introduced in this image is that of identitification. Although we can be fairly confident of the structure of the surface being imaged, we cannot be equally certain how an SFM will image that structure. In general, we cannot be sure whether the bright spots represent the uppermost layer of the surface or some deeper atomic layer, or even a convolution of several layers. For this complex reconstruction, it is possible to assign the bright spots to corner and centre adatoms in the surface,

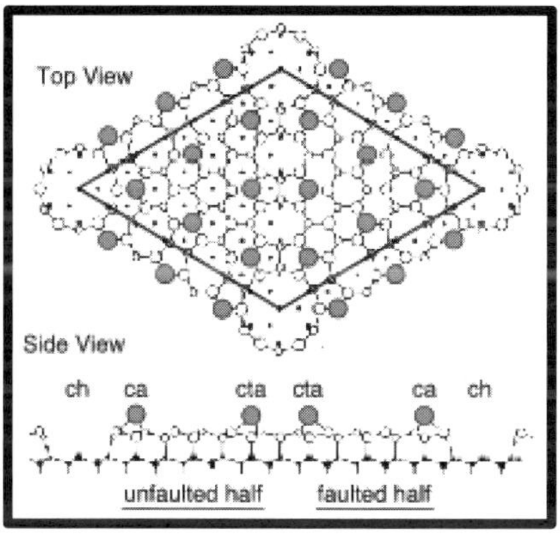

Fig. 2.11. Schematic model of the dimer adatom stacking-fault (DAS) model of the silicon (111) 7×7 surface. The unit cell is shown by the black diamond with the adatoms shown as gray circles. The side view shows the positions of the corner holes (ch), corner adatoms (ca), and centre adatoms (cta). M. A. Lantz and H. J. Hug and R. Hoffman and P. J. A. van Schendel and P. Kappenberger and S. Martin and A. Baratoff and H. -J. Güntherodt, Science 291, 2580 (2001), reprinted with permission.

.

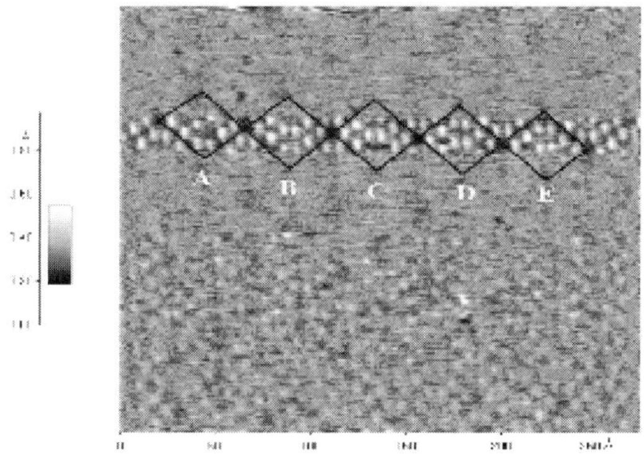

Fig. 2.12. First atomically resolved topographic SFM image of the Si(111) 7×7 surface ($\Delta f = -70$ Hz, $f_0 = 114$ kHz, $k = 17$ N/m, A = 34 nm, $U_{bias} = 0$ V). F. J. Giessibl, Science 267, 68 (1995), reprinted with permission.

but as we shall see, for most other surfaces, assigning identity to the bright spots is a significant challenge.
The development of low-temperature SFM offered a further opportunity to push the limits of resolution on the silicon (111) surface, and also to test some theoretical predictions [76] (see Chapter 9). The first low-temperature images (see Figure 2.13(a)) demonstrate a remarkable improvement from the initial room-temperature studies. The reduction in thermal noise and drift provides a considerable increase in sensitivity, and the resultant images have very high quality and are free from any evidence of tip instability. By increasing the negative frequency shift and scanning a small area at lower speeds, it was also possible to show the interaction of the tip with deeper rest atoms in the surface, as predicted by theory [76]. In Figure 2.13(b) contrast can be clearly seen between the adatom positions at rest atom sites.

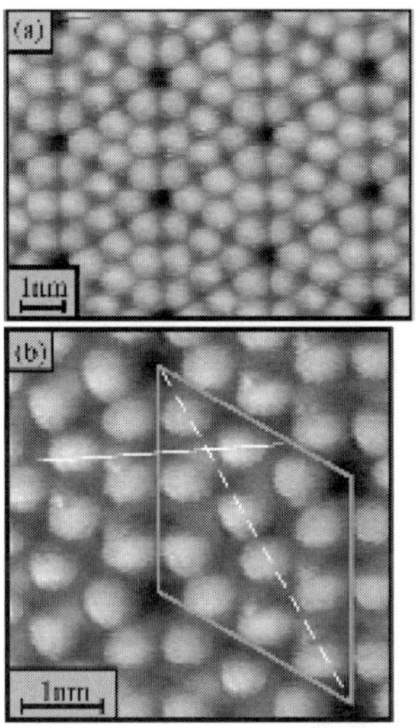

Fig. 2.13. (a) First low-temperature atomically resolved topographic SFM image of the Si(111) 7×7 surface. (b) Image of a smaller area taken with larger frequency change and slower scan speed. ($\Delta f = -27, -31$ Hz, $f_0 = 155$ kHz, Q $= 370{,}000$, $k = 28.6$ N/m). M. A. Lantz and H. J. Hug and P. J. A. van Schendel and R. Hoffmann and S. Martin and A. Baratoff and A. Abdurixit and H. -J. Güntherodt and Ch. Gerber, Phys. Rev. Lett. 84, 2642 (2000). Copyright (2000) by the American Physical Society, reprinted with permission.

Advances in the sensitivity of SFM experiments at room temperature have also produced further evidence of the extreme tip-dependence of contrast patterns. Figure 2.14(a) shows a high-quality image of the Si(111) 7×7 surface, which at first glance appears little different from Figure 2.13(a). However, if we enlarge a single adatom in the image (see 2.14(b)), it is clear that two maxima appear at each adatom site. Since we know there is only one silicon atom at each adatom site, it seems likely that this is an effect of the tip. Theoretical simulations [57, 77] have demonstrated that a silicon tip with two dangling bonds at the apex could provide this kind of double maximum in images.

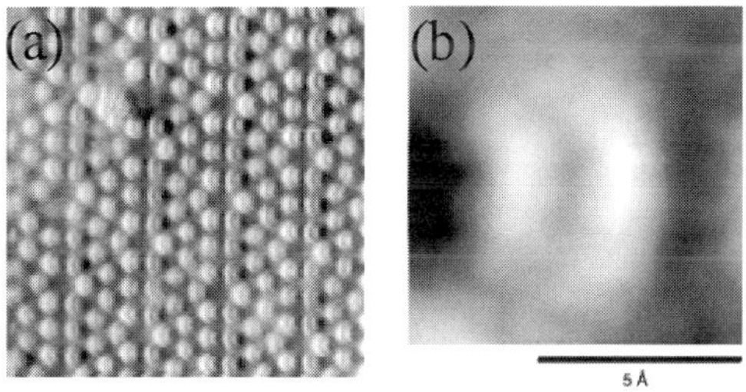

Fig. 2.14. (a) Atomically resolved topographic SFM image of the Si(111) 7×7 surface with a defect in the top left ($\Delta f = -160$ Hz, $f_0 = 17$ kHz, $k = 1800$ N/m, A = 0.8 nm, $U_{\text{bias}} = 1.6$ V). (b) Enlargement of a single adatom image. F. J. Giessibl and S. Hembacher and H. Bielefeldt and J. Mannhart, Science 289, 422 (2000), reprinted with permission.

The tip instabilities mentioned previously are basically a feature of uncontrolled atomic motion on or between the tip and surface; e.g., a surface atom jumps onto the tip, changing its imaging character. This is generally unwanted when one is trying to learn something about the surface, but controlled manipulation of atoms in the surface is an important developmental step for SFM. Again, this was perhaps best realized on the silicon (111) surface [78], where it was possible to pick up an adatom from the surface and then resolve the created vacancy, before finally returning a silicon atom to the defect and imaging the *ideal* surface.
Controlled atomic manipulation is one of the first signs of SFM's progress beyond atomic resolution of surfaces, and it opens the door to controlled chemical reactions and device assembly at the atomic scale. It is highly probable that the silicon (111) 7×7 surface will also feature in these new developments.

2.2.4 SFM case 2: cubic crystals

Cubic ionic materials offer structurally the simplest insulating materials with ideal bulk-terminated surfaces, which are generally inert. The first atomically resolved images of an insulating surface [79, 80] where achieved on alkali halides (specifically NaCl (001) [79]), where preparation of clean surfaces with large atomically flat terraces via cleavage in UHV is reasonably simple. Figure 2.16 demonstrates the types of contrast pattern seen on the different surfaces, and as one would expect, they are quite similar. The separation of bright spots in images corresponds to the bulk lattice positions of like charges in the

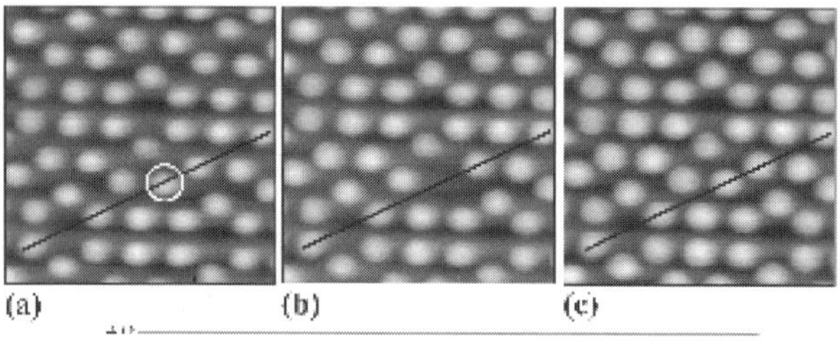

Fig. 2.15. Atomically resolved topographic SFM image of the Si(111)-7 $\times$ 7 surface ($\Delta f = -4$ Hz, $f_0 = 160$ kHz, $k \approx 48$ N/m, A = 26 nm, Q = 170,000 $U_{\text{bias}} = 0$ V) (a) initially, (b) after removal of a single adatom, and (c) after approaching close to the previously created defect and healing defect. N. Oyabu and Ó. Custance and I. Yi and Y. Sugawara and S. Morita, Phys. Rev. Lett. 90, 176102 (2003). Copyright (2003) by the American Physical Society, reprinted with permission.

ionic lattice. Point defects are also visible on NaCl in Figure 2.16, emphasizing the local nature of the experimental data and providing further evidence that true atomic resolution has been achieved. However, beyond this experimental achievement, very little further information can be extracted from the images. The simplicity of the structure and lack of information on the tip means that it is impossible to identify which species, anion or cation, is being imaged as bright, or in fact, whether interstitial regions appear bright. This general problem of SFM remains significant, and only in a few cases (in only one case for cubic crystals; see Chapter 9) has it been resolved. Possible solutions include imaging more complex insulating surfaces, where contrast patterns for each sublattice differ (see CaF_2 (111) in Chapter 7) or using low-temperature force curves over specific atomic sites (see KBr in Chapter 9). These methods are extremely resource intensive, both experimentally and theoretically, and simpler, more general approaches to interpretation are being sought (see Chapter 7).

Experimentally, it would seem natural to move from imaging alkali halides to the more application-rich field of oxide surfaces. For cubic crystals, MgO stands as the obvious example, and theoretical simulations [81], including defects, were performed soon after the initial success on halide surfaces. However, successful atomic resolution on the MgO surface had to wait for half a decade despite the attention of several SFM groups. Unlike the alkali halides, cleavage of MgO usually resulted in a surface covered in nanorubble and localized charged defects, making tip instabilities and crashes much more likely. Atomic resolution was finally achieved using a careful combination of UHV cleavage at room temperature, annealing at 620 K, and minimization of electrostatic forces via applied bias [82], although the significant skill of experimentalists involved should, of course, not be neglected. The resulting image, shown in Figure 2.17, is very similar to those in Figure 2.16, apart from evidence of a tip change. Along with the alumina (Al_2O_3) surface [83], achievement of

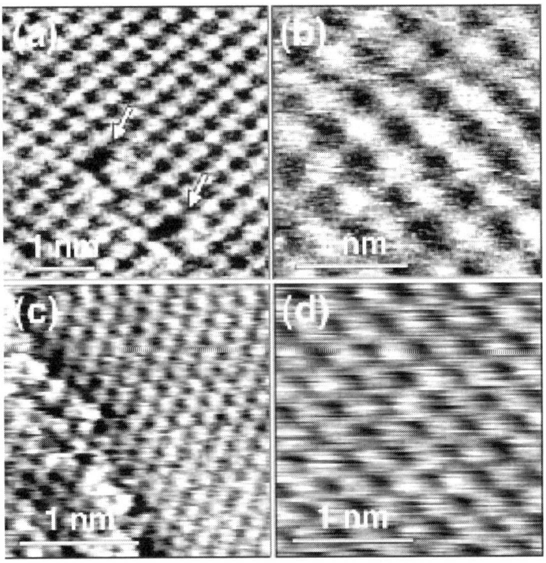

Fig. 2.16. Atomically resolved SFM Δf images of (a) NaCl (001), (b) NaF (001) , (c) LiF (001) , and (d) RbBr (001) ($f_0 \approx 167$ kHz, $k \approx 30$ N/m, A = 13 nm). Point defects are labelled by arrows. M. Bammerlin and R. Lüthi and E. Meyer and J. Lü and M. Guggisberg and C. Loppacher and C. Gerber and H. -J. Güntherodt, Appl. Phys. A 66, S293 (1998), reprinted with permission.

atomic resolution on the MgO surface was perhaps one of the final challenges in imaging bulk insulating surfaces.

One method for circumventing the difficulties of preparing a good insulating sample for imaging is to grow a thin film onto a conducting substrate. This removes the problem of charging and generally allows the preparation of large flat terraces, which can then be imaged via STM or SFM. In SFM, this was most successfully demonstrated for NaCl thin films on Cu (111) [84, 30]. Figure 2.18 shows the so-called Christmas Tree image, with atomic resolution across steps and kinks in the NaCl terrace. The terrace again demonstrates the characteristic cubic crystal contrast pattern, but at stepedges, and especially kink sites, there is a clear increase in brightness. Simulations demonstrated [30] that this is a feature of the reduced coordination of these ions, increasing both the local electrostatic potential gradient and the local atomic displacements.

The improved sensitivity offered by low-temperature SFM has also motivated an effort to measure the exchange force directly, i.e., to measure the different contributions to the tip-surface total force of different local atomic spins. The simple cubic crystal NiO offers the simplest possibility for this type of experiment, since it is antiferromagnetic and its (001) surface presents both spin-up and spin-down Ni ions. Practically, this involves preparing a tip that is spin polarized, such as iron, and then trying to detect the difference in force

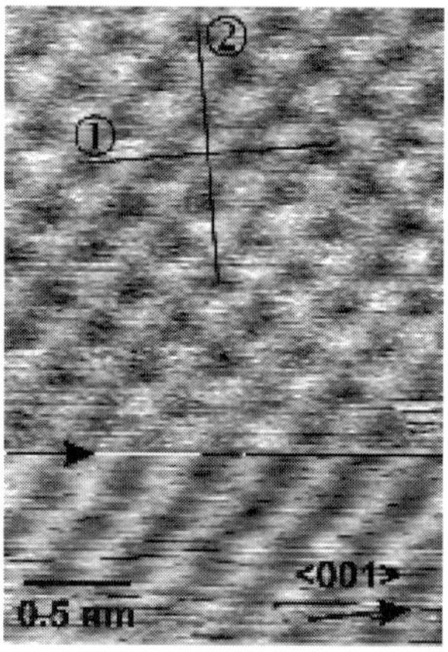

Fig. 2.17. Atomically resolved SFM topographic image of MgO (001) (Δf = -139 Hz, f_0 = 293 kHz, $k \approx 40$ N/m, A = 4 nm, $U_{bias} = -1.4$ V). A tip change in the lower part of the image is marked with an arrow. C. Barth and C. R. Henry, Phys. Rev. Lett. 91, 196102 (2003). Copyright (2003) by the American Physical Society, reprinted with permission.

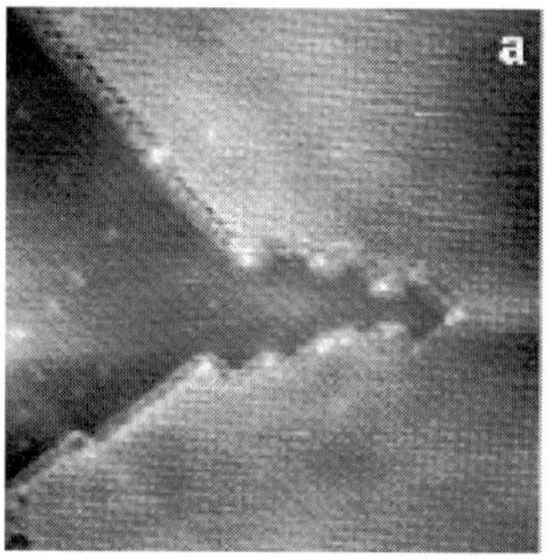

Fig. 2.18. Atomically resolved SFM topographic image of a NaCl thin film on Cu(111) ($\Delta f = -128$ Hz, $f_0 = 158$ kHz, $k = 26$ N/m, A = 1.8 nm, Q = 24,000, U_{bias} = 0 V). R. Bennewitz and A. S. Foster and L. N. Kantorovich and M. Bammerlin and Ch. Loppacher and S. Schär and M. Guggisberg and E. Meyer and A. L. Shluger, Phys. Rev. B 62, 2074 (2000). Copyright (2000) by the American Physical Society, reprinted with permission.

over opposite spin Ni ions. Despite a serious experimental effort [85, 86, 87, 88] this has not yet been achieved, with control of the spin on the tip remaining a difficult problem. However, the efforts have provided a wide selection of experimental images of the surface, and even a full 3D map of the force [88]. Figure 2.19 shows a high-quality image of a step and defect in the surface. As for the previous cubic crystals, interpretation remains difficult [89], although an iron tip would be expected to resolve oxygen as bright. As always, being certain that you have a clean iron tip is difficult.

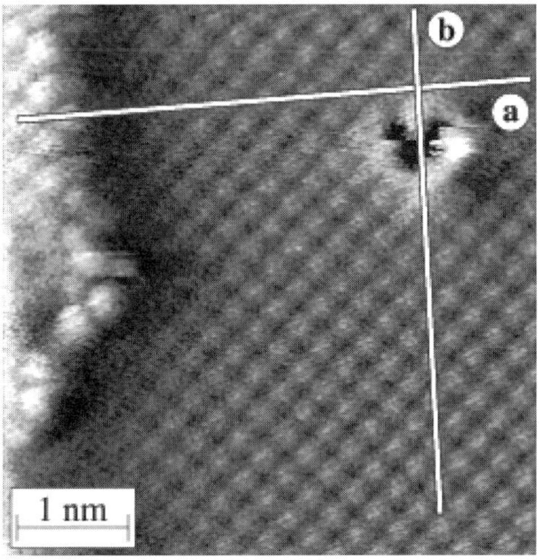

Fig. 2.19. Atomically resolved SFM topographic image of NiO (001) ($\Delta f = -23$ Hz, $f_0 = 201$ kHz, $k \approx 60$ N/m, A = 7.5 nm). W. Allers and S. Langkat and R. Wiesendanger, Appl. Phys. A 72, S27 (2001), reprinted with permission.

References

1. I. Giaever. *Phys. Rev. Lett.*, 5:147, 1960.
2. I. Giaever. *Phys. Rev. Lett.*, 5:464, 1960.
3. G. Binnig and H. Rohrer. *Helv. Phys. Acta*, 55:726, 1982.
4. G. Binnig, H. Rohrer, Ch. Gerber, and E. Weibel. *Appl. Phys. Lett.*, 40:178, 1982.
5. G. Binnig, H. Rohrer, Ch. Gerber, and E. Weibel. *Phys. Rev. Lett.*, 49:57, 1982.
6. G. Binnig, C.F. Quate, and C. Gerber. *Phys. Rev. Lett.*, 56:930, 1986.
7. H-J. Güntherodt, D. Anselmetti, and E. Meyer, editors. *Forces in Scanning Probe Methods.* Kluwer, Dordrecht, 1995.
8. S. Morita, R. Wiesendanger, and E. Meyer, editors. *Noncontact Atomic Force Microscopy.* Springer, Berlin, 2002.

9. D. A. Bonnell. *Probe Microscopy and Spectroscopy: Theory, Techniques and Applications*. Wiley, New York, 2000.
10. R. J. Driscoll, M. G. Youngquist, and J. D. Baledschwieler. *Nature*, 346:294, 1990.
11. F. Ohnesorge and G. Binnig. *Science*, 260:1451, 1993.
12. A. Engel and D. J. Müller. *Nature Structural Biology*, 7:715, 2000.
13. K. D. Jandt. *Surf. Sci.*, 491:303, 2001.
14. P. J. James, M. Antognozzi, J. Tamayo, T. J. McMaster, J. M. Newton, and M. J. Miles. *Langmuir*, 17:349, 2001.
15. T. Aoki, Y. Sowa, H. Yokota, M. Hiroshima, M. Tokunaga, Y. Ishii, and T. Yanagida. *Single Mol.*, 2:183, 2001.
16. A. Philippsen, W. Im, A. Engel, T. Schirmer, B. Roux, and D. J. Mller. *Biophys. J.*, 82:1667, 2002.
17. W. Hebenstreit, M. Schmid, J. Redinger, and P. Varga. *Phys. Rev. Lett.*, 85:5376, 2000.
18. S. Schintke, S. Messerli, M. Pivetta, F. Patthey, L. Libouille, M. Stengel, A. de Vita, and W.-D. Schneider. *Phys. Rev. Lett.*, 87:276801, 2001.
19. R. J. Behm, N. Garcia, and H. Rohrer. *Scanning Tunneling Microscopy and Related Methods*. Kluwer, Dordrecht, 1990.
20. C. J. Chen. *Introduction to Scanning Tunneling Microscopy*. Oxford University Press, Oxford, 1993.
21. S.-W. Hla, L. Bartels, G. Meyer, and K.-H. Rieder. *Phys. Rev. Lett.*, 85:2777, 2000.
22. J. R. Hahn and W. Ho. *Phys. Rev. Lett.*, 87:166102, 2001.
23. K. I. Fukui, H. Onishi, and Y. Iwasawa. *Chem. Phys. Lett.*, 280:296, 1997.
24. A. Sasahara, H. Uetsuka, and H. Onishi. *J. Phys. Chem. B*, 105:1, 2001.
25. T. Nishiguchi, M. Kageshima, N. Ara-Kato, and A. Kawazu. *Phys. Rev. Lett.*, 81:3187, 1998.
26. P. Molinas-Mata, A. J. Mayne, and G. Dujardin. *Phys. Rev. Lett.*, 80:3101, 1998.
27. J. J. Schulz, M. Sturmat, and R. Koch. *Phys. Rev. B*, 62:15402, 2000.
28. L. J. Lauhon and W. Ho. *Phys. Rev. Lett.*, 85:4566, 2000.
29. P. Hoffmann, S. Jeffrey, J. B. Pethica, H. Özgür Özer, and A. Oral. *Phys. Rev. Lett.*, 87:265502, 2001.
30. R. Bennewitz, A. S. Foster, L. N. Kantorovich, M. Bammerlin, Ch. Loppacher, S. Schär, M. Guggisberg, E. Meyer, and A. L. Shluger. *Phys. Rev. B*, 62:2074, 2000.
31. A. Biedermann. *Instrumentelle Optimierung eines Ultrahochvakuum- Rastertunnelmikroskops und Messungen an Graphit- and Silizium- und Platin-Nickel-Einkristall-oberflächen*. Diplomarbeit, Technische Universität, Wien, 1991.
32. M. A. Lantz, H. J. Hug, P. J. A. van Schendel, R. Hoffmann, S. Martin, A. Baratoff, A. Abdurixit, H. J. Güntherodt, and Ch. Gerber. *Phys. Rev. Lett.*, 84:2642, 2000.
33. W. Allers, A. Schwarz, U. D. Schwarz, and R. Wiesendanger. *Rev. Sci. Instrum.*, 69:221, 1998.
34. M. Schmid. *http://www.iap.tuwien.ac.at/www/surface/STM_Gallery*, 1998.
35. Basel. *http://monet.physik.unibas.ch/gue/uhvafm*. 2000.
36. A. L. Shluger, A. I. Livshits, A. S. Foster, and C. R. A. Catlow. *J. Phys.: Condens. Matter*, 11:R295, 1999.

37. F. J. Giessibl. *Jpn. J. Appl. Phys.*, 33:3726, 1994.
38. F. J. Giessibl. *Phys. Rev. B*, 56:16010, 1997.
39. Y. Martin, C. C. Williams, and H. K. Wickramasinghe. *J. Appl. Phys.*, 61:4723, 1987.
40. R. García and R. Pérez. *Surf. Sci. Rep.*, 47:197, 2002.
41. T. R. Albrecht, P. Grütter, D. Horne, and D. Rugar. *J. Appl. Phys.*, 69:668, 1991.
42. F. Moreno-Herrero, J. Colchero, J. Gómez-Herrero, and A. M. Baró. *Phys. Rev. E*, 69:031915, 2004.
43. Courtesy of Omicron Nanotechnology. 2004.
44. M. M. J. Bischoff, C. Konvicka, A. J. Quinn, M. Schmid, J. Redinger, R. Podloucky, P. Varga, and H. van Kempen. *Phys. Rev. Lett.*, 86:2396, 2001.
45. G. A. D. Briggs and A. J. Fisher. *Surf. Sci. Rep.*, 33:1, 1999.
46. F. J. Himpsel, J. E. Ortega, G. J. Mankey, and F. F. Willis. *Adv. Phys.*, 47:511, 1998.
47. K. F. Braun and K. H. Rieder. *Phys. Rev. Lett.*, 88:096801, 2002.
48. R. Wiesendanger and M. Bode. *Solid State Comm.*, 119:341, 2001.
49. J. R. Hahn and W. Ho. *Phys. Rev. Lett.*, 87:196102, 2001.
50. T. Nishino, Ph. Bühlmann, T. Ito, and Y. Umezawa. *Surf. Sci. Lett.*, 490:L579, 2001.
51. M. P. L. Wertz, E. W. van der Vegte, and G. Hadziioannou. *Langmuir*, 13:4939, 1997.
52. B. Capella and G. Dietler. *Surf. Sci. Rep.*, 34:1, 1999.
53. G. Cross, A. Schirmeisen, A. Stalder, P. Grutter, and U. Durig. *Phys. Rev. Lett.*, 80:4685, 1998.
54. M. F. Crommie, C. P. Lutz, and D. M. Eigler. *Phys. Rev. B*, 48:2851, 1993.
55. R. M. Feenstra, J. A. Stroscio, and A. P. Fein. *Surf. Sci.*, 181:295, 1987.
56. T. Eguchi and Y. Hasegawa. *Phys. Rev. Lett.*, 89:266105, 2002.
57. F. J. Giessibl, S. Hembacher, H. Bielefeldt, and J. Mannhart. *Science*, 289:422, 2000.
58. A. S. Foster, W. A. Hofer, and A. L. Shluger. *Curr. Op. Sol. State Mater. Sci.*, 5:427, 2001.
59. A. R. H. Clarke, J. B. Pethica, J. A. Nieminen, F. Besenbacher, E. Laegsgaard, and I. Stensgaard. *Phys. Rev. Lett.*, 76:1276, 1996.
60. W. A. Hofer, J. Redinger, and P. Varga. *Solid State Comm.*, 113:245, 1999.
61. W. A. Hofer, J. Redinger, and R. Podloucky. *Phys. Rev. B*, 64:125108, 2001.
62. W.A. Hofer and J. Redinger. *Surf. Sci.*, 447:51, 2000.
63. D. M. Zehner, J. R. Noolan, H. L. Davis, and C. W. White. *J. Vac. Sci. Tech.*, 18:852, 1981.
64. G. J. R. Jones and B. W. Holland. *Solid State Commun.*, 53:45, 1985.
65. G. Binnig, H. Rohrer, Ch. Gerber, and E. Weibel. *Phys. Rev. Lett.*, 50:120, 1983.
66. V. M. Hallmark, S. Chiang, J. F. Raboli, J. D. Swalen, and R. J. Wilson. *Phys. Rev. Lett.*, 59:2879, 1987.
67. P. Han, E. C. H. Sykes, T. P. Pearl, and P. S. Weiss. *J. Phys. Chem. A*, 107:8124, 2003.
68. D. T. Pierce. *Phys. Scr.*, 38:291, 1988.
69. R. Wiesendanger, H.-J. Güntherodt, G. Güntherodt, R. J. Gambino, and R. Ruf. *Phys. Rev. Lett.*, 65:247, 1990.

70. R. Wiesendanger, D. Bürgler D, G. Tarrach G, A. Wadas, D. Brodbeck, H. J. Güntherodt, G. Güntherodt, R. J. Gambino, and R. Ruf. *J. Vac. Sci. Technol. B*, 9:519, 1991.
71. W. Wulfhekel, H. F. Ding, W. Lutzke, G. Steierl, M. Vazquez, P. Marin, A. Hernando, and J. Kirschner. *J. Appl. Phys.*, 72:463, 2001.
72. M. Bode. *Rep. Progr. Phys.*, 66:523, 2003.
73. M. Bode, R. Pascal, and R. Wiesendanger. *J. Vac. Sci. Technol. A*, 15:1285, 1997.
74. F. J. Giessibl. *Science*, 267:68, 1995.
75. K.Takayanagi, Y.Tanishiro, M.Takahashi, and S.Takahashi. *J. Vac. Sci. Technol.*, A3:1502, 1981.
76. R. Pérez, M. C. Payne, I. Stich, and K. Terakura. *Phys. Rev. Lett.*, 78:678, 1997.
77. M. Huang, M. Čuma, and F. Liu. *Phys. Rev. Lett.*, 90:256101, 2003.
78. N. Oyabu, Ó. Custance, I. Yi, Y. Sugawara, and S. Morita. *Phys. Rev. Lett.*, 90:176102, 2003.
79. M. Bammerlin, R. Lüthi, E. Meyer, A. Baratoff, M. Guggisberg, C. Gerber, L. Howard, and H.-J. Güntherodt. *Probe Microscopy*, 1:3, 1997.
80. M. Bammerlin, R. Lüthi, E. Meyer, J. Lü, M. Guggisberg, C. Loppacher, C. Gerber, and H. J. Güntherodt. *Appl. Phys. A*, 66, 1998.
81. A. I. Livshits, A. L. Shluger, A. L. Rohl, and A. S. Foster. *Phys. Rev. B*, 59:2436, 1999.
82. C. Barth and C. R. Henry. *Phys. Rev. Lett.*, 91:196102, 2003.
83. C. Barth and M. Reichling. *Nature*, 414:54, 2001.
84. R. Bennewitz, M. Bammerlin, M. Guggisberg, C. Loppacher, A. Baratoff, E. Meyer, and H.-J. Güntherodt. *Surf. Interface Anal.*, 27:462, 1999.
85. H. Hosoi, K. Sueoka, K. Hayakawa, and K. Mukasa. *Appl. Surf. Sci.*, 157, 2000.
86. W. Allers, S. Langkat, and R. Wiesendanger. *Appl. Phys. A*, 72:S27, 2001.
87. R. Hoffmann, M. A. Lantz, H. J. Hug, P. J. A van Schendel, P. Kappenberger, S. Martin, A. Baratoff, and H. J. Güntherodt. *Phys. Rev. B*, 67:085402, 2003.
88. S. M. Langkat, H. Hölscher, A. Schwarz, and R. Wiesendanger. *Surf. Sci.*, 527:12, 2003.
89. A. S. Foster and A. L. Shluger. *Surf. Sci.*, 490:211, 2001.

3

Theory of Forces

Here we introduce the theoretical background to the forces important in SPM studies, and try to highlight the particular systems and environments in which certain forces will dominate. In this, we use the standard setups presented in Chapter 2 as a limit on the types of interactions we consider. Hence, for example, we do not discuss the forces important when imaging in liquids [1]. Any separation of forces into categories will be to some extent arbitrary – all forces result from atomic and electronic interactions. However, it is usually convenient to divide the forces according to the length scales on which they are significant: (i) macroscopic forces are those that have a range of at least several nanometers, but are generally chemically independent, (ii) microscopic forces are significant only at ranges of less than 1 nm , but they are much more sensitive to the chemical identity of the atom under the tip.

3.1 Macroscopic forces

3.1.1 Van der Waals force

The *van der Waals (vdW) force* represents the electromagnetic interaction of fluctuating dipoles in the atoms of the tip and surface. On the atomic level, it is one of the weakest interactions, responsible, for example, for bonding in rare gas crystals. It has three main components:

- For neutral atoms, dipoles are created instantaneously due to the fluctuations in the electron charge density, and these induce dipoles in other atoms. The interaction between these instantaneous dipoles is called the *dispersion force* (also London force [2]) and is the most important component.
- In polar molecules, permanent dipoles can induce dipoles in other atoms, and their interaction is called the *induction force* (also Debye force [3, 4]).

- The interaction between permanent dipoles in polar molecules is determined by their orientation, and is termed the *orientation force* (also Keesom force [5]).

This force is nearly always attractive, and therefore small interactions between individual atoms of macroscopic tip and sample sum up to a resulting force on the order of several nanonewtons (nN). Although this force is small by macroscopic standards, it exceeds the chemical forces discussed below and in many cases dominates the tip–surface interaction. The vdW interaction does not vary much as a function of atomic species in comparison to chemical forces, and therefore acts as a long-range macroscopic force.
The full tip contains billions of atoms, and it is impossible to sum all the interactions, but their long-range nature means that it is important to include the full force, and therefore an approximation must be made based on the material and structure of the tip. Assuming that the potential $V(r)$ between two atoms separated by a distance r is known, then the force between them is defined by the gradient of that potential:

$$f(r) = -\nabla V(r). \tag{3.1}$$

For the van der Waals interaction the potential is of the form

$$V(r) = -\frac{C_6}{r^6}, \tag{3.2}$$

where C_6 is the interaction constant as defined by London [2] and is specific to the identity of the interacting atoms. Hamaker [6] then performed the integration of the interaction potential to calculate the total interaction between two macroscopic bodies. Hamaker used the following hypotheses in his derivation:

- additivity: the total interaction can be obtained by the pairwise summation of the individual contributions.
- continuous medium: the summation can be replaced by an integration over the volumes of the interacting bodies assuming that each atom occupies a volume dV with a number density ρ.
- uniform material properties: ρ and C_6 are uniform over the volume of the bodies.

This then allows the total force between two arbitrarily shaped bodies to be given by

$$F_{vdW} = \rho_1 \rho_2 \int_{v_2} \int_{v_1} f(r) dV_1 dV_2, \tag{3.3}$$

where ρ_1 and ρ_2 are the number densities and V_1 and V_2 are the volumes of bodies 1 and 2 respectively. Hamaker then introduced a constant H, known as the "Hamaker constant", which characterizes the resonance interactions between electronic orbitals in two particles and the intervening medium in much the same way as polarizability does in the case of two atoms. The Hamaker constant depends on the properties of both the particles (geometry and material) and the medium. The Hamaker constant for the general interaction is then

$$H = \pi^2 C_6 \rho_1 \rho_2. \tag{3.4}$$

The assumption of additivity is not always appropriate, since the presence of other atoms changes the effective polarizability of a single atom. This problem can be avoided by using the Lifshitz theory [7], where the Hamaker constant is now calculated from the dielectric and optical properties of the materials. This provides a much more accurate estimate of H, but in practice obtaining the necessary dielectric and optical information for a real system is very difficult. Regardless of the method used to determine H, the vdW force is finally determined as a function of the distance for a given tip shape. Many analytical expressions have been derived for different tip shapes [1], and here we will give a few examples that have relevance in the context of SPM. For a sphere of radius R at a distance D from the surface the force is

$$F(D) = -\frac{2HR^3}{3D^2(D+2R)^2}, \tag{3.5}$$

for a pyramidal tip the force is

$$F(D) = -\frac{2H\tan^2\theta}{3\pi D}, \tag{3.6}$$

where θ is the angle between the rotational axis and the edge of the pyramid, and for a conical tip of angle γ and radius R the total force is given by [8]

$$\begin{aligned} F(D) = &\frac{HR^2(1-\sin\gamma)(R\sin\gamma - D\sin\gamma - R - D)}{6D^2(R+D-R\sin\gamma)^2} - \\ &\frac{H\tan\gamma[D\sin\gamma + R\sin\gamma + R\cos(2\gamma)]}{6\cos\gamma(D+R-R\sin\gamma)^2}. \end{aligned} \tag{3.7}$$

Retardation effects

When two atoms are a significant distance apart, the time taken for the electric field due to an instantaneous dipole of the first atom to reach the second atom and return can be greater than the period of the dipole fluctuations. The dispersive interaction can now be repulsive rather than attractive, and in fact beyond a separation of 100 nm the vdW force begins to decay at r^{-7} rather than r^{-6}. This process is known as the retardation effect, and can affect interactions in a vacuum beyond 5 nm (and even closer for interactions in a medium).

3.1.2 Image forces

The image force is the interaction due to the polarization of the conducting electrodes (i.e., of the conducting tip and the substrate) by the charged atoms of the sample. This is important for any tip–surface (or just surface) setup containing conducting materials, e.g., STM, interaction of a conducting tip with an insulating surface in SFM or in studying the properties of an insulating thin film on top of a metal substrate. As for the vdW force, the image force is generally not atom specific, and is therefore important mainly as a contribution to the overall force. However, it has been shown to dominate interactions for certain conditions in SFM [9], and the induced changes in electronic structure are likely to influence STM [10].

Assuming that the potential on conducting electrodes is maintained by external sources (i.e. by the *battery*), then from the point of view of classical electrostatics the polarization of the conductors by external charges is caused by the additional potential on the conductors due to the charges. This extra potential is compensated by a charge flow from one electrode to another to keep the potential on the conductors fixed. This work is done by the battery. As a result, there will be some distribution of the net charge on the surfaces of conductors induced by the point charges situated in the free space between them. The net charge on each conducting electrode will interact with the total charge on other conductors and with the point charges.

Following the derivation in [11], the image interactions introduce an additional energy to the system:

$$U_{el} = -\frac{1}{2}QV + \sum_i q_i\phi(\mathbf{r}_i) + \frac{1}{2}\sum_{i,j} q_i q_j \phi_{ind}(\mathbf{r}_i, \mathbf{r}_j) \qquad (3.8)$$

where q_i is the charge and $\mathbf{r}_i$ the position of atom i, V is the potential difference applied to the metal electrodes, Q is the charge on the tip before polarization, ϕ is the electrostatic potential of the bare electrodes anywhere outside the metals, and ϕ_{ind} is the potential at $\mathbf{r}_i$ due to image charges *induced* on the metals by a unit point charge at $\mathbf{r}_j$. This extra contribution to the total energy can be added self-consistently to calculations [9] to determine its effect on structure and contribution to the total tip-surface force.

3.1.3 Capacitance force

If electrons are allowed to flow between two different conducting materials there will be a contact potential U between them, as the electrons lose energy in the transfer from the material with the smaller work function to the material with the higher one. This effect is exactly the same as that discussed in the previous section for the image force, but now the difference in potential between the tip and surface is due to the contact potential, U, as well as the applied bias V. In effect, in calculating the image force with an applied bias, capacitance force is included as a component of the overall force, and therefore this capacitance force is present in all calculations that include the image force component. However, it is useful to be able to calculate an analytical approximation of the capacitance force for macroscopic systems. The difference in surface potential of the two materials produces an electrostatic energy of the form

$$E_{elec} = \frac{1}{2} C U^2(x, y), \tag{3.9}$$

where C is the tip–sample capacitance. This can be differentiated with respect to tip–surface separation, z, to give the capacitance force between them:

$$F(x, y, z) = \frac{1}{2} \frac{dC}{dz} U^2(x, y). \tag{3.10}$$

The main difficulty in evaluating this expression is in finding a physical expression for $C(z)$ for the real tip–shape. Numerical methods can give an exact value for the force, but they do not allow variations of tip size and curvature to be studied. An approximate analytical method [12] has been developed that allows the capacitance of an axisymmetric tip to be given as [13]

$$C(z) = \frac{1}{U} \int_{tip} 2\pi \rho_s'(z') \sigma_s(z') dz', \tag{3.11}$$

where ρ_s is the analytical surface equation of the tip and σ_s is the surface charge density. For a spherical tip of radius R, the capacitance force is given by

$$F_c(z) = -\frac{\epsilon \epsilon_0 \pi R U^2}{z}, \tag{3.12}$$

where ε_0 is the dielectric constant of vacuum and ϵ is the dielectric constant of the medium.

The importance of the capacitance force due to the tip–surface interaction depends critically on the tip/surface properties and experimental setup. If there is a significant potential difference between the tip and surface, then the capacitance force is an important contribution to the interactions. As stated above, a large potential difference may exist if there is a significant difference between the work functions of the tip and surface material or a large bias is applied in the experiment. However, bias in SFM is normally applied to minimize the effect of work function differences, so capacitance forces due to contact potential and applied bias should in principle cancel each other. This electrostatic minimization process is not well defined, and its success in canceling the capacitance force due to work function differences is not clear. In light of this it is important to understand how the capacitance force compares with other interactions. For a metal surface and a conical silicon tip (e.g., SFM tip), with a potential difference of about 1 V, the capacitance force will dominate tip–surface interactions beyond a separation of about 6 – 7 nm and is comparable to the van der Waals force at about 5 nm.

Work function anisotropies

The discussion of capacitance force above makes an assumption about surfaces that is not always valid. By calculating the capacitance force as a function of z alone, it is assumed that the work function is uniform across the surface. On real surfaces, inequivalencies in the work function across the surface can arise due to surface preparation, adsorbates, crystallographic orientation, and variations in local geometry [14, 15]. Real surfaces of any material are not perfectly smooth; in fact, they are very rough on the micro-scale, and this roughness can lead to inhomogeneities in the surface charge density and work function.

This is especially relevant for the electrostatic minimization procedure used in SFM experiments, since this minimizes the electrostatic forces at a single point on the surface before scanning. Variations in the work function over the region scanned could render the minimization process invalid, or at least approximate. Other studies [14] have already suggested that work function anisotropies are the most likely source of the long-range interactions observed in force microscopy of graphite with diamond tips.

The contribution of work function anisotropies to the tip–surface interaction cannot be calculated explicitly however, they can be represented by increased surface charge density or increased/decreased applied bias in calculating the image force contribution [14].

3.1.4 Forces due to tip and surface charging

Many processes can introduce charge into the tip and surface. Surface preparation by cleavage is known to induce very large charges on insulating surfaces [16, 17, 18, 19], although these can be reduced by annealing. Surface sputtering can also cause charging, as can ion exchange between tip and surface during movement of the tip across the surface (tribocharging). On the microscopic scale, these charge defects can appear as unexpected interactions close to the surface. On the macroscopic scale, tip and surface charging can dominate the interactions. It is commonly assumed that the very large attractive forces that make stable SFM imaging of some insulating oxides difficult, e.g., MgO, are due to significant surface charging after cleavage.

Charging is limited to insulating materials, where the charge density is localized around ion positions and the added charge cannot conduct away. This means that it is not relevant for imaging of metal surfaces, nor for tips that are pure conductors. The charge–charge interaction for a neutral surface, where all the charged defects have been compensated without atomic displacement, decays exponentially into the vacuum and will introduce a contribution to the tip–surface force only at small tip–surface separations. However, charged defects in the surface usually cause atoms to move from their ideal lattice sites, creating dipoles within the surface. Charge–dipole and dipole–dipole interactions [14] have much longer range than charge–charge interactions, and they can introduce electrostatic contributions of very long-range to the tip–surface interactions. It should also be noted that for systems with conducting materials, any charging will also change the image force. Charging of the tip will greatly increase the magnitude of image charges produced in the conductors and hence the image force [9].

3.1.5 Magnetic forces

Magnetic forces are really important only when both the tip and sample demonstrate magnetic behavior, e.g., when both are ferromagnets. For a ferromagnetic tip and sample, the magnetic force contribution can be calculated by first estimating, theoretically or experimentally, the magnetic moment of the tip and then applying

$$[F = \nabla(m.B)], \tag{3.13}$$

where m is the magnetic moment and B the magnetic flux density. For a setup with a ferromagnetic tip and a paramagnetic/diamagnetic sample the force will be due to the interaction of the induced moment in the sample and the diverging field of the tip.

3.1.6 Capillary forces

SPM experiments in air must also consider the role that atmospheric humidity plays in the tip–surface interaction; it is important to realize that in UHV conditions this force component is absent. The presence of liquid water layers on the tip and/or surface can introduce some discontinuous behavior in their interaction. Aside from modifying other interactions, at short–range the liquid layers will "jump into contact", forming a bridge of large meniscus radius between them. This layer will then compress until "hard-contact" between the tip and surface. Further movement of the tip, e.g. removal, will stretch the meniscus until it breaks, with the breaking point determined by the original layer thickness.

If the thickness of the liquid layer is negligible, or the system is in the presence of a condensable vapor, then the effects will be more subtle. For a "dry" tip-surface system, for example, in vacuum or nitrogen, the adhesion force will be due mostly to dispersion forces, and $F_{ad} = 4\pi R\gamma_s$, where γ_s is the surface energy of the solids [1]. If some vapor is introduced, the surface energy will be modified by adsorption and at some relative vapor pressure capillary condensation will occur. The force between the tip and surface is then given by (s, l, v denote solid, liquid, vapor):

$$F_{ad} = 4\pi R\gamma_{lv}\cos\theta + 4\pi\gamma_{sl}, \tag{3.14}$$

where γ_{lv} is the surface tension of the liquid in the condensate, θ is the liquid contact angle, and γ_{sl} is the solid–liquid interfacial tension. The first term is due to the capillary pressure in the liquid bridge, and the second term is associated with the solid–solid interaction across the bridge. Generally, the tip–surface adhesion force is less in vapor than in vacuum, although, as discussed, discontinuous behavior in the interactions is also characteristic of the presence of capillary forces. More advanced treatments of capillary forces have been considered in, for example, [20, 21, 22, 23].

3.2 Microscopic forces

Chemical forces rarely dominate the total force in SPM, yet they remain the most crucial interactions for understanding experimental images. They define the atomic structure of the tip and surface, and are responsible for atomic displacements when the tip is close to the surface. The nature of chemical bonds that form in and between, the tip and surface are intrinsically linked to the chemical forces. In SFM they distinguish atomic identities and are therefore responsible for atomic resolution in images. In STM, if a chemical bond forms between the tip and surface, its energetic level may be lower than the conductance band of the leads (see Chapter 2 STM setup) preventing tunneling.
Due the sensitivity of experiments on these interactions, they are always calculated explicitly and are the only force that cannot usefully be approximated by a continuum model. However, the level of complexity required to calculate the chemical forces depends on the properties and materials in the SPM system. The actual physical components of microscopic forces are basically the interactions between nuclei and electrons in the system, in principle requiring an exact solution of the electronic many-body wavefunction for a given atomic geometry. However, it is generally useful to separate different interaction classes according to the systems in which they dominate:

- Electrostatic forces: Coulomb interaction between ions in the tip and sample. For an ionic surface and an ionic tip the electrostatic force between ions will usually dominate the microscopic forces.
- Polarization forces: polarization of electron-cloud by ions. This is especially relevant when conducting materials, which are highly polarizable, are interacting with insulating materials.
- Van der Waals forces: the microscopic version of the force discussed in the previous section, generally much weaker than the other forces at this scale, but important in imaging inert surfaces like Xenon [24] or in considering the physisorption of inert species.

- Chemical bonding: in the case that the system's materials cannot be well approximated as ideally ionic or inert, it becomes important to take account of chemical bonds that may form between the tip and surface. This is especially important in considering the interactions of reactive tips and surfaces [25], where the need to saturate dangling bonds results in strong tip–surface bonds and correspondingly large microscopic forces.
- Magnetic forces: on the microscopic scale, magnetic forces represent the exchange force between atomic spins in the tip and surface. For a spin polarized tip scanning a magnetic surface, the exchange force will vary according to the spin–state of the atom under the tip.

3.2.1 Theoretical methods for calculating the microscopic forces

Empirical modeling

For highly ionic insulating tips and surfaces (such as CaF_2 in Chapter 7), the chemical forces are dominated by the Coulomb interaction between the ions, and charge transfer processes do not play a significant role. In this case, the chemical forces can be well represented by atomistic simulation (AS) empirical methods , such as the shell model (SM) [26]. In this technique atoms are represented by point charges connected by springs to a massless charged shell. This shell can move independently of the central charge to simulate polarization of the atom. The interactions between cores and shells are controlled by empirical potentials whose parameters are fitted to achieve the best possible comparison with experiment or *ab initio* techniques. The potentials are usually derived from three interactions: (i) electrostatic Coulomb interactions between the atoms (cores and shells), (ii) van der Waals interactions and (iii) short-range repulsive interactions. The charge–charge electrostatic interaction between atoms i and j is given as the sum of four terms:

$$V_i^{elec} = \sum_j^n \frac{q_i q_j}{4\pi\varepsilon_0 \left|\mathbf{r}_{si} - \mathbf{r}_{sj}\right|} + \sum_j^n \frac{Q_i Q_j}{4\pi\varepsilon_0 \left|\mathbf{r}_{ci} - \mathbf{r}_{cj}\right|} + \sum_j^n \frac{Q_i q_j}{4\pi\varepsilon_0 \left|\mathbf{r}_{ci} - \mathbf{r}_{sj}\right|} + \sum_j^n \frac{q_i Q_j}{4\pi\varepsilon_0 \left|\mathbf{r}_{si} - \mathbf{r}_{cj}\right|} \tag{3.15}$$

where $i \neq j$, n is the number of atoms, q_i is the shell charge of atom i, Q_i is the core charge of atom i, $\mathbf{r}_{si}$ is the position vector of the shell of atom i, and $\mathbf{r}_{ci}$ is the position vector of the core of atom i. An example of the non-Coulombic short-range interactions between the shells is the Buckingham two-body potentials. These potentials have the following form:

$$V_i^{short} = \sum_j^n \left(-C \left|\mathbf{r}_{si} - \mathbf{r}_{sj}\right|^{-6} + A e^{-\frac{\left|\mathbf{r}_{si} - \mathbf{r}_{sj}\right|}{\rho}} \right), \tag{3.16}$$

where C, A, and ρ are parametrized constants specific to each pair of shells i and j, and $i \neq j$. Note that for some atoms there is no shell, and all references to distance apply to the position of the core instead. The first term in equation (3.16) represents the attractive van der Waals interaction, and the second term the short-range repulsion due to electron cloud overlap. For shells there is also an additional contribution to the interaction due to the elastic force in the spring connecting core and shell. This force is equal to $k\delta r_i$, where k is the parameterized spring constant between a core–shell pair, and δr_i is the distance between the centers of core and shell for atom i. The spring interaction between the cores and shells is given by

$$V_i^{spring} = \frac{1}{2} k \delta r_i^2 . \tag{3.17}$$

Combining equations (3.15), (3.16), and (3.17) gives the total energy of the system as

$$E = \frac{1}{2} \sum_i^n \left[V_i^{elec} + V_i^{short} + 2V_i^{spring} \right] . \tag{3.18}$$

This can then be minimized with respect to core and shell positions to find the equilibrium geometry of relaxed atoms in the system. Usually certain atoms within the tip–surface unit cell remain frozen to represent the interface between the macroscopic and microscopic features. For infinite systems the unit cell is repeated, according to the system lattice vectors, across space until the atomic interactions converge to the desired accuracy. For bulk samples the cell is repeated in three dimensions, but for surfaces two possibilities exist. One method for calculating surfaces is to cut the infinite bulk system and create a series of slabs that are infinite in two dimensions but separated in the third dimension by a large vacuum gap. These slabs are a good model of a surface if the gap is large enough that there is no significant interaction between the slabs. Another method for calculating surfaces is just to repeat the cell in two dimensions, directly generating a real infinite surface. Note that the electrostatic interaction converges conditionally for an infinite system, and methods such as Ewald summation [27] must be used to calculate this contribution.

Practically, AS calculations are very cheap, and hundreds of atoms can be simulated efficiently on a desktop PC. However, the nature of the parameterization means that the interactions can be somewhat inflexible, and one must be careful when applying them beyond their design. For example, it is often the case that parameters that give excellent results for the bulk properties of a material fail miserably when applied to its surface.

Ab initio modeling

In STM we are always concerned with conducting or semiconducting materials, and the chemical forces cannot be represented in any simple AS method.

The same is true in SFM if we are studying conducting surfaces or tips. In this case more complex theoretical methods that represent the full electronic charge density are required to accurately reflect the chemical bonding and interactions in these strongly covalent or metallic systems. Although intermediate semi-empirical methods exist [27], generally only first-principles techniques can be reliably applied in this regime.

These first principles, or *ab initio*, methods attempt to solve the many-body Schrödinger equation:

$$H\Psi_i(1, 2, ..., N) = E_i\Psi_i(1, 2, ..., N), \tag{3.19}$$

where H is the Hamiltonian of a quantum-mechanical system composed of N particles, Ψ_i is its ith wavefunction, and E_i is the energy eigenvalue of the ith state. The particle coordinates $(1, 2, ..., N)$ are usually associated with a spin and a position coordinate. For electronic systems with nonrelativistic velocities the Hamiltonian for an N-electron system is

$$H = -\frac{1}{2}\sum_{i=1}^{N}\nabla_i^2 + \sum_{i>j}^{N}\frac{1}{|\mathbf{r}_i - \mathbf{r}_j|} + \sum_{i=1}^{N} v(\mathbf{r}_i), \tag{3.20}$$

where the first term of equation (3.20) represents the electron kinetic energy, the second term the electron–electron Coulomb interactions, and the third term the coulomb potential generated by the nuclei. This equation also assumes that the nuclei are effectively stationary with respect to electron motion (Born–Oppenheimer approximation).

Most theoretical approaches to SPM apply the density functional theory (DFT) to solve this problem. In contrast to other methods (such as Hartree-Fock) which try to determine approximations of the electron density or many-electron wavefunction, DFT can "exactly" calculate any ground-state property from the electron density [28]. If we consider the ground state of the electron–gas system in an external potential $v(\mathbf{r})$, the following density functional theorem holds exactly: There is a universal functional $F[\rho(\mathbf{r})]$ of the electron charge density $\rho(\mathbf{r})$ that defines the total energy of the electronic system as

$$E = \int v(\mathbf{r})\rho(\mathbf{r})d\mathbf{r} + F[\rho(\mathbf{r})]. \tag{3.21}$$

The energy of the system can be minimized to find the true electron charge density in the external potential. This theory is exact for a nondegenerate ground state. Unfortunately, as yet an exact general form of the functional $F[\rho(\mathbf{r})]$ has not been found, so approximations, such as the local density approximation (LDA) [29], must be used. From the electronic ground-state solution the forces on the atoms can be calculated, and used to relax the entire atomic configuration until a preset force or energetic convergence limit is reached.

The setup of such calculations is more or less identical to that of the AS simulation, but now no parameterization is required, and accurate calculations require only the atomic number of the elements involved. However, these simulations are orders of magnitude more computationally expensive than AS methods, and calculating a few hundred atoms requires supercomputer resources. In practice, some further approximations and parameterizations can be made to increase efficiency without affecting accuracy, yet full first-principles SPM image simulation remains very resource-consuming.

3.3 Forces due to electron transitions

In the previous sections we considered only those interactions in which the tip and surface are decoupled at the atomic scale, or at most a bond exists between them. However, the situation is different if it is possible for electron transitions between tip and surface; a situation essential for the operation of an STM, but also possible when one is using a conducting tip in SFM. For a brief review of this effect, let us consider the change of physical processes as the tip approaches the surface. A conducting surface and an equally conducting probe will be completely decoupled if the distance between the surface atoms and the foremost atom of the probe (apex atom) is substantially greater than 1 nm. Then the electron states of surface and probe are orthogonal: every product

$$(\chi_\nu^*, V\psi_\mu) := \int d^3r \chi_\nu^*(\mathbf{r})V(\mathbf{r})\psi_\mu(\mathbf{r}) \tag{3.22}$$

will be zero. If the two systems are brought into closer contact, with a distance of about 0.5 – 0.6 nm, the presence of the other lead will have an effect on the electronic structure and electron dynamics on both sides. The two systems in this case are weakly coupled, and the change of the physical situation compared to the long distance range can be described by a perturbation potential V. The two main effects occurring in this range are:
(i) A transition of electrons from one side to the other, the transition rate given by Fermi's golden rule [30]:

$$\Gamma_{\mu\nu} = \frac{2\pi}{\hbar} \left|(\chi_\nu^*, V\psi_\mu)\right|^2 \delta\left(E_\nu - E_\mu\right). \tag{3.23}$$

This relation is equivalent to Bardeen's formulation of the tunneling problem [31, 32]. The reason for using the Bardeen formulation rather than Fermi's golden rule in the calculation of tunneling currents [33] is a technical one: the perturbation potential due to the approach of a surface and a probe is commonly unknown.
For a finite system with a discrete set of eigenvalues, or for nonzero temperatures, the delta function has to be replaced by a smeared-out function, for

example a Gaussian of half-width σ; thus the tunneling current $I_{\mu\nu} = e\Gamma$ for a single transition is described by

$$I_{\mu\nu} = \frac{2\pi e}{\hbar\sigma\sqrt{\pi}} \left|(\chi_\nu^*, V\psi_\mu)\right|^2 e^{-(E_\nu - E_\mu)^2/\sigma^2}. \tag{3.24}$$

(ii) The second effect is a change of the system energy. In a second-order perturbation expansion, the change in the energy of a filled state χ_ν is given by

$$-\Delta E_\nu^{(2)} = (\chi_\nu^*, V\chi_\nu) + \sum_\lambda \frac{\left|(\chi_\nu^*, V\chi_\lambda)\right|^2}{E_\nu - E_\lambda} + \sum_\lambda \frac{\left|(\chi_\nu^*, V\psi_\lambda)\right|^2}{E_\nu - E_\lambda}, \tag{3.25}$$

where the sum goes in principle over all empty states χ_λ and ψ_λ. The first term is the change of the eigenvalue due to the coupling potential V, the second term describes the changes due to transitions between states on one side of the barrier only, while the final term describes transitions across the tunneling barrier between tip and sample states. Only the third type will contribute to the interaction energy between the two surfaces, because the wavefunctions and the potential are exponentially decreasing and the perturbing potential for states of the tip is the potential of the sample surface. Focusing on the energy contribution due to a single pair of states (μ, ν), we obtain within perturbation theory

$$-\Delta E_{\mu\nu} = \frac{\left|(\chi_\nu^*, V\psi_\mu)\right|^2}{E_\nu - E_\mu}. \tag{3.26}$$

Comparing (3.26) and (3.24), we find for the relation between the current $\Delta I_{\mu\nu}$ and interaction energies $\Delta E_{\mu\nu}$ the following expression:

$$\Delta I_{\mu\nu} = \frac{2\pi e}{\hbar\sigma\sqrt{\pi}} (E_\nu - E_\mu)\, e^{-(E_\nu - E_\mu)^2/\sigma^2} \Delta E_{\mu\nu}. \tag{3.27}$$

Note that interaction energies contribute a negative term to the total energy of interacting systems. But to compare with the positive tunneling current we use their absolute values.

In this formulation the tunneling current from a single transition ($\Delta I_{\mu\nu}$) appears to be proportional to its contribution to the interaction energy ($\Delta E_{\mu\nu}$). Due to level broadening, the energy difference $E_\nu - E_\mu$ is of the same scale as σ. Consequently, we set $E_\nu - E_\mu \approx \sigma$. Then the above relation gives

$$\Delta I_{\mu\nu} \approx \frac{2e\sqrt{\pi}}{2.718 \cdot \hbar} \Delta E_{\mu\nu} \approx 4 \cdot \frac{2e}{h} \Delta E_{\mu\nu}. \tag{3.28}$$

Interestingly, the relation then is very similar to the Landauer–Büttiker formulation of the tunneling problem [34]. Although the right-hand side of equation (3.27) does not appear to contain the transition matrix T, it implicitly appears because T, like the interaction energy $\Delta E_{\mu\nu}$, is proportional to

$|(\chi_\nu^*, V\psi_\mu)|^2$. Consequently, as shown by Feuchtwang and others [35], the perturbation treatment is the lowest-order term in a full scattering treatment of the transport. However, numerical results and experimental evidence suggest that the proportionality between net current and total interaction energy holds even beyond the domain of validity of this approximation.
This may be understood by writing the interaction energy as

$$E_{\text{int}} = Tr[\rho V] = -\frac{1}{\pi}\int_{-\infty}^{\mu} G^r(E)VdE, \tag{3.29}$$

where ρ is the one-electron density matrix and G^r the retarded one-electron Green function. The part of E_{int} coming from the mixing of tip and sample states involves the off-diagonal elements of ρ (and hence of G^r) linking tip and sample. These are determined by Dyson's equation, and hence to lowest order in V,

$$G^r = G_0^r + G_0^r V G_0^r. \tag{3.30}$$

Thus the interaction energy to lowest order is

$$E_{\text{int}} = -\frac{1}{\pi}\int_{-\infty}^{\mu} Tr[G_0^r V G_0^r V], \tag{3.31}$$

which (like the total current) is quadratic in V. Every current thus corresponds to interactions between the surface and the tip, and the relation between currents and interaction energies is in general linear. This is in marked contrast to the treatment suggested by Julian Chen [32], which was based on the hydrogen molecule, who predicted that the relation should be quadratic. Extensive numerical calculations showed that Chen's prediction is untenable. However, the ratio between current and interaction energy depends on the detailed electronic structure of surface and tip, since bond formation occurs not only in the small energy window defined by the bias voltage, but across the whole spectrum of eigenvalues. We shall present a unified model of scattering and perturbation based on the Keldysh nonequilibrium formalism in Chapter 5 and show, that the ratio α can be exactly predicted to any order of accuracy. Since this involves a rather technical derivation of current and interaction energy based on electron transport theory, the linearity, which has been verified with great accuracy by simultaneous current and force measurements [36, 37], is only stated, but not fully proved at this point.

Simulating dynamic contours

If α is known, then the *dynamic constant current contour* can be calculated in a rather simple manner. Given that

$$\Delta E(z) = -\alpha \Delta G(z) = -\alpha \frac{\Delta I(z)}{V}, \tag{3.32}$$

z being the vertical distance between tip and surface atoms, $\Delta G(z)$ the conductance, the atoms of the surface will be dislocated by Δz from their ground-state positions due to the onset of chemical bonding. Within a harmonic approximation we may state

$$\Delta E(z) = -k_{harm}\Delta z^2(z). \tag{3.33}$$

Since the distance-dependent current is in general exponentially decaying with the distance

$$I(z) = I_0 \exp(-\kappa z), \tag{3.34}$$

the modified current $I_F(z)$ can be calculated by

$$I_F(z) = I(z) \exp \kappa \sqrt{\frac{\alpha I(z)}{V k_{harm}}}. \tag{3.35}$$

Apart from the constant α, which needs to be determined by model calculations, the necessary input comprises also the elastic constant of surface atoms and the decay constant κ. All of these values are readily available from simulations. The elastic constant, for example, is calculated in the standard manner by shifting the atomic positions of the surface by a few percent of the interlayer spacing and calculating the ensuing forces on the atom with DFT methods. The current decay can be obtained from the calculation of the currents itself. It should be noted that so far, we have considered only one position of the STM tip, the on-top position. If the tip scans the region between surface atoms, the equation needs to be modified. The reason for the modification is the following. Consider a position of the tip slightly off center of a surface atom. In this case the relaxation will no longer be along the direct line of the force, since surface atoms are restricted to vertical motion. Therefore, the actual relaxation will be projected onto the vertical axis and described by a relation close to the cosine of the angle between the vertical axis and the line of force. If the tip is located in a hollow site of the surface, the interaction energy will be statistically distributed between the adjacent atoms. On a (111) surface, it will thus be about one-third, on a (100) surface about one-fourth, of the original value. This feature amounts to a geometric factor, called P for projection, which has to be included in the equation:

$$I_F(z) = I(z) \exp \kappa \sqrt{\frac{P(r_\parallel)\alpha I(z)}{V k_{harm}}}. \tag{3.36}$$

P can be approximated by the following expression:

$$P(r_\parallel) = a\left[1 - b \cdot \tanh\left(4 \cdot r_\parallel/d_0 - 2\right)\right], \tag{3.37}$$

where d_0 is the distance between the on-top site and the nearest hollow site, and a and b are parameters depending on the symmetry of the surface. This expression has the advantage that P varies smoothly with the position $r_{\|}$ of the tip, and that its differential at $r_{\|} = 0$ and $r_{\|} = d_0$ is close to zero. The exact shape of a contour between the two points depends on the arguments of the hyperbolic tangent. In a series of model calculations this shape was compared to the shape of atoms in experimental scans. The arguments given in the above equation reflect these comparisons. However, it should be noted that the suitable range is very small and that experimental scans are in general highly variable with regard to the exact shape of a maximum in a line scan.

3.4 Summary

In this chapter we have tried to establish those forces that are likely to be relevant in an SPM experiment, and discussed those systems where specific forces are especially significant. In order to provide an accessible overview of these different forces, Table 3.1 lists them all, along with the systems and techniques where they are relevant. Note that most of these forces are present in all systems, both in STM and SFM, but in the table we highlight those interactions that may influence the measurement itself, e.g., capacitance forces are present in STM, but do not influence the tunneling current significantly.

Force	STM/SFM	System
Van der Waals	SFM	Present in all systems, but can be damped in other media, e.g., water
Image	Both	Present when both conducting materials and localized charges are interacting , e.g., metal tip and insulating surface
Capacitance	SFM	Present if an electrical connection exists between tip and sample
Due to charging	SFM	Present if tip and surface contain trapped charges, e.g., in insulating materials
Magnetic	Both	Present if tip and/or sample are magnetic
Capillary	Both	Present if there is nonzero humidity, e.g., experiments are not performed in UHV
Microscopic	Both	Always present, although dominance of a particular flavor depends on the properties of tip and surface, e.g., microscopic van der Waals dominates in imaging of graphite
Due to electron transitions	Both	Present if there is direct electron transport between the tip and surface

Table 3.1. Comparison of the technique and system relevance of various forces in SPM.

It is important to emphasize that although STM measures tunneling current and SFM measures forces, the phenomena are not mutually exclusive. Tip–

surface forces will affect STM measurements, especially the microscopic forces between atoms in the tip and surface at close approach. For example, tip-induced displacements of surface species will strongly change the measured current at a given distance. In SFM, electron transitions between tip and surface must also be considered, particularly when one is imaging with a conducting tip.

References

1. J.N. Israelachvili. *Intermolecular and Surface Forces.* Academic Press, London, 1991.
2. F. London. *Trans. Faraday Soc.*, 33:8, 1937.
3. P. J. W. Debye. *Phys. Z.*, 21:178, 1920.
4. P. J. W. Debye. *Phys. Z.*, 22:302, 1921.
5. *Communications Physical Laboratory.* Leyden, Holland, 1912.
6. H. C. Hamaker. *Physica*, 4:1058, 1937.
7. E. M. Lifshitz. *Sov. Phys. JETP*, 2:73, 1956.
8. C. Argento and R. H. French. Parametric tip model and force-distance relation for hamaker constant determination from atomic force microscopy. *J. Appl. Phys.*, 80:6081, 1996.
9. L. N. Kantorovich, A. S. Foster, A. L. Shluger, and A. M. Stoneham. Role of image forces in NC-SFM images of ionic surfaces. *Surf. Sci.*, 445:283, 2000.
10. J. M. Pitarke, P. M. Echenique, and F. Flores. *Surf. Sci.*, 217:267, 1989.
11. L. N. Kantorovich, A. I. Livshits, and A. M. Stoneham. Electrostatic energy calculation for the interpretation of surface probe microscopy experiments. *J. Phys.: Condens. Matter*, 12:795, 2000.
12. S. Hudlet, M. Saint Jean, C. Guthmann, and J. Burger. *Eur. Phys. J.*, 2:5, 1998.
13. M. Saint Jean, S. Hudlet, C. Guthmann, and J. Berger. *J. Appl. Phys.*, 86:5245, 1999.
14. N.A. Burnham, R.J. Colton, and H.M. Pollock. Interpretation of force curves in force microscopy. *Nanotechnology*, 4:64, 1993.
15. N.A. Burnham, R.J. Colton, and H.M. Pollock. *Phys. Rev. Lett.*, 69:144, 1992.
16. J. B. Engelhardt, H. Dabringhaus, and K. Wandelt. *Surf. Sci.*, 448:187, 2000.
17. M. Luna, F. Rieutord, N. A. Melman, Q. Dai, and M. Salmeron. *J. Phys. Chem. A*, 102:6793, 1998.
18. L. B. Harris and J. Fiasson. *J. Phys. C: Solid State Phys.*, 18:4845, 1985.
19. H. J. Wintle. *Meas. Sci. Technol.*, 8:508, 1997.
20. J. Jang, G. C. Schatz, and M. A. Ratner. *J. Chem. Phys.*, 120:1157, 2004.
21. Y. I. Rabinovich, J. J. Adler, M. S. Esayanur, A. Ata, R. K. Singh, and B. M. Moudgil. *Adv. Coll. Int. Sci.*, 96:213, 2002.
22. T. Stifter, O. Marti, and B. Bhushan. *Phys. Rev. B*, 62:13667, 2000.
23. O. H. Pakarinen, A. S. Foster, M. Paajanen, T. Kalinainen, J. Katainen, I. Makkonen, J. Lahtinen, and R. M. Nieminen. *Modelling Simul. Mater. Sci.*, 13:1175, 2005.
24. H. Hölscher, W. Allers, U. D. Schwarz, A. Schwarz, and R. Wiesendanger. *Appl. Phys. A*, 72:S35, 2001.

25. R. Pérez, M. C. Payne, I. Stich, and K. Terakura. *Phys. Rev. Lett.*, 78:678, 1997.
26. B. G. Dick and A. W. Overhauser. *Phys. Rev.*, 112:603, 1958.
27. K. Ohno, K. Esfarjani, and Y. Kawazoe. *Computational Materials Science.* Springer, Berlin, 1999.
28. P. Hohenberg and W. Kohn. *Phys. Rev.*, 136:B864, 1964.
29. W. Kohn and L. J. Sham. *Phys. Rev*, 140:A1133, 1965.
30. C. Kittel. *Quantum Theory of Solids.* John Wiley and Sons, New York, 1963.
31. J. Bardeen. *Phys. Rev. Lett.*, 6:57, 1961.
32. C. J. Chen. *Introduction to Scanning Tunneling Microscopy.* Oxford University Press, Oxford, 1993.
33. W.A. Hofer and J. Redinger. *Surf. Sci.*, 447:51, 2000.
34. M. Bütticker, Y. Imry, R. Landauer, and S. Pinhas. *Phys. Rev. B*, 31:6207, 1985.
35. T. E. Feuchtwang. *Phys. Rev. B*, 13:517, 1976.
36. S. Hembacher, F. J. Giessibl, J. Mannhart, and C. F. Quate. *Phys. Rev. Lett.*, 94:056101, 2005.
37. G. Rubio-Pollinger, P. Joyez, and N. Agrait. *Phys. Rev. Lett.*, 93:116803, 2004.

4

Electron Transport Theory

This chapter introduces the main concepts and mathematical tools currently in use in electron transport theory. Since a detailed exposition of the theoretical framework would require a separate volume, we shall limit the description to the bare essentials.

Readers who wish to gain a more thorough understanding are referred to existing textbooks, e.g., the book on electron transport by Supriyo Datta [1], or the equally exhaustive volume by Hartmut Haug and Antti-Pekka Jauho [2]. On first reading, the chapter can be conveniently omitted and reread once the applications for electron tunneling have been clarified. This will be done in the next chapter, where the different models and their relations are introduced. In fact, this is very much the approach the authors used themselves in developing the numerical methods to calculate tunneling currents in an STM.

From a theoretical point of view a tunneling electron is part of a system comprising two infinite metal leads and an interface consisting of a vacuum barrier and, optionally, a molecule or a cluster of atoms with different properties from those of the infinite leads. The system can be said to be open–the number of charge carriers is not constant–and out of equilibrium: the applied potential and charge transport itself introduce polarizations and excitations within the system. Transport theory has the task to develop the necessary mathematical tools to calculate the number of charge carriers passing through this system in a given interval, depending on the atomic structure of the system and the applied bias voltage.

4.1 Conductance channels

Before embarking on the details of the mathematical framework let us look at transport in a metal lead. Electrons in this case can be conveniently treated as free particles; they are described by plane waves and their dispersion is parabolic. In one dimension,

$$E(k) = \frac{1}{2m}(\hbar k)^2, \qquad \frac{\partial E(k)}{\partial k} = \hbar \frac{\hbar k}{m} = \hbar v_k. \tag{4.1}$$

A number of electrons n per unit length with a velocity v_k will lead to a current equal to env_k. The electron density for a single mode k in a conductor of length L is $1/L$; the total current carried by all k states is consequently

$$I = ne \sum_k v_k f(E) = \frac{e}{L} \sum_k \frac{f(E)}{\hbar} \frac{\partial E(k)}{\partial k}, \tag{4.2}$$

where $f(E)$ in this equation denotes an occupation function, for example a Fermi distribution function. Under periodic boundary conditions the sum can be converted into an integral, taking into account that one mode in k-space corresponds to a length of 2(=spin) $\times$ $L/2\pi$:

$$\sum_k \to 2 \times \frac{L}{2\pi} \int dk, \qquad I = \frac{2e}{h} \int_{E_0}^{\infty} dE f(E). \tag{4.3}$$

The energy E_0 denotes a threshold. For nearly free electrons in a conductor, this energy threshold is given by solutions with $k = 0$. The threshold can be eliminated from the integral with the help of a number distribution $M(E)$, describing the number of modes above the threshold:

$$M(E) = \sum_i \theta(E - E_0^i), \qquad I = \frac{2e}{h} \int_{-\infty}^{\infty} dE f(E) M(E). \tag{4.4}$$

We may conclude from this simple model that the current carried in a conductor depends mainly on the number of nodes provided. For $M(E) \approx f(E) \approx 1$ the current per unit energy is a constant and equal to $2|e|/h$, or about 80 nA/meV.

Now let us consider an energy interval $\Delta E = E_2 - E_1$. If the number of nodes remains constant $M(E) \approx M_0$, and $f(e) \approx 1$, then the total current I_0 and the conductance $G_0 = I_0/((E_2 - E_1)/e)$ will be given by

$$I_0 = \frac{2e^2}{h} M_0 \frac{E_2 - E_1}{e}, \qquad G_0 = \frac{2e^2}{h} M_0. \tag{4.5}$$

The conductance is directly proportional to the number of modes M_0. For a single mode it is a constant, which depends only on the fundamental constants e and h. The minimum resistance in the lead, if only one mode for charge transport is provided, is called the *contact resistance*. It is equal to 12.9 kΩ.

$$R_C = R_0(M_0 = 1) = G_0^{-1}(M_0 = 1) = \frac{h}{2e^2} = 12.9\, k\Omega \tag{4.6}$$

This implies that (i) the resistance is inversely proportional to the number of modes; it thus approaches zero only if the number of modes is close to infinity; and (ii) the the contact resistance is actually quite high and certainly not negligible.

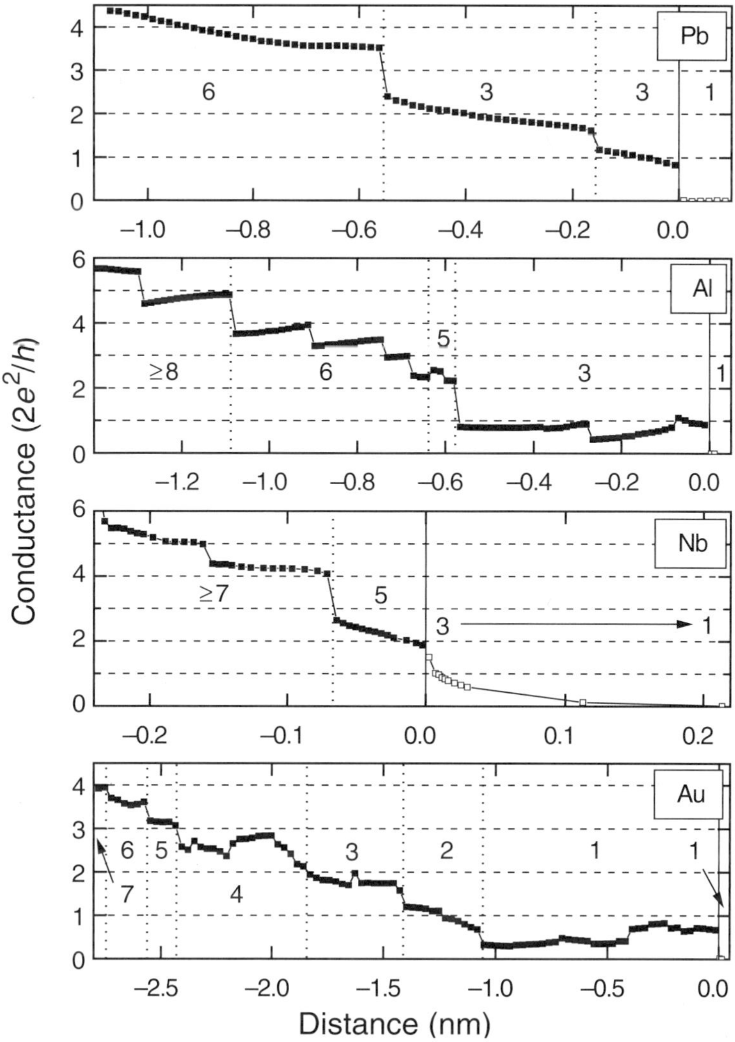

Fig. 4.1. Conductance measurements for four different metals at 1.5 K. The wires were strained up to the breaking point, and the conductance recorded at every distance. The result in all cases is a discrete decrease of the conductance from a multiple of the conductance quantum $G_0 = 2e^2/h$ to G_0, when the wire finally breaks. E. Scheer and N. Agrait and J. C. Cuevas and A. L. Yegati and B. Ludolph and A. Martin-Rodero and G. Rubio Pollinger and J. M. van Ruitenbeek and C. Urbina, Nature 394, 154 (1998). Copyright (1998) Nature Publishing Group, reprinted from [3] with permission.

Contact resistances can be measured in break junctions at close to zero kelvin. In these experiments a thin wire, usually made of gold, is extended until the chemical bonds between the atoms start to break. During the process the wire forms a neck, which ultimately consists of single atoms. The lowest conductance measured in such a wire is found equal to the inverse of the contact resistance (see Figure 4.1). This feature introduces a natural limit to the resistance in scanning tunneling microscopes. The lowest resistance in a metallic lead that can theoretically be measured is the contact resistance R_C. We shall see during the course of this presentation that the minimum resistance in STM measurements is in fact much higher, about 500–1000 kΩ. The two leads, the surface, and the STM tip, are in this situation still far, on a microphysical length scale, from actual contact.

4.2 Elastic transport

Transport through a system of two leads and a conducting interface can be said to be coherent if the electrons retain their phase throughout the whole system. It can be said to be elastic if the electrons also retain their energy. In this section we shall assume that both of these conditions are met. In the following, we shall relax the condition of elasticity and treat also processes, in which the electrons gain or lose energy due to the interactions with other electrons or by exciting lattice or molecular vibrations. It will be seen that the theoretical tools needed in both cases are very similar. In essence, elastic transport can be described by a set of relations based on **scattering matrices** S, phase relations between incoming and outgoing waves in the channels of the two leads; **Green's functions** G, relating amplitudes at the origin of waves and at arbitrary points along their propagation; **self-energies** Σ, changes in the spectrum of eigenfunctions due to the attachment of infinite leads; and **transmission functions** T, probabilities for the transmission of electrons from one lead to the other.
Since the systems we are interested in are very small–the interface in most cases is smaller than a few wavelengths of a Bloch wave or a few nm–we shall consider only coherent transport within our system. We shall also limit the presentation to a two-terminal interface connected to only two leads. Generalizing to multiple leads is rather straightforward, if mathematically somewhat more involved. The formalism to this end can be looked up in the existing literature.

4.2.1 The scattering matrix

A coherent conductor or interface can be seen as a device that connects the properties of incoming and outgoing electron waves in a way that keeps track of their respective phases. Given that the conductor will have physical properties, it is clear that the relation between incoming and outgoing waves will depend

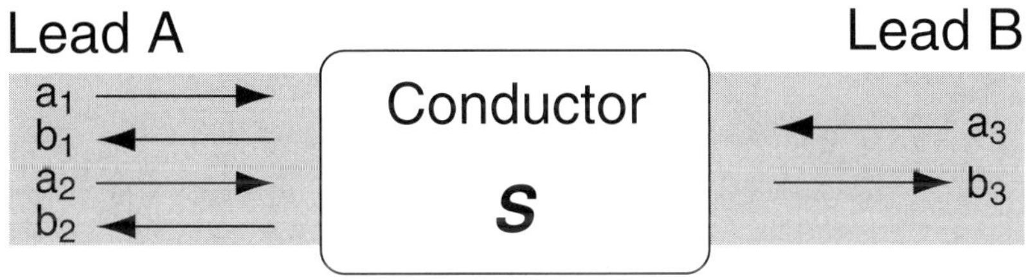

Fig. 4.2. Conductor between two leads. The conductor can be described by a scattering matrix S, which relates the amplitudes of the outgoing waves b to amplitudes of incoming waves a.

on the energy of the electrons as well as their initial phase upon entering the conductor. A convenient way to describe the conductor in an abstract way is by a *scattering matrix* S, which describes the phase relations between incoming and outgoing waves. Let us assume, for the sake of illustration, that at a given energy E the left lead possesses two modes of conductance, and the right lead one (see Figure 4.2). Then the scattering matrix is described by

$$\begin{pmatrix} b_1 \\ b_2 \\ b_3 \end{pmatrix} = \begin{pmatrix} S_{11}(E) & S_{12}(E) & S_{13}(E) \\ S_{21}(E) & S_{22}(E) & S_{23}(E) \\ S_{31}(E) & S_{32}(E) & S_{33}(E) \end{pmatrix} \cdot \begin{pmatrix} a_1 \\ a_2 \\ a_3 \end{pmatrix}. \tag{4.7}$$

Here, the outgoing amplitudes b_i are related to the incoming amplitudes a_i via the matrix, and the total number of modes summed up over both leads equals three, the dimension of the scattering matrix for the conductor. In general, the number of modes will differ from one lead to the other. The dimension of the scattering matrix M_S is given by the total number of (incoming or outgoing) modes for both channels

$$M_S(E) = \sum_{\text{leads}} M_{\text{lead}}(E). \tag{4.8}$$

As well as the scattering matrix, the number of propagating channels or modes will depend on energy.

An easy example

An easy example for a coherent conductor is a potential barrier in an infinite one-dimensional lead (See Figure 4.3). The solution of the problem is part of every undergraduate textbook in quantum mechanics. It is given by applying the Schrödinger equation and accounting for the boundary conditions at the two ends of the potential barrier. The scattering matrix, in this problem, is given by the relation between the amplitudes of incoming and outgoing waves. For a potential barrier of width d, the scattering matrix is given by

$$S(E) = \begin{pmatrix} S_{11}(E) & S_{12}(E) \\ S_{21}(E) & S_{22}(E) \end{pmatrix} = \begin{pmatrix} r(d,E) & t(d,E) \\ t^{+}(d,E) & r(d,E) \end{pmatrix}, \tag{4.9}$$

where $t(d,E)$ and $r(d,E)$ are the transmission and reflection coefficients of the barrier at a given energy. In general, the transmission coefficient is a complex numbers, while r, due to the symmetry of the problem, must be real. Both coefficients depend on the constant potential V_0 in the lead and the energy E or wavenumber k of the single electron solutions. The example is very simple because we have to account only for one propagating mode at a given energy, the two leads are identical, and the potential barrier is symmetric. In general, none of these conditions hold, and the scattering matrix is much more complicated to evaluate. However, we can, for the time being, assume that it can still be calculated by an application of the Schrödinger equation. If one asks what the probability would be in the situation that an electron entering the conductor via mode a_1 would leave it via mode b_2, then the answer for the potential barrier is very simple: it is the absolute value of the matrix element of S relating a_1 to b_2, or $|t^{+}(d,E)|^2 = |t(d,E)|^2$.

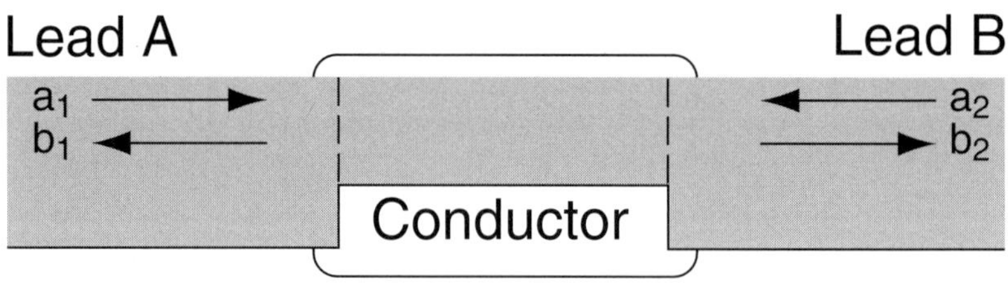

Fig. 4.3. Potential barrier in a one dimensional lead. The potential barrier is equal to a coherent conductor; the transmission probability depends on the energy E.

4.2.2 Transmission functions

In general, the number of modes in each lead will be greater than one, and the numbers will be different for each lead, say lead A and lead B. However, the transmission probability T for the transfer of electrons from one mode in lead A, say n, to a mode in lead B, say m, is generally given by the square of the corresponding element of the S-matrix:

$$T_{m\leftarrow n}(E) = |S_{m\leftarrow n}(E)|^2. \tag{4.10}$$

Let us evaluate the transmission probability for the initial example. In this case we have two incoming and outgoing modes on the left, and one incoming and outgoing mode on the right (Figure 4.2). The three lines of the matrix equation read in this case

$$
\begin{aligned}
b_1 &= S_{11}(E)a_1 + S_{12}(E)a_2 + S_{13}(E)a_3, \\
b_2 &= S_{21}(E)a_1 + S_{22}(E)a_2 + S_{23}(E)a_3, \\
b_3 &= S_{31}(E)a_1 + S_{32}(E)a_2 + S_{33}(E)a_3.
\end{aligned} \tag{4.11}
$$

If we are interested in the current from the left lead to the right, we have to sum over all contributions connecting a_1 and a_2 with b_3. The sum of all transmission probabilities, or the *transmission function* $\overline{T}$, is in this case

$$\overline{T}_{B\leftarrow A}(E) = |S_{31}(E)|^2 + |S_{32}(E)|^2. \tag{4.12}$$

This means that in general we have to sum up over all modes of lead A, and all modes of lead B propagating in the right direction. Symbolically, we can write this summation as

$$\overline{T}_{B\leftarrow A}(E) = \sum_{m(A)} \sum_{n(B)} T_{n\leftarrow m}(E) = \sum_{m(A)} \sum_{n(B)} |S_{n\leftarrow m}(E)|^2. \tag{4.13}$$

We could also have reversed the direction of current and estimated the transmission function from lead B to lead A. In this case we get (now we sum up all contributions from a_3 into b_1 and b_2)

$$\overline{T}_{A\leftarrow B}(E) = |S_{13}(E)|^2 + |S_{23}(E)|^2. \tag{4.14}$$

As one can prove, the two transmission functions (4.12) and (4.14) are actually equal. To this end, we consider the total current passing through all incoming and outgoing channels. Charge conservation requires that the two currents be equal; thus

$$\sum_m |a_m|^2 = \sum_m |b_m|^2. \tag{4.15}$$

The sums can be interpreted as scalar products of two vectors, $\mathbf{a}$ and $\mathbf{a}^+$, or $\mathbf{b}$ and $\mathbf{b}^+$, respectively

$$\mathbf{a}^+\mathbf{a} = \mathbf{b}^+\mathbf{b}. \tag{4.16}$$

Inserting this relation into the definition of the S-matrix, we obtain

$$\mathbf{a}^+\mathbf{a} = (S\mathbf{a})^+(S\mathbf{a}) = \mathbf{a}^+(S^+S)\mathbf{a} \quad \Rightarrow \quad S^+S = SS^+ = I. \tag{4.17}$$

The S matrix, we find, must be unitary. Its Hermitian conjugate (transposed and the complex conjugate taken of its elements) is equal to the inverse matrix. But this means that for two elements S_{ij} and S_{ji} we may write

$$S_{ij} = S_{ji}^*. \tag{4.18}$$

Inspecting now (4.12) and (4.14), we see that the transmission function in both cases is composed of the same contributions, because, via the unitarity of the S-matrix, $S_{13} = S^*_{31}$ and $S_{23} = S^*_{32}$. The two functions are therefore identical. A further consequence of the unitarity is that the transmission probability over all modes is equal to unity

$$S^+ S = S S^+ = I \quad \Rightarrow \quad \sum_{m=1}^{M_S(E)} |S_{mn}(E)|^2 = \sum_{m=1}^{M_S(E)} |S_{nm}(E)|^2 = 1. \tag{4.19}$$

As a reminder, we have inserted the explicit dependency of S and M_S on the energy E. The reader is asked to keep this explicit dependency on energy in mind, even if it is in the following, for brevity of notation, sometimes omitted.

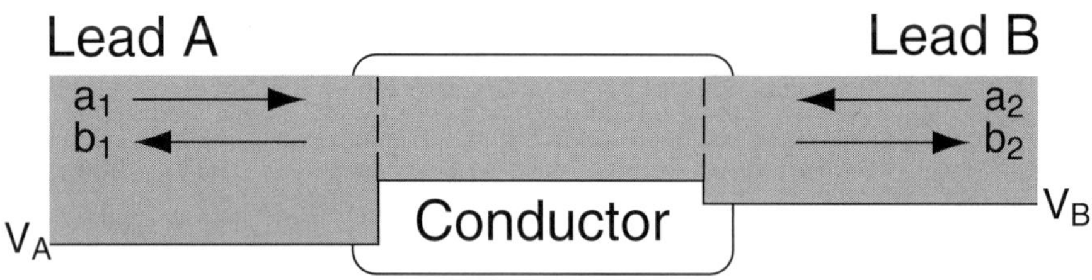

Fig. 4.4. Modified potential barrier in a one-dimensional lead. The potentials within the two leads A and B are now different. The transition probability $A \rightarrow B$ is no longer equal to the probability $B \rightarrow A$.

Scattering in time

It seems that this definition of S does not include the most important feature of current transport: the dependency on time. The sequence of physical events is clear: an electron wave impinges on the conductor from a lead; it is partly transmitted within the conductor and leaves it via the other lead. S, it seems, does not include this feature. So how does the scattering matrix account for current flow? To see clearly how this works, let us modify the potential barrier. Let us say it has a different potential for lead A, given by V_A, and lead B, described by V_B (see Figure 4.4). Clearly, then, the transition probability $A \rightarrow B$ is different than the probability $B \rightarrow A$. The matrix then has the form

$$S(d,E) = \begin{pmatrix} r_A(d,E) & t_{B\leftarrow A}(d,E) \\ t_{A\leftarrow B}(d,E) & r_B(d,E) \end{pmatrix}. \tag{4.20}$$

In fact, the matrix is no longer unitary, unless the different propagation velocities, proportional to the wavevector k, are included in the matrix definition.

The reason is that we have based the calculation on the amplitudes of the wavefunctions, and not the current flow between the leads. To account for the different propagation velocities, we have to modify the matrix S by a factor $\sqrt{k_A/k_B}$ for the transmission amplitudes. Since k vectors are proportional to electron velocities, we can also choose the ratio $\sqrt{v_A/v_B}$ to this end:

- The reader may calculate the necessary ratio himself, by simply using a potential step. In this case the two transmission amplitudes will comply with the relation $t_{B\leftarrow A} = k_A/k_B \cdot t_{A\leftarrow B}$. In order to make the scattering matrix unitary, the two matrix elements have to be multiplied by $\sqrt{k_B/k_A}$ and $\sqrt{k_A/k_B}$, respectively.

Using velocities instead of k, the unitary matrix of our modified example then reads

$$S(d,E) = \begin{pmatrix} r_A(d,E) & \sqrt{v_B/v_A} \cdot t_{B\leftarrow A}(d,E) \\ \sqrt{v_A/v_B} \cdot t_{A\leftarrow B}(d,E) & r_B(d,E) \end{pmatrix}. \tag{4.21}$$

As a brief inspection shows, the definition of the scattering matrix in (4.11) already incorporates this feature, since the condition of current conservation between incoming and outgoing current leads to the result that the matrix is unitary. We therefore conclude from this analysis one important feature of scattering matrices: they must be modified if the theoretical model is based on wavefunction amplitudes rather than currents. We shall need this conclusion in our relation between Green's functions and scattering matrices derived in the following sections.

4.2.3 A brief introduction to Green's functions

It seems that so far we have not gained much ground. We know that the transmission function, which gives us the probability for the transfer of electrons from incoming to outgoing modes and vice versa, is related to the S-matrix; and we know, in simple cases, that the S-matrix can be calculated by solving the Schrödinger equation. However, that does not tell us how we should calculate the conductance for an energy-dependent number of nodes and a complicated system composed of chemically different atoms. To this end we have to relate the S-matrix to some quantity that can be routinely calculated in solid-state theory, even if these calculations are rather involved and based on numerical self-consistency cycles, as, for instance, in electronic structure calculations. This quantity, as we shall see, is the Green's function of a system.

What is a Green's function?

Due to their wide range of applications, Green's functions are today used in many parts of physics. For the purpose of transport theory, we may focus on their importance for the solution of a very fundamental problem. The problem can be stated as follows:

- What is the effect of a unit excitation at some point of our system, say $\mathbf{r}'$, at another point, say $\mathbf{r}$, if the excitation is propagated by waves?

We may think, initially, that the solution to the problem, the Green's function, must have the same general form as a wave (function) itself, since the propagation is governed by the same physical relations. Thus it must be a solution to some sort of wave equation, either the optical wave equation for photon transport or the Schrödinger equation for electron transport. If this wave equation can be written as an operator equation, where $\hat{D}$ stands for a differential operator like

$$\hat{D}_{\text{ph}} = \nabla^2 - \frac{1}{c^2}\frac{\partial^2}{\partial t^2} \qquad \text{Photons,} \tag{4.22}$$

$$\hat{D}_{\text{el}} = -\frac{\hbar^2}{2m}\nabla^2 + U - E \qquad \text{Electrons,} \tag{4.23}$$

then the Green's function $G(\mathbf{r}, \mathbf{r}')$ of the problem for most of the system will be described by

$$\hat{D}_{\text{ph}}\, G_{ph}(\mathbf{r}, \mathbf{r}') = 0 \qquad \text{Photons,} \tag{4.24}$$

$$\hat{D}_{\text{el}}\, G_{el}(\mathbf{r}, \mathbf{r}') = 0 \qquad \text{Electrons.} \tag{4.25}$$

However, this formulation does not account for a change of the wave due to an excitation at $\mathbf{r}'$. To include this feature in our description of the system we have to add a term that enforces a unit change at our point of origin, $\mathbf{r}'$. This is done by a δ-functional $\delta(\mathbf{r} - \mathbf{r}')$, so that the Green's function for electrons is described by

$$\hat{D}_{\text{el}}\, G(\mathbf{r}, \mathbf{r}') = -\delta(\mathbf{r} - \mathbf{r}'). \tag{4.26}$$

A point on locality

If we assume that the inverse operator $\hat{D}_{\text{el}}^{-1}$ exists, so that the successive operation of $\hat{D}_{\text{el}}^{-1}$ and $\hat{D}_{\text{el}}$ leaves us only with the excitation at $\mathbf{r} = \mathbf{r}'$, then we may write

$$\hat{D}_{\text{el}}^{-1} \cdot \hat{D}_{\text{el}} = -\delta(\mathbf{r} - \mathbf{r}'). \tag{4.27}$$

Inserting into (4.26), and accounting for the fact that $G(\mathbf{r}, \mathbf{r}')$ is a function, so that we can exchange the order of multiplication, we get

$$\hat{D}_{\text{el}}^{-1}\hat{D}_{\text{el}}\, G(\mathbf{r}, \mathbf{r}') = -G(\mathbf{r}, \mathbf{r}')\,\delta(\mathbf{r} - \mathbf{r}') = -\hat{D}_{\text{el}}^{-1}\delta(\mathbf{r} - \mathbf{r}'),$$

$$G(\mathbf{r}, \mathbf{r}') \equiv \hat{D}_{\text{el}}^{-1}. \tag{4.28}$$

The Green's function of the problem is therefore equivalent to the inverse operator $\hat{D}_{\text{el}}^{-1}$. However, this equivalence has to be treated with caution. The operator itself, $\hat{D}_{\text{el}}$, is local, since the local derivatives and the potential U both depend only on the coordinate $\mathbf{r}$. This is not the case for the inverse operator $\hat{D}_{\text{el}}^{-1}$ or the Green's function $G(\mathbf{r}, \mathbf{r}')$, which both depend on $\mathbf{r}$ *and* $\mathbf{r}'$. Both are therefore intrinsically nonlocal. If we write the equivalence for the Schrödinger operator in the form

$$G(\mathbf{r}, \mathbf{r}') = \left[-\frac{\hbar^2}{2m} \nabla^2 + U(\mathbf{r}) - E \right]^{-1}, \tag{4.29}$$

then we should remember that the inverse operation also makes the expression explicitly dependent on $\mathbf{r}'$, which covers the rest of our system. It is clear that this feature introduces a dependency on the boundaries, since the inverse operation will introduce wave propagation from the boundaries to our point of excitation. To show how we deal with this problem, let us consider an example, a one-dimensional wire.

The one-dimensional Green's function

The potential can be assumed constant U_0 throughout the wire. In this case the Green's function is described by the relation

$$\left(E - U_0 + \frac{\hbar^2}{2m} \frac{\partial^2}{\partial z^2} \right) G(z, z') = \delta(z - z'). \tag{4.30}$$

Apart from the point $z = z'$, the relation is equal to the Schrödinger equation for a one-dimensional wavefunction $\psi(z)$:

$$\left(E - U_0 + \frac{\hbar^2}{2m} \frac{\partial^2}{\partial z^2} \right) \psi(z) = 0 \tag{4.31}$$

We expect it therefore to have the same solutions, plane waves, throughout the wire. For outgoing waves from $z = z'$ we therefore can write

$$G(z, z') = \begin{cases} A_1 \exp[ik(z - z')] & z > z', \\ A_2 \exp[-ik(z - z')] & z < z'. \end{cases} \tag{4.32}$$

The amplitudes A_i are found, as is customary, by considering the boundary conditions. At $z = z'$ the Green's function must be continuous to comply with current conservation, thus we obtain

$$[G(z > z') - G(z < z')]_{z \to z'} = 0 \quad \Longrightarrow \quad A_1 = A_2. \tag{4.33}$$

The difference of the second derivatives at this point must be equal to a delta functional. Since the delta functional is a derivative of the step function $\theta(z - z')$, the first derivatives at this point must change by a discrete amount:

$$\left[\frac{\partial}{\partial z}G(z>z') - \frac{\partial}{\partial z}G(z<z')\right]_{z\to z'} = \frac{2m}{\hbar^2} \quad \Longrightarrow \quad ikA_1 = \frac{m}{\hbar^2}. \tag{4.34}$$

Hence the Green's function for outgoing waves is given by

$$G_{\text{out}}(z,z') = -\frac{i}{\hbar v}\exp(ik|z-z'|), \qquad \text{where} \qquad v = \frac{\hbar k}{m}. \tag{4.35}$$

We could also have sought solutions for incoming waves; in this case the signs of the exponents are reversed, and applying the same boundary conditions leads to a Green's function with the opposite sign for its amplitude. The Green's function of incoming waves is consequently

$$G_{\text{in}}(z,z') = +\frac{i}{\hbar v}\exp(-ik|z-z'|), \qquad \text{where} \qquad v = \frac{\hbar k}{m}. \tag{4.36}$$

It is customary to refer to these solutions of the problem, which correspond to different boundary conditions at infinity, by a different name: the outgoing solutions are usually called the *retarded*, the incoming the *advanced* Green's functions:

$$\begin{aligned} G^{\text{R}}(z,z') &= -\frac{i}{\hbar v}\exp(ik|z-z'|), \\ G^{\text{A}}(z,z') &= +\frac{i}{\hbar v}\exp(-ik|z-z'|), \end{aligned} \tag{4.37}$$

where

$$k \equiv \frac{\sqrt{2m(E-U_0)}}{\hbar}, \qquad v \equiv \frac{\hbar k}{m}.$$

The two solutions present only one difficulty: they are singular for $k \to 0$. This feature makes it difficult to apply them in transport calculations, because the result will necessarily diverge. To circumvent the problem, one usually introduces a small imaginary energy component into the Schrödinger equation.

Imaginary energy components $i\eta$

If we introduce an imaginary energy component into the equation for the one dimensional Green's function (4.30),

$$\left(E + i\eta - U_0 + \frac{\hbar^2}{2m}\frac{\partial^2}{\partial z^2}\right)G(z,z') = \delta(z-z') \tag{4.38}$$

then the wavevector k changes to

$$k' = \frac{\sqrt{2m(E+i\eta-U_0)}}{\hbar} \approx \frac{\sqrt{2m(E-U_0)}}{\hbar}\left(1+\frac{i\eta}{2(E-U_0)}\right) = k(1+i\epsilon). \tag{4.39}$$

Inserting this result into the retarded and advanced Green's functions, we see that only one, the retarded Green's function G^{R}, remains finite at infinity, while G^{A} will grow beyond all limits. We also see that G^{R} in this case remains well behaved also for $k = 0$. The infinitesimal energy component thus solves our problems quite elegantly: (i) It makes one Green's function the only possible solution, and the function is therefore unique; and (ii) it avoids the pole at $k = 0$. By starting from a different equation (4.38), with $-i\eta$, we arrive at the advanced Green's function as the only valid solution. Generalizing the results so far, and with the shortcut for the Hamilton operator

$$\hat{H}_0 = -\frac{\hbar^2}{2m}\nabla^2 + U(\mathbf{r}), \tag{4.40}$$

we can write for the two Green's functions the following relations:

$$G^{\mathrm{R}} = \left(E - \hat{H}_0 + i\eta\right)^{-1}, \qquad \eta \to 0^+, \tag{4.41}$$

$$G^{\mathrm{A}} = \left(E - \hat{H}_0 - i\eta\right)^{-1}, \qquad \eta \to 0^+. \tag{4.42}$$

Eigenvector expansions

In solid state physics we are often confronted with the situation that we know the eigenvalues and eigenvectors of our system, since we have calculated them, for example, by density functional theory. In this case it is very simple to derive the Green's function of the system. Let us consider the solution of a system described by one particle electron states. The Schrödinger equation of the system gives

$$\hat{H}_0\psi_i(\mathbf{r}) = \epsilon_i\psi_i(\mathbf{r}), \tag{4.43}$$

where ϵ_i and $\psi_i(\mathbf{r})$ are the solutions with band index i. The complete set of solutions forms a complete orthonormal set of functions, so that every function may be expanded in eigenfunctions $\psi_i(\mathbf{r})$:

$$\int d^3r\psi_j^*(\mathbf{r})\psi_i(\mathbf{r}) = \delta_{ij}. \tag{4.44}$$

The retarded Green's function can therefore be expanded in the eigenfunctions $\psi_i(\mathbf{r})$ in the following way:

$$G^{\mathrm{R}}(\mathbf{r},\mathbf{r}') = \sum_i C_i(\mathbf{r}')\psi_i(\mathbf{r}). \tag{4.45}$$

Substituting the expansion into the relation for the retarded Green's function,

$$\left(E - \hat{H}_0 + i\eta\right) G^{\mathrm{R}}(\mathbf{r}, \mathbf{r}') = \sum_j \left(E - \epsilon_j + i\eta\right) C_j(\mathbf{r}')\psi_j(\mathbf{r}) = \delta(\mathbf{r} - \mathbf{r}'), \quad (4.46)$$

multiplying by $\psi_i^*(\mathbf{r})$, and integrating over the whole system,

$$\sum_j \left(E - \epsilon_j + i\eta\right) C_j(\mathbf{r}') \int d^3r\, \psi_i^*(\mathbf{r})\psi_j(\mathbf{r}) = \int d^3r \psi_i^*(\mathbf{r})\delta(\mathbf{r} - \mathbf{r}'), \quad (4.47)$$

making use of (4.44), we arrive at the following result for $C_i(\mathbf{r}')$:

$$\sum_j \left(E - \epsilon_j + i\eta\right) C_j(\mathbf{r}')\delta_{ij} = \psi_i^*(\mathbf{r}'),$$

$$C_i(\mathbf{r}') = \frac{\psi_i^*(\mathbf{r}')}{E - \epsilon_i + i\eta} \quad (4.48)$$

The retarded Green's function is consequently

$$G^{\mathrm{R}}(\mathbf{r}, \mathbf{r}') = \sum_i \frac{\psi_i^*(\mathbf{r}')\psi_i(\mathbf{r})}{E - \epsilon_i + i\eta}. \quad (4.49)$$

By an identical derivation starting from the relations for the advanced Green's function, we arrive at a similar expansion:

$$G^{\mathrm{A}}(\mathbf{r}, \mathbf{r}') = \sum_i \frac{\psi_i^*(\mathbf{r}')\psi_i(\mathbf{r})}{E - \epsilon_i - i\eta}. \quad (4.50)$$

Exchanging the coordinates $\mathbf{r}$ and $\mathbf{r}'$ in (4.50) and computing the complex conjugate, we realize, by inspection, that

$$\left[G^{\mathrm{A}}(\mathbf{r}', \mathbf{r})\right]^* = \sum_i \frac{\psi_i(\mathbf{r})\psi_i^*(\mathbf{r}')}{E - \epsilon_i + i\eta} = G^{\mathrm{R}}(\mathbf{r}, \mathbf{r}'). \quad (4.51)$$

Since the only assumption that went into our derivation is the existence of a complete orthonormal set of eigenfunctions for our Hamiltonian, the result is quite generally valid; that is transposing and taking the complex conjugate transforms one (retarded) Green's function into the other (advanced) one:

$$\left[G^{\mathrm{R}}\right]^+ = G^{\mathrm{A}}, \qquad \left[G^{\mathrm{A}}\right]^+ = G^{\mathrm{R}}. \quad (4.52)$$

From a practical point of view this means that one of them is actually redundant: we shall therefore limit the presentation in the following to the retarded Green's function when we talk of *the* Green's function of a system.
It should be noted that this derivation is valid only if we can represent eigenfunctions of the system as solutions of the Schrödinger equation with a unique

potential. This is generally the case in mean field approximations to electron interactions, as in density functional theory. In a many-body context, the same conclusion is not necessarily true. In fact, as will be seen later, the only solutions we can find at present for our transport problem are based on perturbation theory, where interactions are seen as slight perturbations of the noninteracting Hamiltonian.

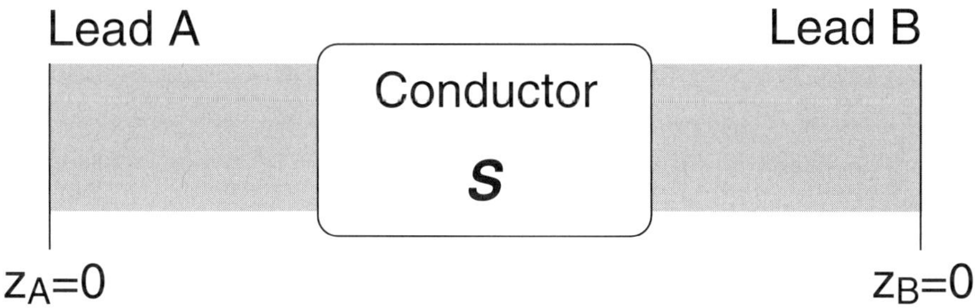

Fig. 4.5. Potential barrier in a one-dimensional lead. The potential barrier is equal to a coherent conductor, the transmission probability depends on the energy E.

4.2.4 Green's functions and scattering matrices

After this digression and introduction of Green's functions, we are now in a position to take a second look at the scattering matrix S. First, let us remember that the scattering matrix is defined based on currents, rather than amplitudes. If we base our theoretical model on amplitudes, we have to modify the transmission amplitudes $t_{A(B)\leftarrow B(A)}$ by $\sqrt{v_{A(B)}/v_{B(A)}}$ (see (4.21)).
We want to calculate the Green's function connecting the points $z_A = 0$ and $z_B = 0$, which are within the two leads A and B at opposite sides of the conductor (see Figure 4.5):

$$G_{B\leftarrow A}(z_A, z_B) \equiv G^{\mathrm{R}}_{B\leftarrow A}(z_A, z_B). \tag{4.53}$$

From the previous results we know that a unit excitation at $z_A = 0$ has two consequences: it leads to a wave of amplitude $A_1 = +i/\hbar v$ away from the conductor, and a wave of amplitude $A_2 = -i/\hbar v$ in the direction of the conductor (see (4.35) and (4.36)). The wave toward the conductor is scattered by the conductor, its transmission amplitude to propagate into lead B is $t_{B\leftarrow A}$. Hence we may write for the Green's function

$$G^{\mathrm{R}}_{B\leftarrow A}(z_A, z_B) = \delta_{BA} \cdot A_1 + t_{B\leftarrow A} A_2. \tag{4.54}$$

The Kronecker delta describes that the excitation arises in lead A, and will be zero if we are within lead B, but one if we are within A. Since the scattering matrix element S_{BA} is related to $t_{B\leftarrow A}$ via

$$S_{BA} = \sqrt{\frac{v_A}{v_B}} \cdot t_{B\leftarrow A}, \tag{4.55}$$

we can express the scattering matrix in terms of the retarded Green's function by

$$S_{BA} = -\delta_{BA} + i\hbar\sqrt{v_A v_B} \cdot G^{\mathrm{R}}_{BA}. \tag{4.56}$$

This relation, which was derived in the 1980s by Fisher and Lee [4], relates transport properties through an interface to the Green's function of this interface. Since Green's functions can be calculated by existing electronic structure methods, it is one of the theoretical bases of modern simulation techniques in transport theory.

4.2.5 Scattering matrices for multiple channels

In a wire with multiple propagation modes, extending along z we may write the Green's function as a plane wave along z modulated by amplitudes $A^{\pm}_m$ and transverse wavefunctions $\chi_m(\mathbf{r}_2)$ with $\mathbf{r}_2 = (x, y)$. Without loss of generality these transverse wavefunctions are assumed to be real. Thus

$$G^{\mathrm{R}}(z, z') = \sum_m A^{\pm}_m \chi_m(\mathbf{r}_2) \exp\left[ik_m|z - z'|\right]. \tag{4.57}$$

The transverse wavefunction must satisfy a two-dimensional Schrödinger equation, or

$$\left[-\frac{\hbar^2}{2m}\nabla_2^2 + U(\mathbf{r}_2)\right] \chi_m(\mathbf{r}_2) = \epsilon_{m,0}\chi_m(\mathbf{r}_2). \tag{4.58}$$

The potential $U(\mathbf{r}_2)$ confines electron motion in the $\mathbf{r}_2$-direction; the wavefunctions $\chi_m(\mathbf{r}_2)$ are orthogonal, since they represent the spectrum of the two-dimensional Hamiltonian for a discrete set of eigenvalues. For the purpose of demonstration we also assume that they are real:

$$\int d^2r \chi_m(\mathbf{r}_2)\chi_n(\mathbf{r}_2) = \delta_{mn}. \tag{4.59}$$

To obtain the amplitudes $A^{\pm}_m$ we use the boundary conditions at $z = z'$, which amount to

$$\left[G^{\mathrm{R}}(z,z')\right]_{z=z'+} = \left[G^{\mathrm{R}}(z,z')\right]_{z=z'-},$$
$$\left[\frac{\partial G^{\mathrm{R}}(z,z')}{\partial z}\right]_{z=z'+} - \left[\frac{\partial G^{\mathrm{R}}(z,z')}{\partial z}\right]_{z=z'-} = \frac{2m}{\hbar^2}\,\delta\left(\mathbf{r}_2 - \mathbf{r}'_2\right). \tag{4.60}$$

Inserting the definition of $G^{\mathrm{R}}(\mathbf{r}_2)$ into the boundary conditions we arrive, as before, at two characteristic equations:

$$\sum_m A_m^+ \chi_m(\mathbf{r}_2) = \sum_m A_m^- \chi_m(\mathbf{r}_2), \tag{4.61}$$

$$\sum_m ik_m \left[A_m^+ + A_m^-\right]\chi_m(\mathbf{r}_2) = \frac{2m}{\hbar^2}\,\delta(\mathbf{r}_2 - \mathbf{r}'_2). \tag{4.62}$$

Multiplying by $\chi_n(\mathbf{r}_2)$ and integrating over d^2r gives

$$A_m^+ = A_m^-, \qquad ik_m\left(A_m^+ + A_m^-\right) = \frac{2m}{\hbar^2}\,\chi_m(\mathbf{r}'_2). \tag{4.63}$$

It follows that the amplitude A_m is proportional to the wavefunction χ_m at the point of excitation $\mathbf{r}'_2$:

$$A_m^+ = A_m^- = -\frac{i}{\hbar v_m}\chi_m(\mathbf{r}'_2). \tag{4.64}$$

The Green's function for the infinite wire is consequently described by

$$G^{\mathrm{R}}(\mathbf{r},\mathbf{r}') = -\sum_m \frac{i}{\hbar v_m}\chi_m(\mathbf{r}_2)\chi_m(\mathbf{r}'_2)\exp\left[ik_m|z-z'|\right]. \tag{4.65}$$

Consider now that we have two leads: lead A with m and lead B with n conducting channels. Then the Green's function between two points at opposite sides of the conductor has to be modified. Instead of (4.54) we now have

$$G^{\mathrm{R}}(\mathbf{r}_2(B),\mathbf{r}_2(A)) = \sum_{m(A)}\sum_{n(B)}\left[\delta_{nm}A_m^- + t_{nm}A_m^+\right]\chi_n(\mathbf{r}_2(B)). \tag{4.66}$$

Using the results for $A_m^\pm$ and accounting for propagation velocities, we may write for $G^{\mathrm{R}}(\mathbf{r}_2(B),\mathbf{r}_2(A))$,

$$\begin{aligned} G^{\mathrm{R}}&\left(\mathbf{r}_2(B),\mathbf{r}_2(A)\right) \\ &= -\sum_{m'(A)}\sum_{n'(B)}\frac{i}{\hbar\sqrt{v_{m'}v_{n'}}}\chi_{n'}(\mathbf{r}_2(B))\left[\delta_{n'm'} + S_{n'm'}\right]\chi_{m'}(\mathbf{r}_2(A)). \end{aligned} \tag{4.67}$$

Generally, one is interested in the elements of the scattering matrix, which give the transmission function and thus the transmission probability for electron propagation through a device. These elements are obtained from the above equation by multiplying by $\chi_m(\mathbf{r}_2(A))\chi_n(\mathbf{r}_2(B))$ and integrating over d^2r, making use of the orthogonality condition for the wavefunctions χ:

$$S_{nm} = -\delta_{nm} + i\hbar\sqrt{v_n v_m} \iint d^2r d^2r' \chi_n(\mathbf{r'}_2)\, G^R\,(\mathbf{r'}_2, \mathbf{r}_2)\;\chi_m(\mathbf{r}_2). \quad (4.68)$$

Here, we have dropped the explicit notation that $\mathbf{r}_2$ is within lead A and $\mathbf{r'}_2$ within lead B. Note also that all components of this equation retain their energy dependency.

4.2.6 Self-energies Σ

So far we have only described the propagation of electrons from one side of a conductor to the other. The key quantities needed to calculate the transmission probability were found to be the Green's functions G^{R} or the propagator for a given transition between two points at opposite sides of the conductor. These, in turn, are related to the scattering matrix, and a summation of all the elements of the scattering matrix gives us the desired transmission probability via the Fisher–Lee equation (4.68).
However, it is generally impossible to calculate the Green's functions of a system comprising a conductor and two infinite leads for the simple reason that the leads will have an effect on the electronic structure and propagation within the conductor. This effect needs to be included in the calculation. The general strategy to this end is to determine the effect of the infinite leads at the boundaries of the conductor itself.

The method of finite differences: infinite wire

To understand the procedures involved we consider initially a simple one-dimensional system with a discrete set of lattice points, equally spaced at a distance $z_j - z_{j-1} = a$, where a is a constant. The Hamilton operator of the system is then given by

$$H = -\frac{\hbar^2}{2m}\frac{d^2}{dz^2} + V(z). \quad (4.69)$$

The potential $V(z)$ is also described at every lattice point, we write V_j for $V(z = ja)$. Multiplying H by a function $f(z)$, equally discretized, we obtain for a point $z = ja$ the following relation:

$$[HF]_{z=ja} = \left[-\frac{\hbar^2}{2m}\frac{d^2 f}{dz^2}\right]_{z=ja} + V_j f_j. \quad (4.70)$$

The method of finite differences derives its name from the treatment of differentials. For the first derivative of $f(z = ja)$ we may consider the finite difference between the values $f(z = (j + 1/2)a)$ and $f(z = (j - 1/2)a)$, thus

$$\left[\frac{df}{dz}\right]_j = \frac{1}{a}\left[f_{j+1/2} - f_{j-1/2}\right]. \tag{4.71}$$

The second derivative is consequently

$$\left[\frac{d^2f}{dz^2}\right]_j = \frac{1}{a}\left\{\left[\frac{df}{dz}\right]_{j+1/2} - \left[\frac{df}{dz}\right]_{j-1/2}\right\} = \frac{1}{a^2}\left[f_{j+1} - 2f_j + f_{j-1}\right]. \tag{4.72}$$

The one-dimensional Schrödinger equation is then transformed into a matrix equation, where every row of the matrix contains only three elements: the diagonal element and its two neighbors. Within an atomic orbital representation of the Hamiltonian such a matrix is known as the "tight-binding" matrix, since it contains only the diagonal and the "hopping" parameters to the nearest neighbors. The matrix relation reads

$$[Hf]_j = (V_j + 2t)\, f_j - t f_{j-1} - t f_{j+1} = \sum_i H_{ji} f_i, \tag{4.73}$$

$$t \equiv \frac{\hbar^2}{2ma^2}. \tag{4.74}$$

Since we consider an infinite lead, the matrix H is infinite-dimensional. Considering the matrix near the point $z = ja = 0$, we can write the components explicitly as

$$H = \begin{pmatrix} V_{-2} + 2t & -t & 0 & 0 & 0 \\ -t & V_{-1} + 2t & -t & 0 & 0 \\ 0 & -t & V_0 + 2t & -t & 0 \\ 0 & 0 & -t & V_1 + 2t & -t \\ 0 & 0 & 0 & -t & V_2 + 2t \end{pmatrix}. \tag{4.75}$$

Once we have the Hamiltonian matrix of a system, the Green's function is simply the inverse matrix according to the prescription (see previous sections)

$$G_A^R = \left[(E + i\eta)I - H\right]^{-1}. \tag{4.76}$$

Conductor and infinite wire

However, there is a slight problem: the matrix is infinite-dimensional, since we are considering the Hamiltonian of an infinite wire. Inverting an infinite matrix is not feasible, so we have to think of an indirect approach to deal with the properties of the lead. To this end we first separate a system containing a conductor and an infinite lead into two subsystems described by separate Hamiltonians (see Figure 4.6). The overall Green's function of the system can then be partitioned into submatrices in the following way:

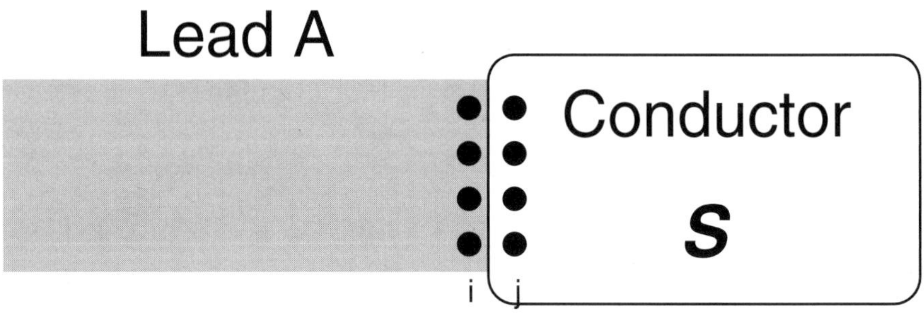

Fig. 4.6. A system comprising a conductor and an infinite lead.

$$\begin{pmatrix} G_A & G_{AC} \\ G_{CA} & G_C \end{pmatrix} \equiv \begin{pmatrix} (E-i\eta)I - H_A & T_A \\ T_A^+ & EI - H_C \end{pmatrix}^{-1} =: M^{-1}. \tag{4.77}$$

The coupling matrix T_A will be nonzero only for adjacent points of conductor and wire, labeled by indices (j, i) (see Figure 4.6). Multiplying (4.77) by M we obtain the following conditions:

$$[(E+i\eta)I - H_A]\, G_{AC} + T_A G_C = 0, \tag{4.78}$$

$$[EI - H_C]\, G_C + T_A^+ G_{AC} = I. \tag{4.79}$$

The matrix G_{AC}, which describes propagation at the interface between conductor and wire, is then given by

$$G_{AC} = -G_A^{\mathrm{R}} T_A G_C, \tag{4.80}$$

$$G_A^{\mathrm{R}} = [(E+i\eta)I - H]^{-1}. \tag{4.81}$$

Plugging this result into the previous conditions we see that the Green's function of the conductor in the presence of an infinite wire is given by

$$G_C = \left[EI - H - T_A^+ G_A^{\mathrm{R}} T_A\right]^{-1}. \tag{4.82}$$

In contrast to G_A^{R} the matrix G_C of the conductor in the presence of an infinite lead is finite. To obtain the Green's function the last term has to be evaluated only for points at the interface between conductor and lead.

What is self-energy?

The last term in the Green's function matrix, containing the effect of the lead on the propagation in the conductor, is usually called the "self energy". The self-energy in itself is an additional energy term, usually complex, which has an analogue in electron–electron and electron–phonon interactions. Here, we follow Datta [1] and recent work in transport theory by considering it as just another Hamiltonian-like entity, which arises naturally as soon as we consider open instead of closed systems. Using the symbol Σ, we can write for the Green's function of the conductor coupled to an infinite lead the result

$$G_C^{\mathrm{R}} = \left[EI - H_C - \Sigma^{\mathrm{R}} \right]^{-1}, \tag{4.83}$$

$$\left[\Sigma^{\mathrm{R}} \right]_{i,i'} = t^2 G_A^{\mathrm{R}}(i,i'), \tag{4.84}$$

where t is the hopping parameter between adjacent lattice points (see above) and (i,i') are points at the interface from the wire to the conductor.

Wires with multiple modes

The concept can be generalized to wires with multiple modes and a discrete cross section described by a transversal potential $U(\mathbf{r}_2)$. The procedure is similar to the one used in deriving the Fisher–Lee equation. However, it is somewhat modified in the case of a discrete Hamiltonian and under the condition of a semi-infinite wire that terminates at $z = 0$. For details please refer to Datta [1], Chapter 3. The Green's function of a semi-infinite wire on a discrete lattice is described in the plane $z = a$ at the conductor by

$$G_A^{\mathrm{R}}(i,i') = -\frac{1}{t} \sum_m \chi_m(i) \exp(ik_m a) \chi(i'). \tag{4.85}$$

The self-energy in this case is consequently

$$\Sigma_A^{\mathrm{R}}(i,i') = -t \sum_{m(A)} \chi_m(i) \exp(ik_m a) \chi_m(i'). \tag{4.86}$$

Calculating the advanced self energy Σ_A^{A} we likewise obtain

$$\Sigma_A^{\mathrm{A}}(i,i') = -t \sum_{m(A)} \chi_m(i) \exp(-ik_m a) \chi_m(i'). \tag{4.87}$$

Taking into account that the momentum $\hbar v_m$ in a discrete system is equal to ([1], Chapter 3),

$$\hbar v_m = \frac{\partial E_m}{\partial k_m} = 2at \sin(k_m a), \tag{4.88}$$

we can define for the difference between retarded and advanced self energy of lead A a new quantity, labeled Γ:

$$i\left[\Sigma_A^{\mathrm{R}}(i,i') - \Sigma_A^{\mathrm{A}}(i,i')\right] = \sum_{m(A)} \chi_m(i)\left[2t\sin(k_m a) = \frac{\hbar v_m}{a}\right]\chi_m(i') \equiv \Gamma_A, \tag{4.89}$$

$$\Gamma_A = i\left[\Sigma_A^{\mathrm{R}} - \Sigma_A^{\mathrm{A}}\right]. \tag{4.90}$$

One may call this new quantity a *contact*. It provides a very compact notation for calculating the transmission through a conductor with two leads. From (4.68),

$$S_{nm} = -\delta_{nm} + i\hbar\sqrt{v_n v_m}\iint d^2r d^2r' \chi_n(\mathbf{r'}_2)\, G^R\left(\mathbf{r'}_2, \mathbf{r}_2\right)\, \chi_m(\mathbf{r}_2), \tag{4.91}$$

we get for the square of S_{nm}, under the condition that $m \neq n$,

$$|S_{nm}|^2 = \frac{\hbar^2 v_n v_m}{a^2}\sum_{i,i',j,j'} \chi_n(j)\chi_n(j')\, G^{\mathrm{R}}(j,i)\, \chi_m(i)\chi_m(i')\, G^{\mathrm{A}}(i',j'), \tag{4.92}$$

where we have used the usual prescription for the transformation between integrals and summations, and also the fact that the transposed and conjugated retarded Green's function is equal to the advanced one:

$$\int d(xy) \to \frac{1}{a}\sum_{i(xy)} \qquad G^{\mathrm{A}}(i',j') = G^{\mathrm{R}}(j',i')^* \tag{4.93}$$

Substituting the contacts Γ_A and Γ_B,

$$\Gamma_A(i,i') = \chi_m(i)\frac{\hbar v_m}{a}\chi_m(i'), \tag{4.94}$$

$$\Gamma_B(j,j') = \chi_n(j)\frac{\hbar v_n}{a}\chi_n(j'), \tag{4.95}$$

into the equation, we get for the transmission

$$\overline{T}_{BA} = \sum_{m(A)}\sum_{n(B)} |S_{nm}|^2 = \sum_{i,i',j,j'} \Gamma_B(j',j) G^{\mathrm{R}}(j,i)\Gamma_A(i,i')G^{\mathrm{A}}(i',j'). \tag{4.96}$$

The expression to the right is just the sum over all the diagonal elements of the resulting matrix, or the *trace* (Tr) of the matrix:

$$\overline{T}_{BA}(E) = Tr\left[\Gamma_B(E)G^{\mathrm{R}}(E)\Gamma_A(E)G^{\mathrm{A}}(E)\right]. \tag{4.97}$$

Landauer–Büttiker equation

A transmission probability of $\overline{T} = 1$ contributes a conductance quantum $2e^2/h$ to the total conductance through the conductor. The conductance $\sigma(E)$ is thus

$$\sigma(E) = \frac{dI(E)}{dV} = \frac{2e^2}{h} Tr \left[\Gamma_B(E) G^{\mathrm{R}}(E) \Gamma_A(E) G^{\mathrm{A}}(E) \right]. \tag{4.98}$$

Integrating over an energy range E_0 to E_1 we obtain the current through the device:

$$I = \frac{1}{e} \int_{E_0}^{E_1} \sigma(E) dE = \frac{2e}{h} \int_{E_0}^{E_1} dE \, Tr \left[\Gamma_B(E) G^{\mathrm{R}}(E) \Gamma_A(E) G^{\mathrm{A}}(E) \right]. \tag{4.99}$$

Apart from the electron distribution functions, which are strictly speaking due to the chemical potentials of the leads and thus do not enter the picture here, this is one formulation of the Landauer–Büttiker equation [5], widely used in current transport codes. It should be noted that the relation is valid only in the limit of zero bias. This fact, which bears on the inclusion of distribution functions from the outset, is related to the omission of the change of electron properties in the conductor due to finite bias voltages in the leads. It will be analyzed in more detail in the following sections.

Physically speaking, the transmission through the conductor is thus the product of all the different pathways described by the contacts to the leads A, B and the propagation through the conductor. The expression is generally valid for elastic transport through an interface, even though, as seen in the following sections, the Green's functions and self energies in solid systems are generally calculated somewhat differently.

The main result of this section, which should be remembered in the following, is the elimination of infinite leads from the resulting transport equations, even though these leads are, at least within the finite differences method, exactly accounted for. This feature is quite astonishing; it seems to be due to the fact that the changes to conductance properties can actually be localized at the interface, the contacts of the conductor. Since it is clear, from the preceding exposition, that the Green's functions of the leads enter the description of transport properties only in a very limited sense, in fact only through their properties at the interface to the conductor, we shall drop the explicit notation and generally refer to the Green's function of the conductor as *the* Green's function of the system.

4.3 Nonequilibrium conditions

So far, we have treated a system in equilibrium. Even though there is no limitation on the atomic structure of the leads of the conducting interface, we have neglected two essential ingredients of electron transport in most situations: the effect of finite bias voltage, and the effect of thermal conditions.

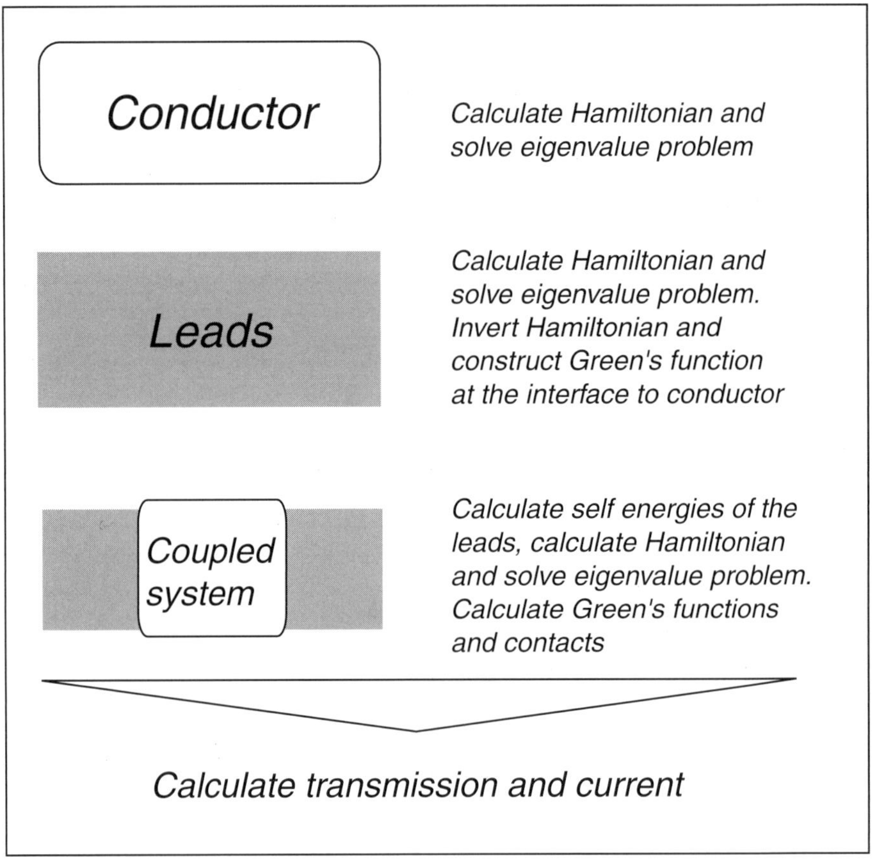

Fig. 4.7. Computational scheme to calculate the current through a conductor with two infinite leads based on elastic transport theory.

4.3.1 Finite-bias voltage

Finite-bias voltage can be included by a scheme suggested by Taylor [6]. In this scheme the conductor interface is chosen in such a way that it includes a few layers of the infinite leads. The Hartree potential of the lead (the Coulomb term) is shifted by a finite value $\Delta V_H = eV_{\text{bias}}$. If the two leads of the conductor are changed by different values of ΔV_H, then the net effect will be a voltage drop across the conductor, which can be calculated self-consistently using the changed values of the Hartree potentials at the leads as boundary conditions for the solution of the Poisson equation. This, in turn, makes it possible to continue the conductor potential smoothly into the lead (see Figure 4.8).

A solution of the Poisson equation is part of every DFT self-consistency cycle, since the effective potential, used to solve the Kohn–Sham equations [7],

$$V_{\text{eff}}(\mathbf{r}) = V_H[\rho(\mathbf{r})] + V_{XC}[\rho(\mathbf{r})] \tag{4.100}$$

contains the Hartree term and the exchange–correlation potential V_{XC}. In principle, including finite-bias voltages should thus be accessible to every standard DFT method. Convergence of the method can be tested numerically by convergence of the charge density at the boundaries between the leads and the conductor. For a supercell geometry, which is used in many state-of-the-art DFT codes, the procedure needs to be slightly modified. In this case the external Hartree potential will be linear with the position within the lead, and reach its maximum at the interface, while it is zero at the limit of the supercell (see Figure 4.8). In this case it also seems infeasible to choose two different leads, since the interface between the two leads at the supercell boundary will introduce artificial scattering effects.

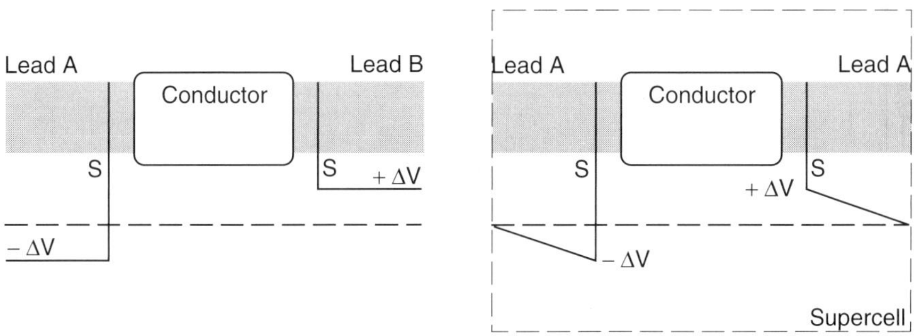

Fig. 4.8. Finite-bias voltage for calculating the electronic structure within the conductor. Additional Hartree potential for a cluster approach (left) and a supercell geometry (right). Screening within the metal leads has the effect that the lead bulk conditions are already obtained within a few atomic layers from the conductor interface (S).

The treatment rests on the assumption that the effect of an additional potential in a metal will be screened within a few atomic layers. However, for long range effects, bound to occur, for example, in semiconductors, the assumption is not generally justified.

4.3.2 Spectral functions and charge density

Applying a bias voltage to the two leads raises a problem not accounted for in standard DFT methods. There, a self-consistency cycle usually consists in solving the Kohn–Sham equations for the effective potential of a given charge distribution, and calculating the updated charge density by filling all electron states up to the chemical potential μ of the system. For a system containing N electrons this means that

$$N = \int_{V_S} d^3r\, n(\mathbf{r}) = \int_{V_S} d^3r \sum_i^{\epsilon_i=\mu} \psi_i^*(\mathbf{r})\psi_i(\mathbf{r}). \tag{4.101}$$

Here, V_S is the volume of the system, ϵ_i the energy eigenvalue of state i, and $\psi_i(\mathbf{r})$ its Kohn–Sham orbital. In a system that involves a finite potential between the two leads, no such Fermi level can be defined. A different route to tackle the problem is to calculate the charge density from the Green's function of a system. From the spectral decomposition of the retarded Green's function under the condition that $\mathbf{r} = \mathbf{r}'$ (See (4.49)),

$$G^{\mathrm{R}}(\mathbf{r}, \mathbf{r}, E) = \sum_i \frac{\psi_i^*(\mathbf{r})\psi_i(\mathbf{r})}{E - \epsilon_i + i\eta}, \tag{4.102}$$

we get for the real and imaginary parts, by multiplying by $E - \epsilon_i - i\eta$:

$$G^{\mathrm{R}}(\mathbf{r}, \mathbf{r}, E) = \underbrace{\sum_i \frac{(E - \epsilon_i)\psi_i^*(\mathbf{r})\psi_i(\mathbf{r})}{(E - \epsilon_i)^2 + \eta^2}}_{=\Re(G^{\mathrm{R}})} - i \underbrace{\sum_i \frac{\eta\psi_i^*(\mathbf{r})\psi_i(\mathbf{r})}{(E - \epsilon_i)^2 + \eta^2}}_{=\Im(G^{\mathrm{R}})}. \tag{4.103}$$

In the limit $\eta \to 0$, the factor containing the energy-dependency transforms into a delta functional:

$$\lim_{\eta \to 0} \frac{\eta}{(E - \epsilon_i)^2 + \eta^2} = \pi\delta(E - \epsilon_i). \tag{4.104}$$

We obtain therefore for the imaginary part

$$\Im\left[G^{\mathrm{R}}(\mathbf{r}, \mathbf{r}, E)\right] = -\pi \sum_i |\psi_i^*(\mathbf{r})|^2\, \delta(E - \epsilon_i) = -\, \pi n(\mathbf{r}, E). \tag{4.105}$$

The charge density ρ at a given location and energy can therefore be obtained from the Green's function at this particular location:

$$n(\mathbf{r}, E) = -\frac{1}{\pi} \Im\left[G^{\mathrm{R}}(\mathbf{r}, \mathbf{r}, E)\right]. \tag{4.106}$$

An identical deduction could start from the advanced, instead of the retarded, Green's function. In this case the result will be

$$n(\mathbf{r}, E) = +\frac{1}{\pi} \Im\left[G^{\mathrm{A}}(\mathbf{r}, \mathbf{r}, E)\right]. \tag{4.107}$$

Combining the two results, we may write the charge density as a difference between G^{R} and G^{A} with

$$n(\mathbf{r}, E) = \frac{i}{2\pi}\left[G^{\mathrm{R}}(\mathbf{r}, \mathbf{r}, E) - G^{\mathrm{A}}(\mathbf{r}, \mathbf{r}, E)\right]. \tag{4.108}$$

The charge density in this case can be seen as the diagonal element (due to $\mathbf{r} = \mathbf{r}'$) of a more general structure, which is called the *spectral function*

$A(\mathbf{r}, \mathbf{r}', E)$. The trace of this spectral function, i.e., the sum over its diagonal elements, gives the number of electron states:

$$A(\mathbf{r}, \mathbf{r}', E) \equiv i\left[G^{\mathrm{R}}(\mathbf{r}, \mathbf{r}', E) - G^{\mathrm{A}}(\mathbf{r}, \mathbf{r}', E)\right], \tag{4.109}$$

$$A \equiv i\left[G^{\mathrm{R}} - G^{\mathrm{A}}\right] \qquad n(E) = \frac{1}{2\pi} Tr[A(E)]. \tag{4.110}$$

This means that once we know the Green's functions of our system, we can compute the total charge by simply integrating the trace of the spectral function over energy:

$$N = \int_{-\infty}^{E_0} \frac{dE}{2\pi} Tr[A(E)] = N_0. \tag{4.111}$$

This, in turn, tells us the energy level E_0, which forms the upper limit of the integration, if we require that the total number of electrons in the system remain constant, N_0. In this way the problem of defining the Fermi level for different parts of the coupled lead–conductor–lead circuit is avoided. Every eigenvalue then corresponds either to an occupied (if $\epsilon_i < E_0$) or unoccupied (if $\epsilon_i > E_0$) state of the electrons. From this information, the charge density throughout the system can be computed.

4.3.3 Spectral functions and contacts

The spectral function is in fact a more general version of the Green's function. Green's functions relate a wavefunction of unit amplitude at one point of the system $\mathbf{r}$ to the amplitude of the wavefunction at a point $\mathbf{r}'$, while the spectral function relates an arbitrary amplitude at $\mathbf{r}$ to the amplitude at $\mathbf{r}'$. To demonstrate this feature let us consider $A(\mathbf{r}, \mathbf{r}', E)$, where $E = \epsilon_k$. The energy is thus equal to one of the eigenvalues of the system. If η is chosen sufficiently small, the summation can be limited to one term only, k:

$$G^{\mathrm{R}}(\mathbf{r}, \mathbf{r}', \epsilon_k) \approx \frac{\psi_k^*(\mathbf{r}')\psi_k(\mathbf{r})}{i\eta}, \qquad G^{\mathrm{A}}(\mathbf{r}, \mathbf{r}', \epsilon_k) \approx -\frac{\psi_k^*(\mathbf{r}')\psi_k(\mathbf{r})}{i\eta}. \tag{4.112}$$

The spectral function is consequently

$$A(\mathbf{r}, \mathbf{r}', \epsilon_k) = i[G^R(\mathbf{r}, \mathbf{r}', \epsilon_k) - G^A(\mathbf{r}, \mathbf{r}', \epsilon_k)] \approx \frac{2}{\eta}\psi_k^*(\mathbf{r}')\psi_k(\mathbf{r}). \tag{4.113}$$

It describes the correlation of amplitudes between two different points of the system. For this reason it is sometimes called a correlation function. It is related not only to the Green's function, but also to the contacts of a system. To show this feature we use the matrix definition of the Green's function in the presence of a lead, and the definition of a contact (see (4.90)):

$$G^{\mathrm{A}} = \left[EI - H - \Sigma^{\mathrm{A}}\right]^{-1},$$
$$G^{\mathrm{R}} = \left[EI - H - \Sigma^{\mathrm{R}}\right]^{-1}, \qquad i\Gamma = \Sigma^{\mathrm{A}} - \Sigma^{\mathrm{R}}. \tag{4.114}$$

For the difference between the inverse retarded and advanced Green's functions we get consequently

$$\left[G^{\mathrm{R}}\right]^{-1} - \left[G^{\mathrm{A}}\right]^{-1} = EI - H - \Sigma^{\mathrm{R}} - EI + H + \Sigma^{\mathrm{A}} = i\Gamma. \tag{4.115}$$

Multiplying from the left by G^{R} and from the right by G^{A} gives

$$G^{\mathrm{A}} - G^{\mathrm{R}} = iA = iG^{\mathrm{R}}\Gamma G^{\mathrm{A}} \quad \Rightarrow \quad A = G^{\mathrm{R}}\Gamma G^{\mathrm{A}}. \tag{4.116}$$

Exchanging G^{R} and G^{A} in the multiplication gives us a second relation so that

$$A = G^{\mathrm{R}}\Gamma G^{\mathrm{A}} = G^{\mathrm{A}}\Gamma G^{\mathrm{R}}. \tag{4.117}$$

Consider now the Landauer–Büttiker equation again, where the transmission channels between two leads A, B are described by the trace of the matrix product:

$$\overline{T}_{BA}(E) = Tr\left[\Gamma_B(E)G^{\mathrm{R}}(E)\Gamma_A(E)G^{\mathrm{A}}(E)\right] = Tr\left[\Gamma_B(E)A(E)\right]. \tag{4.118}$$

Comparing with the standard formulations in quantum statistics, where the average value of an operator $\hat{O}$ is given by the trace of the product of the operator and the density of states ρ, $\langle\hat{O}\rangle = Tr\left[\hat{O}\rho\right]$, we see that the spectral function plays essentially the role of a generalized density of states. The equation can then be interpreted as describing the transmission of contact Γ_A in the presence of the system comprising the conductor and lead B.

The definition of the spectral function and the derived relations might seem somewhat academic at this point, but we will see presently that they allow us to understand the essential concepts of electron propagation when systems are driven out of equilibrium. To analyze nonequilibrium situations in more detail we first have to turn again to self-energies.

4.3.4 Self-energy Σ again

So far we have considered the self-energy as an additional complex term to the Hamiltonian, which describes the changes to electron states in a conductor, if it is coupled to one or more leads. Within the conductor we have assumed that the electron waves are coherent, i.e., they are in phase over the length scale of the conductor. But this is true only, if the electrons do not interact with each other or with phonons of the conductor lattice. Interactions induce additional phase breaking within the conductor itself, which needs to be included to obtain the transmission from one lead to the other.

A way to include phase breaking for different processes was shown by Keldysh in 1964 [8]. The main achievement of the method was to relate the equilibrium state Green's functions G^{R} and G^{A} to the Green's functions of a system under nonequilibrium conditions. Let us first consider the effect of self energies Σ on the current flow within the conductor. If we rewrite the matrix definition of the retarded Green's function (4.114) to a local representation, we have to replace matrix multiplications by integrals over space. Thus we get

$$(E-H)\,G^{\mathrm{R}}(\mathbf{r},\mathbf{r}')-\int d^3r_1\,\Sigma^{\mathrm{R}}(\mathbf{r},\mathbf{r}_1)\,G^{\mathrm{R}}(\mathbf{r},\mathbf{r}_1)=\delta(\mathbf{r}-\mathbf{r}'). \tag{4.119}$$

The source term on the right represents the unit excitation at point $\mathbf{r}'$. If we omit this term, then we arrive at a Schrödinger equation including the self-energy term:

$$E\Psi(\mathbf{r})=H\Psi(\mathbf{r})-\int d^3r_1\,\Sigma^{R}(\mathbf{r},\mathbf{r}_1)\,\Psi(\mathbf{r}_1). \tag{4.120}$$

The self-energy term in this case shows up as an additional energy component in the Hamiltonian. Previously, we considered only the effect of infinite leads on the conductance properties, by relating self-energy to transmission probabilities via the Green's function G^{R} of the conductor. In a more general picture we may consider self-energies as a potential, which not only signifies leads, but also phase-altering processes within the conductor itself. The reason such a view is justified can be seen if we consider the sources and sinks of currents in the conductor. From the current operator for wavefunctions $\Psi(\mathbf{r})$,

$$J(\mathbf{r})=\frac{i\hbar e}{2m}\left[\Psi(\mathbf{r})\nabla\Psi^*(\mathbf{r})-\Psi^*(\mathbf{r})\nabla\Psi(\mathbf{r})\right], \tag{4.121}$$

we get for the divergence

$$\nabla J(\mathbf{r})=\frac{ie}{\hbar}\left[\Psi^*(\mathbf{r})(H\Psi(\mathbf{r}))-\Psi(\mathbf{r})(H\Psi(\mathbf{r}))^*\right]. \tag{4.122}$$

Using the expressions from (4.120), we obtain for the source of current

$$\begin{aligned}\nabla J(\mathbf{r})=&\frac{ie}{\hbar}\left[\int d^3r_1\Psi(\mathbf{r}_1)\Sigma^{\mathrm{R}}(\mathbf{r},\mathbf{r}_1)\Psi^*(\mathbf{r})-\int d^3r_1\Psi^*(\mathbf{r}_1)\Sigma^{\mathrm{A}}(\mathbf{r}_1,\mathbf{r})\Psi(\mathbf{r})\right]\\&-\frac{e}{\hbar}\left\{\int d^3r_1\Psi(\mathbf{r}_1)\Psi^*(\mathbf{r})i\left[\Sigma^{\mathrm{R}}(\mathbf{r},\mathbf{r}_1)-\Sigma^{\mathrm{A}}(\mathbf{r},\mathbf{r}_1)\right]\right\}.\end{aligned} \tag{4.123}$$

Here, we have used the fact that the advanced self energy is the conjugate of the retarded one and exchanged the variables in the second integral. Integrating over $\mathbf{r}$ and remembering that a contact is defined by $i\Gamma=\Sigma^{\mathrm{A}}-\Sigma^{\mathrm{R}}$, we obtain for the current sinks

$$\int d^3r\, \nabla J(\mathbf{r}) = \frac{e}{\hbar} \int d^3r\, \Psi^*(\mathbf{r}) \int d^3r_1 \Gamma(\mathbf{r}, \mathbf{r}_1) \Psi(\mathbf{r}_1). \tag{4.124}$$

At every point of the conductor where the self-energies are not zero, we encounter either a current source or a sink since coherent propagation of electron waves terminates at this point or another coherent trajectory starts. Essentially, we deal at this point with the transition from one state of electrons in phase space to another one.

The physical properties of the contact depend on the interactions considered, as does the exact form of the self-energy accounting for it. In the formalism of nonequilibrium Green's functions, this feature of a system is included by two new variables, symbolized by $\Sigma^<$ and $\Sigma^>$. There is some variability about the definition and the name of these functions. Traditionally, they are called Σ^{R} and Σ^{A}, "self energies", e.g. in the original Keldysh publications [8], in the papers by Appelbaum and Brinkman [9], and Caroli et al. [10]. In the book by Datta, as well as in some more recent publications, they are referred to as "scattering functions" [1]. We shall retain the name self-energy as well as the traditional notation throughout this book. As before, the difference between the self- energies $\Sigma^>$ and $\Sigma^<$ defines a contact, i.e., a point, where the current either has a source or a sink. In matrix notation,

$$\Gamma(E) = i\left[\Sigma^>(E) - \Sigma^<(E)\right]. \tag{4.125}$$

We have included the energy dependency of the matrices as a reminder to the reader that all quantities depend on the energy considered. "Sigma greater than" and "sigma less than", $\Sigma^>(E)$ and $\Sigma^<(E)$, are not equal to $\Sigma^{\mathrm{R}}(E)$ and $\Sigma^{\mathrm{A}}(E)$. While the latter describe only the existence of electron states at a specific energy, the former also describe whether these states are occupied.

Self energies $\Sigma^<$ and $\Sigma^>$ of leads

The difference can be shown, for example, for a lead that may have a different chemical potential μ due to a changed Hartree potential within the lead, but is otherwise thought to be in thermal equilibrium. In this case "sigma less than" and "sigma greater than" comply with the following relations [1]:

$$\Sigma^<(E) = i f(E,\mu)\Gamma(E), \qquad \Sigma^>(E) = i\left[f(E,\mu) - 1\right]\Gamma(E), \tag{4.126}$$

where f is the Fermi distribution function for a given chemical potential μ. Their difference in this case can also be stated in terms of the self-energies $\Sigma^{\mathrm{A}}(E)$ and $\Sigma^{\mathrm{R}}(E)$, since

$$i\left[\Sigma^>(E) - \Sigma^<(E)\right] = \Gamma(E) = i\left[\Sigma^{\mathrm{R}}(E) - \Sigma^{\mathrm{A}}(E)\right]. \tag{4.127}$$

At present, there exists no theoretical treatment of leads out of equilibrium. However, this is not decisive, since the conducting interface can always be assumed large enough to include all relevant inelastic and nonequilibrium effects.

Self-energy due to electron–electron interactions

In the lowest order of a perturbation expansion, described by the Hartree–Fock approximation, the contact due to electron–electron interactions is zero (see [1], p. 307). Thus

$$i\left[{}^{ee}\Sigma^{>}(E) - {}^{ee}\Sigma^{<}(E)\right] = {}^{ee}\Gamma(E) = 0. \tag{4.128}$$

Since we shall be concerned only with the lowest-order expansions, we can safely neglect electron–electron processes in the description of nonequilibrium processes.

Self-energy due to electron–phonon interactions

The same does not hold for electron–phonon interactions. Physically speaking, the process is quite clear: an electron excites a phonon along its trajectory; it loses energy and continues its path along a different trajectory. The formalism to describe the processes has to account not only for the loss of energy due to phonon excitation, but also for a potential gain if energy is transferred back from the phonons to the propagating electrons. In this case, and in a local representation, the self-energies are described by (see [11]):

$${}^{eph}\Sigma^{<}(\mathbf{r}, \mathbf{r}', E) = \int d(\hbar\omega) D(\mathbf{r}, \mathbf{r}', \hbar\omega) G^{<}(\mathbf{r}, \mathbf{r}', E - \hbar\omega), \tag{4.129}$$

$${}^{eph}\Sigma^{>}(\mathbf{r}, \mathbf{r}', E) = \int d(\hbar\omega) D(\mathbf{r}, \mathbf{r}', \hbar\omega) G^{>}(\mathbf{r}, \mathbf{r}', E + \hbar\omega). \tag{4.130}$$

The functions D describe the correlation and energy spectrum of the phonon adsorption and emission processes. $\hbar\omega < 0$ corresponds to emission, and $\hbar\omega > 0$ to adsorption of a phonon. Here, we have also introduced the nonequilibrium Green's functions $G^{<}$ and $G^{>}$, which will be defined in terms of the retarded and advanced Green's functions in the next section. It is not straightforward to connect either of these equations to a specific process, either the adsorption of energy from a propagating electron by a phonon, or the emission of phonon energy and the termination of a phonon excitation. The ultimate reason for this ambiguity is the structure of correlations: processes that may have a specific order in time and a corresponding unique transfer of energy, may possess opposite features if they are considered along reversed time evolution. For correlations, both pathways are permissible: a strict order of events is therefore no longer described by the correlations. The Green's functions in both cases reflect the response of the system to this process. This ambiguity becomes especially clear if we look at the explicit form of D in the integral equations, which describes the phonon correlation function:

$$D(\mathbf{r},\mathbf{r}',\hbar\omega) = \sum_q |U_q|^2 \left[e^{-iq(\mathbf{r}-\mathbf{r}')}N_q\delta(\omega-\omega_q) + e^{+iq(\mathbf{r}-\mathbf{r}')}(N_q+1)\delta(\omega+\omega_q)\right]. \tag{4.131}$$

In this relation U_q is the interaction potential between electrons and a phonon mode of wavevector q, the delta functions describe energy conservation, and the phonon distribution function N_q is the Bose–Einstein distribution function

$$N_q = \left[\exp(\hbar\omega/k_B T) - 1\right]^{-1}. \tag{4.132}$$

The functional form of D implies that it accounts for two separate processes described simultaneously by the phonon correlation function. If we consider one specific phonon mode, the self-energy $\Sigma^<$ will be

$$\begin{aligned} {}^{\text{eph}}\Sigma^<(\mathbf{r},\mathbf{r}',E) = |U_q|^{\,2} \Big[& e^{-iq(\mathbf{r}-\mathbf{r}')}N_q G^<(\mathbf{r},\mathbf{r}',E-\hbar\omega_q) \\ & + \; e^{+iq(\mathbf{r}-\mathbf{r}')}(N_q+1)G^<(\mathbf{r},\mathbf{r}',E+\hbar\omega_q)\Big]. \end{aligned} \tag{4.133}$$

Here, the first line describes the adsorption of a phonon by an electron (which is part of the system of propagating electrons and thus encoded in $G^<$) with energy $(E - \hbar\omega_q)$, while the second gives the emission of a phonon by an electron with energy $(E + \hbar\omega_q)$. Both processes contribute to the number of phonons with $E_q = \hbar\omega_q$; therefore both processes have to be included in the self-energy. This feature, and the corresponding ambiguity in the formulations, which is essentially due to an accounting problem, makes self-energies due to inelastic processes rather difficult to understand or to visualize.

Sum rules for self-energies

In first-order perturbation theory, which we used for the self-energies of all interactions so far, it is always assumed that the process, or the perturbation, is small enough so that the rest of the system remains in equilibrium. For the existence of a number of different origins of self-energy terms this means that we can treat every term on its own and consider the total effect on the system of propagating electrons as a sum of partial effects described, individually, by a self-energy term. We have seen that electron–electron interactions can remain unconsidered, since the first-order perturbation result makes self-energy terms vanish. If we consider a system containing two leads A, B and inelastic effects in the conducting interface, then the total self-energy will be a sum of all contributions, or

$$\Sigma^{<}(\mathbf{r},\mathbf{r}',E) = \Sigma^{<}_{A}(\mathbf{r},\mathbf{r}',E) + \Sigma^{<}_{B}(\mathbf{r},\mathbf{r}',E) + {}^{\mathrm{eph}}\Sigma^{<}(\mathbf{r},\mathbf{r}',E), \quad (4.134)$$

$$\Sigma^{>}(\mathbf{r},\mathbf{r}',E) = \Sigma^{>}_{A}(\mathbf{r},\mathbf{r}',E) + \Sigma^{>}_{B}(\mathbf{r},\mathbf{r}',E) + {}^{\mathrm{eph}}\Sigma^{>}(\mathbf{r},\mathbf{r}',E). \quad (4.135)$$

Considering a system composed of a conductor and two leads, taking into account electron–phonon interactions within the conductor, and assuming that the leads A, B are in thermal equilibrium, we obtain for the self-energies (in matrix notation)

$$\begin{aligned}\Sigma^{<}(E) = &\ [if(E,\mu_A)\Gamma_A(E) + if(E,\mu_B)\Gamma_B(E)] \\ &+ D(\hbar\omega)G^{<}(E-\hbar\omega), \qquad (4.136)\\ \Sigma^{>}(E) = &\ [i\left(f(E,\mu_A)-1\right)\Gamma_A(E) + i\left(f(E,\mu_B)-1\right)\Gamma_B(E)] \\ &+ D(\hbar\omega)G^{>}(E+\hbar\omega). \qquad (4.137)\end{aligned}$$

Apart from the nonequilibrium Green's functions, treated presently, the main problem of electron transport in open systems is to find the self-energies of the infinite leads. Surveying the literature and considering the self-consistency procedure, e.g., due to the change of Hartree potentials at the two leads [6], this is indeed the main obstacle for a wide application of the formalism.

A note on supercells

In a supercell approach, the problem might actually be easier to solve. Given a linear increase of the Hartree potential over the length of the coupled leads (see Figure 4.8), the reaction of the system will be to develop two surface dipoles at the interfaces with the conductor. In the bulk region of the coupled leads, the electron distribution can thus be expected to be close to the equilibrium distribution without any applied bias voltage. In this case the equations will reduce to

$$\Sigma^{<}(E) = if(E,\mu_0)\left[\Gamma^{\mathrm{L}}_{A}(E) + \Gamma^{\mathrm{R}}_{A}(E)\right] + D(\hbar\omega)G^{<}(E-\hbar\omega), \quad (4.138)$$

$$\Sigma^{>}(E) = i\left[f(E,\mu_0)-1\right]\left[\Gamma^{\mathrm{L}}_{A}(E) + \Gamma^{\mathrm{R}}_{A}(E)\right] + D(\hbar\omega)G^{>}(E+\hbar\omega).$$

Here, $\Gamma^{\mathrm{L(R)}}_{A}(E)$ denotes the contact at the left (right) of the conductor. If the contacts are calculated far inside the leads, so that the surface dipoles are part of the conductor interface, then they need to be calculated only for the equilibrium state (with the chemical potential μ_0) of the leads. However, so far, transport simulations of open systems are performed exclusively within a local basis set, and no way has been found to include the treatment into our most precise electronic structure methods, i.e., plane wave and full potential density functional calculations.

4.3.5 Nonequilibrium Green's functions

The nonequilibrium Green's function of the system depends on the advanced and retarded Green's functions as defined above, and the self-energies including inelastic effects. Within the formalism developed by Keldysh [8], they are given by the following matrices:

$$G^{<}(E) = G^{\mathrm{R}}(E)\Sigma^{<}(E)G^{\mathrm{A}}(E), \qquad G^{>}(E) = G^{\mathrm{R}}(E)\Sigma^{>}(E)G^{\mathrm{A}}(E). \tag{4.139}$$

In a real space representation the same relations read

$$G^{<}(\mathbf{r},\mathbf{r}',E) = \iint d^3r_1 d^3r_2 G^{\mathrm{R}}(\mathbf{r},\mathbf{r}_1,E)\Sigma^{<}(\mathbf{r}_1,\mathbf{r}_2,E)G^{\mathrm{A}}(\mathbf{r}_2,\mathbf{r}',E),$$
$$G^{>}(\mathbf{r},\mathbf{r}',E) = \iint d^3r_1 d^3r_2 G^{\mathrm{R}}(\mathbf{r},\mathbf{r}_1,E)\Sigma^{>}(\mathbf{r}_1,\mathbf{r}_2,E)G^{\mathrm{A}}(\mathbf{r}_2,\mathbf{r}',E). \tag{4.140}$$

While it is possible to relate charge density and total charge to the retarded and advanced Green's function of a system in equilibrium (see (4.108)) this is valid only in the nonequilibrium case for an energy level smaller than the Fermi level of both leads [6]. To calculate the total charge of a system, the energy integration of section to the charge density of the system (see Section 4.3.2) must be split into two parts: the first part relies on the retarded Green's function $G^{\mathrm{R}}(E)$, it will give a value n_1:

$$n_1 = -\frac{i}{2\pi}\int_{-\infty}^{E_1} dE Tr[G^{\mathrm{R}}(E) - G^{\mathrm{A}}(E)]. \tag{4.141}$$

Here E_1 is the minimum value of $(\mu_A - eV, \mu_B + eV)$. In the intermediate range the charge within the system must be determined from the nonequilibrium function. Since G^{R} is nonanalytic below the real axis and G^{A} is nonanalytic above, the integration over energy for $G^{<}$ has to be performed along the real axis. Thus

$$n_2 = \int_{E_1}^{E_2} \frac{dE}{2\pi} Tr[G^{<}(E)], \tag{4.142}$$

where E_2 is the maximum value of $(\mu_A - eV, \mu_B + eV)$. The total charge in the system can now be set constant, so that $n_1 + n_2 = N_0$. We note at this place that the nonequilibrium Green's function is related to charge density in the energy range between the effective potentials of the two leads. Since this is also the range in which electron transport occurs, we may relate $G^{<}(E)$ to the square of a hypothetical many-body wavefunction and use the relation to link transport properties to the nonequilibrium Green's function. Exact derivations of current under nonequilibrium conditions are given in the literature cited at the end of this chapter. However, they are commonly cast in the symbols and concepts of second quantization, which makes them quite difficult to comprehend for nonspecialists; the essential relations can, moreover, be obtained in the heuristic fashion described in the next subsection [1].

4.3.6 Electron transport in nonequilibrium systems

We saw that the diagonal elements of the nonequilibrium Green's function are equal to the charge density. We therefore interpret the function as the product of amplitudes of a hypothetical many-body wavefunction at two different locations $\mathbf{r}$ and $\mathbf{r}'$:

$$\frac{1}{2\pi}\left[G^{<}(\mathbf{r},\mathbf{r},E)\right] = n(\mathbf{r},E) \quad \longrightarrow \quad G^{<}(\mathbf{r},\mathbf{r}',E) = 2\pi\, i\, \Psi^*(\mathbf{r}')\Psi(\mathbf{r}). \tag{4.143}$$

The current density $J(\mathbf{r},E)$ is given by the derivative of the density, it complies with

$$J(\mathbf{r},E) = -\frac{ie\hbar}{2m}\left[(\nabla - \nabla')\,\Psi(\mathbf{r})\Psi^*(\mathbf{r}')\right]_{\mathbf{r}=\mathbf{r}'}, \tag{4.144}$$

where the differential ∇ acts on the coordinate $\mathbf{r}$, while ∇' acts on $\mathbf{r}'$. Using the transformation (4.143), this leads to

$$J(\mathbf{r},E) = -\frac{1}{2\pi}\frac{e\hbar}{2m}\left[(\nabla - \nabla')\,G^{<}(\mathbf{r},\mathbf{r}',E)\right]_{\mathbf{r}=\mathbf{r}'}. \tag{4.145}$$

With the same transformation we obtain for the sources of current the following relation:

$$\nabla J(\mathbf{r},E) = \frac{e}{h}\left[H(\mathbf{r})G^{<}(\mathbf{r},\mathbf{r}',E) - G^{<}(\mathbf{r},\mathbf{r}',E)H^*(\mathbf{r}')\right]_{\mathbf{r}=\mathbf{r}'}. \tag{4.146}$$

Here we have reordered the second term to comply with the order of multiplication used for matrices, which shall be introduced presently. Given that the source is the diagonal expression of the term in square brackets, we use the same method as before for the charge density and define a general source function S by

$$S(\mathbf{r},\mathbf{r},E) \equiv \nabla J(\mathbf{r},E). \tag{4.147}$$

The general relation between source function $S(\mathbf{r},\mathbf{r}',E)$ and nonequilibrium Green's function $G^{<}(\mathbf{r},\mathbf{r}',E)$ then reads

$$S(\mathbf{r},\mathbf{r}',E) = \frac{e}{h}\left[H(\mathbf{r})G^{<}(\mathbf{r},\mathbf{r}',E) - G^{<}(\mathbf{r},\mathbf{r}',E)H^*(\mathbf{r}')\right]. \tag{4.148}$$

In matrix notation the same equation states

$$S(E) = \frac{e}{h}\left[HG^{<}(E) - G^{<}(E)H\right]. \tag{4.149}$$

To relate this result to the self-energies and retarded and advanced Green's functions of the conductor, we use the kinetic equation (see previous sections)

$$G^< = G^{\mathrm{R}} \Sigma^< G^{\mathrm{A}}, \tag{4.150}$$

which gives

$$S(E) = \frac{e}{h} \left[H G^{\mathrm{R}}(E) \Sigma^<(E) G^{\mathrm{A}}(E) - G^{\mathrm{R}}(E) \Sigma^<(E) G^{\mathrm{A}}(E) H \right]. \tag{4.151}$$

Using now the definitions of the retarded and advanced Green's functions,

$$(EI - H - \Sigma^{\mathrm{R}}) G^{\mathrm{R}} = I \quad \longrightarrow H G^{\mathrm{R}} = E G^{\mathrm{R}} - \Sigma^{\mathrm{R}} G^{\mathrm{R}} - I, \tag{4.152}$$
$$G^{\mathrm{A}} (EI - H - \Sigma^{\mathrm{A}}) = I \quad \longrightarrow G^{\mathrm{A}} H = E G^{\mathrm{A}} - G^{\mathrm{A}} \Sigma^{\mathrm{A}} - I, \tag{4.153}$$

we arrive at the following expression for the source matrix $S(E)$:

$$S(E) = \frac{e}{h} \left[G^{\mathrm{R}}(E) \Sigma^<(E) - \Sigma^<(E) G^{\mathrm{A}}(E) - \Sigma^{\mathrm{R}}(E) G^<(E) + G^<(E) \Sigma^{\mathrm{A}}(E) \right]. \tag{4.154}$$

The trace of this matrix, which is equal to an integration over space, yields the current flow through the surface of the conductor. The expression can be simplified if we consider that we are interested only in the trace of the matrix expression. An exchange of the order of multiplication leaves the trace constant:

$$G^{\mathrm{R}} \Sigma^< - \Sigma^< G^{\mathrm{A}} - \Sigma^{\mathrm{R}} G^< + G^< \Sigma^{\mathrm{A}} = \Sigma^< (G^{\mathrm{R}} - G^{\mathrm{A}}) - (\Sigma^{\mathrm{R}} - \Sigma^{\mathrm{A}}) G^<.$$

And since the spectral function $G^{\mathrm{R}} - G^{\mathrm{A}}$ and the $\Sigma^{\mathrm{R}} - \Sigma^{\mathrm{A}}$ are equal to

$$G^{\mathrm{R}} - G^{\mathrm{A}} = G^> - G^< \qquad \Sigma^{\mathrm{R}} - \Sigma^{\mathrm{A}} = \Sigma^> - \Sigma^<, \tag{4.155}$$

we get for the current flow through the surface of our system

$$Tr[S(E)] = \frac{e}{h} Tr \left[\Sigma^<(E) G^>(E) - \Sigma^>(E) G^<(E) \right]. \tag{4.156}$$

This equation, describing the current flow through the surface of a conductor under nonequilibrium conditions, is the central result of transport theory within the Keldysh formalism. However, it gives the total current and is therefore not restricted to the current flowing in and out of the conductor interface via the leads. To isolate the components passing through the leads, we remember the sum rules for the self-energies, and take only a single component, say the component of one lead A:

$$Tr[S(E)] = \frac{e}{h} Tr\left[\Sigma_A^<(E) G^>(E) - \Sigma_A^>(E) G^<(E)\right]. \tag{4.157}$$

And the current passing through the interface is then given by

$$I = \frac{e}{h} \int_{\mu_A - eV}^{\mu_B + eV} dE\, Tr\left[\Sigma_A^<(E) G^>(E) - \Sigma_A^>(E) G^<(E)\right]. \tag{4.158}$$

The required information for the calculation of the current through the interface under nonequilibrium conditions is thus the self energy of the leads and the nonequilibrium Green's function. In general, the main problem in actual calculations (see next section) is posed by the self-energy terms, which have to be calculated *inside* the conductor interface.

Relation to the Landauer–Büttiker relation

Even though the result looks quite different from the one we obtained for elastic tunneling and the Landauer–Büttiker approach, the nonequilibrium result can actually be reduced to the elastic formulation. This will also take care of the Fermi distribution functions, which in the elastic relation are introduced somewhat arbitrarily, but can be shown to arise under the condition of leads in thermal equilibrium. To show the equivalence we first expand the functions $G^{<(>)}$ in the usual form, and get for the above expression

$$\Sigma_A^< G^> - \Sigma_A^> G^< = \Sigma_A^< G^{\mathrm{R}} (\Sigma^< - i\Gamma) G^{\mathrm{A}} - (\Sigma_A^< - i\Gamma_A) G^{\mathrm{R}} \Sigma^< G^{\mathrm{A}}. \tag{4.159}$$

Next we split the self-energy into terms due to lead B and inelastic terms, which are omitted:

$$\Sigma^< = \Sigma_{\mathrm{ine}}^< + \Sigma_B^< =: \Sigma_B^< \qquad \Gamma = \Gamma_{\mathrm{ine}} + \Gamma_B =: \Gamma_B, \tag{4.160}$$

which leaves us with the following expression:

$$\Sigma_A^< G^> - \Sigma_A^> G^< = -i\left[\Sigma_A^< G^{\mathrm{R}} \Gamma_B G^{\mathrm{A}} - \Gamma_A G^{\mathrm{R}} \Sigma_B^< G^{\mathrm{A}}\right]. \tag{4.161}$$

And finally we consider two leads in thermal equilibrium, so that the self energies are given by (see previous sections)

$$\Sigma_A^< = i f(\mu_A) \Gamma_A, \qquad \Sigma_B^< = i f(\mu_B) \Gamma_B. \tag{4.162}$$

The current passing through the interface is therefore equal to the expression we already obtained, but completed by the two Fermi distributions of the leads (the factor of 2 is added due to two spin directions):

$$\begin{aligned} I(V) = {} & \frac{2e}{h} \int_{\mu_A - eV}^{\mu_B + eV} dE\, \left[f(\mu_A, E) - f(\mu_B, E)\right] \\ & \times Tr\left[\Gamma_A(E) G^R(E) \Gamma_B(E) G^A(E)\right]. \end{aligned} \tag{4.163}$$

It seems therefore justified to say that the only differences between the Landauer–Büttiker relation and the nonequilibrium relations for the conductance across an interface are due to only two features:

- finite bias, which is via the self-consistent charge distributions of the interface part of the nonequilibrium treatment.
- effect of electron–phonon interactions in the interface itself.

Apart from these two features the treatments within the two separate frameworks are equivalent. Considering that electron–phonon interactions, e.g., in tunneling conditions, change the current values by less than 5%, it seems generally safe to rely on elastic transport models.

4.4 Transport within standard DFT methods

In this last section on transport theory we shall undertake to sketch available methods for calculating currents in a two-terminal device. On the one hand, we shall describe how the problem is solved within a basis set of local orbitals [12], on the other hand, we shall describe how such a solution can be obtained within standard methods of density functional theory. At present, the most efficient methods to this end use three-dimensional repeat units, called *supercells*. Moreover, they are based on a plane wave expansion of electron eigenvectors. The problem we have to address is thus (i) the inequivalence of boundary conditions at the conductor–lead interfaces as soon as bias voltages are applied; (ii) the localization of eigenfunctions based on plane wave basis sets; (iii) the calculation of nonequilibrium Green's functions, contacts, and self-energy terms under these conditions; (iv) the self-consistency of the resulting solutions with respect to the charge density and the chemical potentials throughout the interface; (v) the calculation of transport properties. We start with a detailed description of a transformation that transforms a Green's function into a matrix, which can be formulated in any basis set.

4.4.1 Green's function matrix

Within a given basis set, the equation for the Green's function of a system can generally be written in matrix form. The point of departure is the real-space definition of the Green's function:

$$[H(\mathbf{r}) - Z]\, G^{\mathrm{R}}(\mathbf{r}, \mathbf{r}', E) = -\delta(\mathbf{r} - \mathbf{r}'). \tag{4.164}$$

Here, $Z = E + i\eta$ is the complex eigenvalue associated with the retarded Green's function G^{R}. The corresponding eigenvalue for G^{A} is Z^*. We now rewrite the Green's function in a given basis set (for atomic orbitals the functions ϕ_m are centered at atomic positions of the system):

$$G^{\mathrm{R}}(\mathbf{r},\mathbf{r}',E) = \sum_{mn} \phi_m(\mathbf{r})\, G^{\mathrm{R}}_{mn}(E)\, \phi^*_n(\mathbf{r}'). \tag{4.165}$$

Multiplying (4.164) by $\phi_{n'}(\mathbf{r}')$ and integrating over d^3r' gives

$$\sum_{mn} [H(\mathbf{r}) - Z]\, \phi_m(\mathbf{r})\, G^{\mathrm{R}}_{mn}(E)\, S_{nn'} = -\phi_{n'}(\mathbf{r}). \tag{4.166}$$

The overlap matrix $S_{nn'}$ is defined by

$$S_{nn'} = \int d^3r \phi^*_n(\mathbf{r})\phi_{n'}(\mathbf{r}). \tag{4.167}$$

Multiplying by $\phi^*_{m'}(\mathbf{r})$ and integrating over d^3r leads finally to the matrix equation

$$\textstyle\sum_{mn} \left(H_{m'm} - ZS_{m'm}\right) G^{R}_{mn}(E) S_{nn'} = -S_{m'n'}, \tag{4.168}$$

$$H_{m'm} \equiv \int d^3r \phi^*_{m'}(\mathbf{r})\, H(\mathbf{r})\, \phi_m(\mathbf{r}). \tag{4.169}$$

Multiplying by the inverse overlap matrix S^{-1}, where $S\,S^{-1} = I$, we arrive at the matrix expression for the Green's function in a local basis set:

$$\sum_{n'} \left(ZS_{mn'} - H_{mn'}\right) G^{R}_{n'n}(E) = \delta_{mn}. \tag{4.170}$$

Inverting the matrix $\left(ZS_{nn'} - H_{nn'}\right)^{-1} \equiv \left(ZS - H\right)^{-1}_{nn'}$ and multiplying by the inverted matrix we get

$$\delta_{nm} G^{R}_{nn'}(E) = \left(ZS - H\right)^{-1}_{nn'} \delta_{nm} \quad \Longrightarrow \quad G^{R}_{mn'}(E) = \left(ZS - H\right)^{-1}_{mn'}. \tag{4.171}$$

By an identical procedure using Z^* instead of Z we arrive at the advanced instead of the retarded Green's function, so that both Green's functions in matrix form are given by

$$G^{\mathrm{R}}_{mn}(E) = \left(ZS - H\right)^{-1}_{mn}, \qquad G^{\mathrm{A}}_{mn}(E) = \left(Z^*S - H\right)^{-1}_{mn}. \tag{4.172}$$

It may seem at this point that we are left with an unknown parameter η, since in principle the Green's functions are defined as the limits for $\eta \to \pm 0$. Numerically, however, we may assign a definite value to η depending on our system and under the condition that

$$\int_{-\infty}^{E_F} dE \sum_n \left[i\left(G^{\mathrm{R}}(E) - G^{\mathrm{A}}(E)\right)\right]_{nn} = 2\pi N_0, \tag{4.173}$$

where N_0 is the number of electrons in our system. From a technical point of view it has to be considered that matrix inversion, which is essential to the calculation of the system response with Green's functions, is an $O(N^2)$ routine and therefore computationally expensive. The solution to this problem may be the extensive use of iterative schemes to construct the Green's function without relying on matrix inversion [13, 14]. The result of this expansion is thus the Green's function matrix of the system in a given basis set.

4.4.2 General self-consistency cycle

In standard DFT, a self-consistency cycle begins with a spatial distribution of electron charge, which is then used to construct the effective potential. The solutions of the Schrödinger equation based on the effective potential are finally filled with electrons until the total charge is equal to N, the number of electrons in the system, and the new charge distribution is again used to calculate the effective potential for the next self consistency cycle. This straightforward and well-established scheme does not work under nonequilibrium conditions for the simple reason that the electrons under these conditions do not obey a common Fermi distribution. Instead, one has to iterate self consistent solutions by a different cycle, like the following:

1. Calculate the self-energies of the leads.
2. Calculate the Hartree potential and the Hamiltonian between the two leads for an applied bias voltage.
3. Calculate the self-energies of the interface.
4. Calculate the nonequilibrium Green's functions.
5. Find the charge density distribution by integrating the Green's functions over energy.
6. Begin the next iteration.

Numerically, the problem of the sketched self-consistency cycle are the matrix inversions related to the self-energies and the Green's functions of the interface.

4.4.3 Self-energy of the leads

The leads can be thought to consist of a few layers of metal, with an applied Hartree potential to account for varying bias voltages. The setup will either be periodic, as in the supercell geometries of most DFT codes, or it will implement a cluster description of the leads attached to one side of the conductor. In the following we shall focus on periodic supercells; the only difference to cluster calculations is the change of the Fermi functions at the leads, as will be seen presently. The setup for the separate calculation of the leads is shown in Figure 4.9.

The retarded and advanced Green's function matrices can be calculated by the procedure described above. Once they are known, we can construct the spectral function by

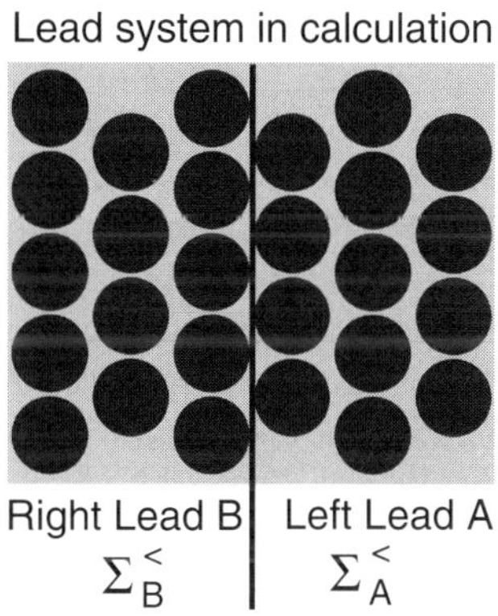

Fig. 4.9. Calculation of self-energies of metal leads. The leads are represented by six metal layers; the self-energies are calculated from the Green's function for the three left (lead A) and the three right (lead B) layers.

$$A_{mn} = i\left(G^{\mathrm{R}}_{mn} - G^{\mathrm{A}}_{mn}\right). \tag{4.174}$$

The spectral function is related to the contact Γ of the lead by (see (4.117))

$$A = G^{\mathrm{R}}\Gamma G^{\mathrm{A}}. \tag{4.175}$$

The contact of the lead is consequently

$$\Gamma = i\left[G^{\mathrm{R}}\right]^{-1}\left[G^{\mathrm{R}} - G^{\mathrm{A}}\right]\left[G^{\mathrm{A}}\right]^{-1} = \left[ZI - H\right]A\left[Z^{*}I - H\right]. \tag{4.176}$$

In matrix notation and including the indices of the matrices, the same equation reads

$$\Gamma_{mn} = |Z|^2 A_{mn} - ZA_{mr}H_{rn} - Z^{*}H_{mr}A_{rn} + H_{mr}A_{rs}H_{sn}. \tag{4.177}$$

The self-energy of the lead can then be calculated assuming that the lead is in thermal equilibrium. The self-energy is

$$\Sigma^{<}_{mn}(E) = if(\mu, E)\Gamma_{mn}(E). \tag{4.178}$$

Given the decay length in metals of about three atomic layers, the self-energy term of the lead will have to be calculated only for three layers adjacent to the interface. If the lead system is composed of six atomic layers, then the three left layers describe the self-energy of the right lead, the three right layers the self-energy of the left lead. In a periodic setup the Fermi distribution functions of both leads will be the ground-state Fermi distributions ($\mu = \mu_0$), while for a cluster approach the chemical potentials will be shifted by the bias voltage $\mu = \mu_0 \pm eV$.

4.4.4 Hartree potential and Hamiltonian of the interface

The leads are attached to the conductor including three additional metal layers to account for the surface dipoles due to the applied bias voltage (see Figure 4.10). Given the changed Hartree potentials at the transition from the lead interface into the conductor, the Hartree potential and the effective potential throughout the interface have to be calculated by the self-consistent procedure sketched in Section 4.3.1. The Hamiltonian matrix of the system then contains matrix elements for the L layers of the interface, and six additional elements for the self-energies of the leads. Writing down the resulting Hamiltonian matrix in a generic way, which means that we represent each layer in the matrix by only one matrix element, we arrive at

$$H_{nm} = \begin{pmatrix} (\Sigma_A^<)_{4-6} & 0 & 0 \\ 0 & (H_{\text{int}})_{1-L} & 0 \\ 0 & 0 & (\Sigma_B^<)_{1-3} \end{pmatrix}. \qquad (4.179)$$

In this notation the self-energies of the leads are represented by 3×3 matrices, where lead A corresponds to layers $4-6$ and lead B to layers $1-3$ of the independent calculation. The interface is described by $L \times L$ matrices. Since individual layers are generally composed of more than one orbital, the atoms as well as the local orbitals centered at each atom will enter the description as additional indices.

4.4.5 Self-energies of the interface

Once the Hamiltonian of the interface including the self-energy terms of the leads is constructed, we may compute the retarded and advanced Green's functions of the interface by matrix inversion:

$$G^{\text{R}}(E) = [Z - H]^{-1}, \qquad G^{\text{A}}(E) = [Z^* - H]^{-1}. \qquad (4.180)$$

In this case there is no immediate solution to determine the numerically appropriate value of the infinitesimal complex constant η; this parameter has to be updated at the end of a self-consistency cycle by using the condition of charge continuity at the positions of the metal leads.
Including electron–phonon coupling for a phonon mode of frequency ω in the interface, the self-energy of the interface is the solution of the following equation:

$$\begin{aligned} \Sigma^<(E) = &\; if(\mu_0, E)\left[\Gamma_A(E) + \Gamma_B(E)\right] \\ &+ D(\hbar\omega)G^{\text{R}}(E-\hbar\omega)\Sigma^<(E)G^{\text{A}}(E+\hbar\omega). \end{aligned} \qquad (4.181)$$

An identical equation exists for $\Sigma^>(E)$; in this case $f(\mu_0, E)$ has to be replaced by $1-f(\mu_0, E)$. This equation has to be solved iteratively, since the same term,

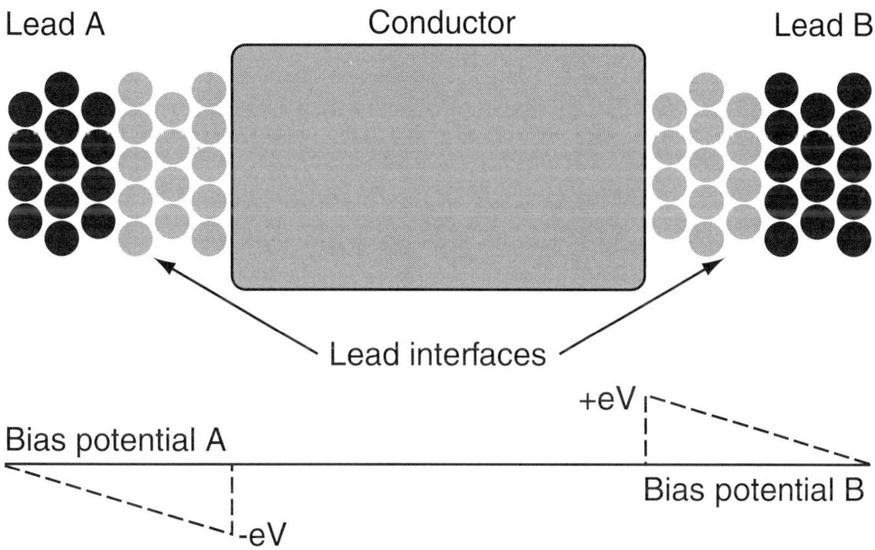

Fig. 4.10. System setup for DFT calculations of transport properties. The leads are represented by three metal layers of a separate calculation; the three layers of the lead interface account for charge polarization due to finite-bias voltages. The whole interface of leads and conductor is repeated in a supercell approach.

the self-energy, appears on the left- and on the right-hand sides. The usual procedure for an iterative solution is to start with the zero approximation given by

$$\Sigma_0^<(E) = if(\mu_0, E)\left[\Gamma_A(E) + \Gamma_B(E)\right], \tag{4.182}$$

which in matrix form reads

$$\left[\Sigma_0^<(E)\right]_{mn} = if(\mu_0, E) \cdot \begin{pmatrix} (\Gamma_A(E))_{4-6} & 0 & 0 \\ 0 & 0 & 0 \\ 0 & 0 & (\Gamma_B(E))_{1-3} \end{pmatrix}. \tag{4.183}$$

The zero-order approximation is then used to find the approximation of first order, which is described by

$$\begin{aligned} \Sigma_1^<(E) = \; & if(\mu_0, E)\left[\Gamma_A(E) + \Gamma_B(E)\right] \\ & + D(\hbar\omega) G^{\mathrm{R}}(E - \hbar\omega) \Sigma_0^<(E) G^{\mathrm{A}}(E + \hbar\omega). \end{aligned} \tag{4.184}$$

Repeating the iteration and checking for convergence of the obtained result, the true self energy of the interface will be found after a sufficient number of iterations.

4.4.6 Nonequilibrium Green's functions of the interface

Once the self-energies and the Green's functions of the system are known the nonequilibrium Green's functions can be obtained by matrix multiplication from

$$G^{<(>)}(E) = G^{\mathrm{R}}(E)\Sigma^{<(>)}(E)G^{\mathrm{A}}(E). \tag{4.185}$$

Calculation of the Green's functions of the interface $G^{<(>)}(E)$ allows one to calculate the total charge within the interface *and* the local charge at the leads. The energy integration of a section of the charge density of the system (see Section 4.3.2) must be split into two components: the first component relies only on the retarded and advanced Green's functions $G^{\mathrm{R}}(E)$ and $G^{\mathrm{A}}(E)$, it will give a value n_1:

$$n_1 = -\frac{i}{2\pi}\int_{-\infty}^{E_1} dE Tr[G^{\mathrm{R}}(E) - G^{\mathrm{A}}(E)]. \tag{4.186}$$

Here E_1 is the minimum value of $(\mu_0 - eV/2, \mu_0 + eV/2)$. In the intermediate range the charge within the system must be determined from the nonequilibrium function. Since G^{R} is nonanalytic below the real axis and G^{A} is nonanalytic above, the integration over energy for $G^<$ has to be performed along the real axis. Thus

$$n_2 = -\frac{i}{2\pi}\int_{E_1}^{E_2} dE Tr[G^<(E)], \tag{4.187}$$

where E_2 is the maximum value of $(\mu_0 - eV/2, \mu_0 + eV/2)$. The total charge in the interface can now be set constant, so that $n_1 + n_2 = N_{\mathrm{int}}$. Given that the number of electrons in the interface is known, we may estimate the numerical value of η from the result obtained in the integration.
In addition, the charge density can be calculated locally, since the projection onto the orbitals at the leads is accessible. This allows one to estimate the level of consistency obtained in the calculation. It is required, if the system is fully converged, that the charge density at the leads matches the charge density obtained from the separate calculation. In case the two charge densities do not match to a sufficient degree, the whole cycle is repeated.

4.4.7 Calculation of nonequilibrium transport properties

Once the calculation is sufficiently converged, we may calculate any physical property with the help of the nonequilibrium Green's functions. However, one is generally interested only in actual transport quantities. The current through the leads is given by (see previous sections, (4.158))

$$I(V) = \frac{e}{h}\int_{\mu_0 - eV/2}^{\mu_0 + eV/2} dE\, Tr\left[\Sigma_A^<(E)G^>(E) - \Sigma_A^>(E)G^<(E)\right]. \tag{4.188}$$

Since the system in a supercell geometry is based on equilibrium conditions at the leads A, B, we cannot calculate the current through the whole system, since it will vanish. This can be directly seen from the Landauer–Büttiker equation, which states, under the condition that $\mu_A = \mu_B = \mu_0$ (see (4.163)),

$$\begin{aligned} I(V) = \tfrac{2e}{h} & \int_{\mu_0 - eV/2}^{\mu_0 + eV/2} dE \; [f(\mu_0, E) - f(\mu_0, E)] \\ & \times Tr \left[\Gamma_A(E) G^{\mathrm{R}}(E) \Gamma_B(E) G^{\mathrm{A}}(E) \right] = 0. \end{aligned} \tag{4.189}$$

However, both of the lead interfaces are out of equilibrium due to the applied Hartree potential. Therefore, one has to calculate the current directly from the lead interfaces, using the nonequilibrium formulation:

$$I(V) = \frac{e}{h} \int_{\mu_0 - eV/2}^{\mu_0 + eV/2} dE \, Tr \left[\Sigma^<_{\mathrm{A,int}}(E) G^>_{\mathrm{int}}(E) - \Sigma^>_{\mathrm{A,int}}(E) G^<_{\mathrm{int}}(E) \right], \tag{4.190}$$

where the matrices have the following form

$$\left[\Sigma^<_{\mathrm{A,int}}(E) \right]_{mn} = \begin{pmatrix} (\Sigma^<(E))_{11} & (\Sigma^<(E))_{12} & (\Sigma^<(E))_{13} & ... \, 0 \\ (\Sigma^<(E))_{21} & (\Sigma^<(E))_{22} & (\Sigma^<(E))_{23} & ... \, 0 \\ (\Sigma^<(E))_{31} & (\Sigma^<(E))_{32} & (\Sigma^<(E))_{33} & ... \, 0 \\ ... & ... & ... & ... \\ 0 & 0 & 0 & ... \, 0 \end{pmatrix}, \tag{4.191}$$

$$\left[G^<_{\mathrm{int}}(E) \right]_{mn} = \begin{pmatrix} (G^<(E))_{11} & (G^<(E))_{12} & (G^<(E))_{13} & ... & (G^<(E))_{1L} \\ (G^<(E))_{21} & (G^<(E))_{22} & (G^<(E))_{23} & ... & (G^<(E))_{2L} \\ (G^<(E))_{31} & (G^<(E))_{32} & (G^<(E))_{33} & ... & (G^<(E))_{3L} \\ ... & ... & ... & ... & ... \\ (G^<(E))_{L1} & (G^<(E))_{L2} & (G^<(E))_{L3} & ... & (G^<(E))_{LL} \end{pmatrix}. \tag{4.192}$$

Given that the self-consistency cycle ensures that the leads A, B themselves are in equilibrium, the only sources of charge transfer through the interface must be located in the lead interfaces. Since the lead interfaces consist of only a few (typically about three) layers, the trace is easy to evaluate. It will be, for an interface of three layers,

$$Tr \left[\Sigma^<_{\mathrm{A,int}}(E) G^>_{\mathrm{int}}(E) - \Sigma^>_{\mathrm{A,int}}(E) G^<_{\mathrm{int}}(E) \right] = \sum_{i,j=1,3} \Sigma^<_{ij} G^>_{ji} - \sum_{i,j=1,3} \Sigma^>_{ij} G^<_{ji}. \tag{4.193}$$

The current through the interface under nonequilibrium conditions is consequently given by the following integral:

$$I(V) = \frac{e}{h} \int_{\mu_0 - eV/2}^{\mu_0 + eV/2} dE \left[\sum_{i,j=1,3} \Sigma_{ij}^{<}(E) G_{ji}^{>}(E) - \sum_{i,j=1,3} \Sigma_{ij}^{>}(E) G_{ji}^{<}(E) \right] . \tag{4.194}$$

It should be noted that at present, calculations of the nonequilibrium transport properties of interfaces are with a few exceptions performed with atomic orbital-like basis sets and pseudopotential approximations for the atomic cores [6, 15]. Given that these methods are not all that reliable for the calculation of, for example, the magnetic transport properties in multilayers, it seems that the approach sketched in this section may provide a blueprint for more accurate simulations in the future.

4.5 Summary

We have given in this chapter an introduction to the current state of electron transport theory, in view of applications to tunneling problems. The theoretical framework, based on Green's functions of open systems, was shown to be adaptable, via its perturbative extension into nonequilibrium environments, to treat all relevant physical processes at the atomic scale, essentially from first principles. The present implementations rely on tight-binding schemes or local orbital geometry; within these limits the theory can cope with finite-bias potentials and inelastic effects due to electron–electron or electron–phonon interactions. It can be foreseen that the framework, once it is extended to cover also plane-wave and full-potential methods, will provide the backbone of transport simulations on the atomic scale, whenever high accuracy is the decisive issue.

Further reading

As in every expanding field, the literature on the topic covers by now a large number of publications. It would be impossible to list all the relevant articles, we therefore want to present only a short list of publications, which we think were either fundamental for the development of the field, or which give an overview clear enough to be understood also by nonspecialist readers. STM experimenters and theorists alike work, after all, on a very different problem, that of low conductance across a vacuum barrier. Their take on the theory and the general problem of electron transport will be explored in detail in the next chapter.

Introduction

S. Datta, *Electronic Transport in Mesoscopic Systems*, Cambridge University Press, Cambridge (1995).

H. Haug and A.-P. Jauho, *Quantum Kinetics in Transport and Optics of Semiconductors*, Springer Series in solid State Sciences, Vol. 123, Springer Berlin (1996).

Intermediate

J. H. Davies, S. Hershfield, P. Hyldgaard, J. W. Wilkins, Physical Review B 47, 4603 (1993).
J. Taylor, H. Guo, and J. Wang, Physical Review B 63, 245407 (2001).
M. Brandbyge, J.-L. Mozos, P. Ordejon, J. Taylor, and K. Stokbro, Physical Review B 65, 165401 (2002).
F. Michael and M. D. Johnson, Physica B 339, 31 (2003).

In depth

J. Rammer and H. Smith, Reviews of Modern Physics 58,323 (1994).
T. E. Feuchtwang, Physical Review B 13, 517 (1976).
D. C. Langreth, *1975 NATO Advanced Study Institute on Linear and Nonlinear Electron Transport in Solids, Antwerpen 1975*, Vol B17, Plenum, New York (1976).
C. Caroli, R. Combescot, P. Nozieres, D. Saint-James, Journal of Physics C 5, 21 (1972).
L. P. Kadanoff and G. Baym, *Quantum Statistical Mechanics*, Benjamin, New York (1962).
L. V. Keldysh, Soviet Physical Journal 20, 1018 (1965).

References

1. S. Datta. *Transport in Mesoscopic Systems.* Cambridge University Press, Cambridge UK, 1995.
2. H. Haug and A.-P. Jauho. *Quantum Kinetics in Transport and Optics of Semiconductors.* Springer Series in solid State Sciences, Vol. 123, Springer Berlin, 1996.
3. E. Scheer, N. Agrait, J. C. Cuevas, A. L. Yegati, B. Ludolph, A. Martin-Rodero, G. Rubio Pollinger, J. M. van Ruitenbeek, and C. Urbina. *Nature*, 394:154, 1998.
4. D. S. Fisher and P. A. Lee. *Phys. Rev. B*, 23:6851, 1981.
5. M. Bütticker, Y. Imry, R. Landauer, and S. Pinhas. *Phys. Rev. B*, 31:6207, 1985.
6. J. Taylor. *Ab-initio Modelling of Transport in Atomic Scale Devices.* PhD thesis, McGill University, Montreal, Canada, 2000.
7. W. Kohn and L. J. Sham. *Phys. Rev.*, 140:A1133, 1965.
8. L. V. Keldysh. *Sov. Phys. JETP*, 20:1018, 1965.
9. J. A. Appelbaum and W. F. Brinkman. *Phys. Rev.*, 186:464, 1969.
10. C. Caroli, R. Combescot, P. Nozieres, and D. Saint-James. *J. Phys. C: Solid State Phys.*, 5:21, 1972.

11. J. Rammer and H. Smith. *Rev. Mod. Phys.*, 58, 1994.
12. D. Ordejon, D. Drabold, R. Martin, and M. P. Grumbach. *Phys. Rev. B*, 51:1456, 1996.
13. D. G. Pettifor and D. L. Weaire. *The Recursion Method and Its Applications.* Springer Verlag, Berlin, 1985.
14. S. Y. Wu, J. Cocks, and C. S. Jahanthi. *Phys. Rev. B*, 49:7957, 1994.
15. M. Brandbyge, J.-L. Mozos, P. Ordejon, J. Taylor, and K. Stokbro. Theory of single molecule vibrational spectroscopy and microscopy. *Phys. Rev. B*, 65:165401, 2002.

5

Transport in the Low Conductance Regime

For practical purposes the framework introduced in the last chapter is commonly too wide a frame of reference, in particular if the experimental environment itself limits the probability of specific processes, as in scanning probe instruments. There, the main limitation to the transport of electrons is the vacuum barrier between the surface and the probe tip. In this case, changes of the conductance across the tunneling barrier due to electron interactions can be considered small and may be conveniently treated within perturbation models. The main task is then reduced to describing the transport of electrons across the barrier, based on the physical properties of the two leads, and to incorporate additional effects like the onset of chemical bonding in high-resolution scans or electron–phonon excitations when crossing the threshold of a specific vibration mode by suitable extensions of the basic model. These effects, together with variations of the tunneling current due to the magnetic properties of the systems, account for the bulk of observations in experiments. At present, four theoretical models are used in nearly all simulations of scanning tunneling experiments. In increasing order of theoretical difficulty, these models are the following:

- The Tersoff–Hamann approach [1, 2], where constant-current contours are modelled from the electronic structure of the surface alone. The approach is based on perturbation theory and one decisive assumption about electron states of the tip.
- The transfer Hamiltonian or the Bardeen approach [3], where the tip electronic structure is explicitly included in the calculation. This is the original perturbation model; it assumes that for every electron only very few pathways exist in its transition between the two leads.
- The scattering or Landauer–Büttiker approach [4], which includes multiple pathways of tunneling electrons from their initial to their final crystalline states. Apart from that, it is equivalent to the previous method.

- The Keldysh or nonequilibrium Green's function approach [5, 6], which also incorporates inelastic effects such as electron–electron and electron–phonon scattering.

5.1 Tersoff–Hamann(TH) approach

This method is today incorporated in nearly every state-of-the-art DFT code. Despite an extension of existing simulation methods, especially with respect to quantitative comparisons between experiments and theory, it continues to be the "workhorse" [7] of STM theory. In this method the tunneling current is proportional to the local density of states at the position of the STM tip [1, 2]:

$$I(\mathbf{R}) \propto \sum_{E_n > E_F - eV_{\text{bias}}}^{E_n < E_F} |\psi(\mathbf{R}, E_n)|^2 =: n\left(\mathbf{R}, V_{\text{bias}}\right). \tag{5.1}$$

Here, I is the tunneling current, E_n the eigenstates of the crystal electrons, E_F the Fermi level, V_{bias} the bias voltage, and n the electron density. In many standard situations, e.g., in the research of molecular adsorption or surface reconstructions, the model provides a reliable qualitative picture of the surface topography, even though it does not generally reproduce the observed corrugation values.

5.1.1 Easy modeling: applying the Tersoff–Hamann model

Numerical methods to compute the tunneling currents from first principles by the Bardeen method, used to elucidate even subtle features of experimental images, are already well advanced (see the following sections). However, from a practical point of view it is frequently desirable to gain an understanding of experiments without highly demanding and thus very time-consuming model calculations. In principle, the TH model, which is based on the electronic structure of the analyzed surface alone, provides just such an easy method. In particular, since advanced codes in DFT generally come with an interface to compute constant-density contours, simulated images can be calculated in a straightforward manner. In this spirit, one seeks to determine the limits within which the method is reliable, and to estimate the density contour value, which roughly corresponds to a given tunneling current. Both of these objectives are attainable if the Bardeen method of calculating the currents is simplified by suitable approximations, which we shall demonstrate presently. Concerning the reliability of the TH model, the following criteria seem sufficient:

- No substantial chemical interactions between surface and tip.

This condition is not trivial to quantify, since the experimental measure of tip–sample separation, the tunneling resistance R, which is given by the ratio of applied bias voltage and tunneling current $R = V_{\text{bias}}/I$, differs strongly for different systems and experimental conditions. On metal surfaces, the distance has to be larger than 5–6 Å. This corresponds, for an ambient environment and very low bias voltages (less than 80 meV) to a tunneling resistance R of 10–100 MΩ. Temperature enters this estimate, because a thermal environment of transiting electrons allows them to reach a substantially higher number of final states even under the condition of elastic tunneling: under ambient conditions the energy difference between the initial and final states can differ by about 80 meV. For this reason the experimental tunneling resistance (e.g., for a bias voltage of -1 mV and a current of 1 nA), and thus the estimate of the distance, can be quite misleading. Comparing with the actual values obtained from explicit calculations of all possible transitions, the estimate under these conditions is too low by one or two orders of magnitude. Thermal excitations, in short, have a similar effect as increased bias voltages.
On semiconductors, the corresponding problem is the exact location of the Fermi level with respect to the upper band edge of the valence band. In this case the same condition (100 MΩ tunneling resistance) can lead to very small distances if the chosen bias voltage includes only very few states of the semiconductor surface. However, if the bias voltage is high enough (above 2 V, say), then this condition is usually sufficient.

- A feature size of surface structures that is well above the typical length scale of electron states of the STM tip.

This condition is far easier to quantify, since the typical length scale of a tip state is about half the interatomic distance of the tip metal; it is therefore between one and two Å. For feature sizes well above this value, the exact geometry of tip states will not enter the shape of the current contour in a decisive way. It is evident that this condition is in general not fulfilled in high-resolution scans, i.e., scans with atomic resolution. In all other cases it is quite safe to omit the explicit structure of the STM tip in a simulation.
Under these conditions the constant-current contour can be related to the charge density contour of the surface in a unique way. This can be done in the following way:
Most DFT codes contain a feature to sum up the charge density within a given energy interval. For a bias voltage of $-V_{\text{bias}}$ one starts by computing the density for the interval $E_F - V_{\text{bias}}$ to E_F. The appropriate contour for a given current value can then be estimated with the following approximations:

1. The bulk of the tunneling current passes through a small cross section of about 2 Å radius.
2. The decay length of surface states (and tip states) is equal to the decay length of an electron state at the Fermi level of a metal surface with a

workfunction $\Phi \approx 4$ eV. The wavefunction of the state is thus (in atomic units)

$$\psi(z) = \psi_0 e^{-kz}, \qquad k = \sqrt{2\Phi}. \tag{5.2}$$

3. The convolution of surface and tip states in the Bardeen integral is simplified by setting $\psi_{\text{sample}} \approx \chi_{\text{tip}}$ and by assuming that the first term in the integral is of the same order of magnitude as the difference:

$$\begin{aligned} I &= C \cdot \left| \int_S dS \left(\chi_{\text{tip}}^* \frac{\partial \psi_{\text{sample}}}{\partial z} - \psi_{\text{sample}} \frac{\partial \chi_{\text{tip}}^*}{\partial z} \right) \right|^2 \\ &\approx C \cdot \Delta S^2 k^2 n^2(\text{sample}). \end{aligned} \tag{5.3}$$

Here, C denotes a constant, ΔS the area of wavefunction overlap, and n the electron density. Since all the constants are known [8], the estimate is straightforward and yields

$$n(I) \quad \left[\text{Å}^{-3}\right] \approx 2 \cdot 10^{-4} \sqrt{I} \quad [\text{I in nA}] \tag{5.4}$$

For a current value of 1 nA, e.g., on a metal surface, the appropriate charge density contour will thus be at $2 \cdot 10^{-4}$Å^{-3}.

5.2 Perturbation approach

Within a transfer Hamiltonian approach the two subsystems of sample and tip are treated as separate entities. This approach is also known as the Bardeen approach [3]. The tunneling current is then described by the equation

$$I = \frac{4\pi e}{\hbar} \sum_{\mu,\nu} \left| \int_S \left(\chi_\nu^* \nabla \psi_\mu - \psi_\mu \nabla \chi_\nu^* \right) \right|^2 \delta\left(E_\nu - E_\mu - eV_{\text{bias}} \right). \tag{5.5}$$

Here, χ_ν are the eigenstates with energy E_ν of the STM tip, ψ_μ the eigenstates of the surface with energy E_μ. The integral extends over the separation surface S between sample and tip, the summation includes all eigenstates within a given interval from the Fermi level. This interval is determined by experimental conditions, e.g., the temperature within the STM.
It is clear from this expression that the quality of the wavefunctions in the vacuum range above the surface is decisive for good agreement between experiments and simulations. In fact, the most suitable expansion is a 2-dimensional Fourier expansion in the lateral direction. Linear combinations of atomic orbitals (LCAO) in this respect have the disadvantage that they decay too rapidly into the vacuum, which in turn renders the currents and corrugations at a given distance unreliable.

5.2.1 Explicit derivation of the tunneling current

The original paper by Bardeen, in which this method to calculate the current through a metal–insulator–metal junction was first derived [3], proves to be quite difficult for theory students and experimentalists. In fact, the essential step is based on a clever integration by parts, within the framework of many-body theory. This makes its explicit relation to the wavefunctions, calculated within density functional theory, quite unclear. For this reason we add the derivation based on time-dependent perturbation theory at this point. It was developed by Julian Chen [9]. The derivation is based on two explicit assumptions:

1. Without any interaction or current flow between the two leads of the tunneling junction, the whole system comprises a discrete set of orthogonal eigenstates, which are conveniently split into eigenstates located at the surface, ψ_μ, and eigenstates located at the tip, χ_ν. Under this condition the two Hamiltonians of the subsystems differ only by their potentials.
2. The total potential under the condition of tunneling is a sum of two potentials: one for the surface, U_S, and one for the tip, U_T. Since both of these potentials are exponentially decaying in the vacuum range, their overlap at a surface in the vacuum range between these two barriers will be very small and can be neglected.

At $t < 0$ the tip potential U_T is turned off. The Schrödinger equation for the sample system then reads

$$\left(-\frac{\hbar^2}{2m}\nabla^2 + U_S\right)\psi_\mu = E_\mu \psi_\mu. \tag{5.6}$$

At $t = 0$ the tip potential is turned on and the sample system starts to evolve according to the time dependent Schrödinger equation

$$\left(-\frac{\hbar^2}{2m}\nabla^2 + U_S + U_T\right)\Psi = i\hbar\frac{\partial \Psi}{\partial t}. \tag{5.7}$$

The tip wavefunctions, on the other hand, are described by an equivalent Schrödinger equation:

$$\left(-\frac{\hbar^2}{2m}\nabla^2 + U_T\right)\chi_\nu = E_\nu \chi_\nu. \tag{5.8}$$

The essential step in the derivation is developing the wavefunction Ψ in terms of the unperturbed tip states χ_ν,

$$\Psi = \sum_\nu a_\nu(t)\chi_\nu e^{-iE_\nu t/\hbar}, \tag{5.9}$$

and to make the ansatz for the coefficients $a_\nu(t)$,

$$a_\nu(t) = (\chi_\nu, \psi_\mu) e^{-i(E_\mu - E_\nu)t/\hbar} + c_\nu(t), \tag{5.10}$$

where $c_\nu(0) = 0$. Now the wavefunction Ψ is a linear combination of the original state ψ_μ and all the tip states χ_ν described by

$$\Psi = \psi_\mu e^{-iE_\mu t/\hbar} + \sum_\nu c_\nu(t) \chi_\nu e^{-E_\nu t/\hbar}. \tag{5.11}$$

By inspection one sees that the transition amplitude is then given by the conventional expression in first-order perturbation theory:

$$c_\nu(t) = \frac{1}{i\hbar} \int_0^t dt' e^{i(E_\nu - E_\mu)t'/\hbar} (\chi_\nu, U_T \psi_\mu). \tag{5.12}$$

For a close to continuous spectrum as in metallic systems the integration in the limit of infinite time yields a delta functional because of the following relations:

$$\left| \int_0^t dt' e^{i\omega t} \right|^2 = \left| \frac{\sin(\omega t/2)}{\omega/2} \right|^2, \qquad \lim_{t \to \infty} \frac{\sin^2 \alpha t}{\alpha^2 t} = \pi \delta(\alpha). \tag{5.13}$$

And since the transition probability is the square of the transition amplitude, the transition rate (or the rate of transiting electrons per unit time) will be

$$\omega_{\mu\nu}^1 = \frac{|c_\nu|^2}{t} = \frac{2\pi}{\hbar} \delta(E_\nu - E_\mu) |(\chi_\nu, U_T \psi_\mu)|^2. \tag{5.14}$$

With the help of the Schrödinger equation the decisive matrix element $M_{\mu\nu}$ for the transition from state ψ_μ of the surface to χ_ν of the tip

$$M_{\mu\nu} = \int_{\Omega_T} d\tau \chi_\nu^* U_T \psi_\mu. \tag{5.15}$$

can be rewritten to a Bardeen-like form. The integral encompasses only the region of the tip, because the potential U_T is zero outside. With Schrödinger's equation for the tip states we may write

$$M_{\mu\nu} = \int_{\Omega_T} d\tau \psi_\mu U_T \chi_\nu^* = \int_{\Omega_T} d\tau \psi_\mu \left(E_\nu + \frac{\hbar^2}{2m} \nabla^2 \right) \chi_\nu^*. \tag{5.16}$$

The energies E_μ and E_ν must be equal due to the delta functional; therefore the matrix element can also be written in the following form (note that the surface potential is zero in the region of integration):

$$\begin{aligned} E_\mu &= E_\nu : \\ &\Rightarrow M_{\mu\nu} = \frac{\hbar^2}{2m} \int_{\Omega_T} d\tau \left(\psi_\mu \nabla^2 \chi_\nu^* - \chi_\nu^* \nabla^2 \psi_\mu \right). \end{aligned} \tag{5.17}$$

And with the help of Gauss's theorem the integral is transformed into a surface integral over the separation surface S, while the operator of kinetic energy becomes a gradient:

$$M_{\mu\nu} = -\frac{\hbar^2}{2m}\int_S d\mathbf{S}\,(\chi_\nu^* \nabla \psi_\mu - \psi_\mu \nabla \chi_\nu^*)\,. \tag{5.18}$$

The matrix element has the dimension of *energy*. Integrating over all the states of the tip and the sample, taking into account the occupation probabilities, the tunneling current is

$$\begin{aligned} I = \; & \frac{4\pi e}{\hbar}\int_{-\infty}^{+\infty} d\epsilon\,[f(E_F - eV + \epsilon) - f(E_F + \epsilon)] \\ & \times \rho_S(E_F - eV + \epsilon)\,\rho_T(E_F + \epsilon)|M_{\mu\nu}|^2, \end{aligned} \tag{5.19}$$

where $f(E) = [1 + \exp(E - E_F)/k_B T]^{-1}$ is the Fermi distribution function, $\rho_S(E_F)$ is the density of states (DOS) of the sample, and $\rho_T(E_F)$ is the DOS of the tip. The result is essentially the one used by Tersoff and Hamann as a basis of their calculation (see the previous section).

5.2.2 Tip states of spherical symmetry

It is tempting to simplify the result by a reasonable assumption about the tip states χ_ν, which reduces the problem further, ideally in such a way that the tip system does not have to be explicitly included in the theoretical model. How this can be done was shown first by Tersoff and Hamann, and later by Chen [1, 2, 10]. The essential step is to consider tip states that are Green's functions of the vacuum Schrödinger equation. These Green's functions are described by

$$\left(\nabla^2 - \kappa^2\right) G\left(\mathbf{r} - \mathbf{R}\right) = -\delta\left(\mathbf{r} - \mathbf{R}\right), \tag{5.20}$$

where $\kappa = \sqrt{2m\phi}/\hbar$. Apart from the point $\mathbf{r} = \mathbf{R}$, $G\left(\mathbf{r} - \mathbf{R}\right)$ is also a solution of the Schrödinger equation for free electrons:

$$\left(\nabla^2 - \kappa^2\right) \chi_\nu\left(\mathbf{r}\right) = 0. \tag{5.21}$$

Explicitly, it is given by a radial function, centered at the tip apex $\mathbf{R}$:

$$G\left(\mathbf{r} - \mathbf{R}\right) = \left|\mathbf{r} - \mathbf{R}\right|^{-1} \exp -\kappa\left|\mathbf{r} - \mathbf{R}\right|. \tag{5.22}$$

Inserting (5.20) into (5.17), taking into account that $\kappa^2 = 2mE_\mu/\hbar^2$, we can integrate the tunneling matrix element over the tip volume and obtain

$$M_{\mu\nu} = \frac{\hbar^2}{2m} \int_{\Omega_T} d\tau \left(\psi_\mu \nabla^2 G\left(\mathbf{r} - \mathbf{R}\right) - G\left(\mathbf{r} - \mathbf{R}\right) \nabla^2 \psi_\mu \right)$$

$$= -\frac{\hbar^2}{2m} \int_{\Omega_T} d\tau \delta\left(\mathbf{r} - \mathbf{R}\right) \psi_\mu = -\frac{\hbar^2}{2m} \psi_\mu(\mathbf{R}), \tag{5.23}$$

$$|M_{\mu\nu}|^2 = \left| \frac{\hbar^2}{2m} \psi_\mu(\mathbf{R}) \right|^2 = \left(\frac{\hbar^2}{2m} \right)^2 n_\mu(\mathbf{R}), \tag{5.24}$$

where $n_\mu(\mathbf{R})$ is the electron density of states μ at the center of the STM tip apex. This is the result Tersoff and Hamann obtained by a slightly different route [1, 2].

5.2.3 Magnetic tunneling junctions

In magnetic systems the transport properties of electrons depend not only on their wavefunctions and eigenvalues, but also on their spin state. For an explicit calculation of electron propagation in a magnetic system, let us consider the situation in a tunneling junction between a crystal surface and an STM tip in real space. Magnetic anisotropy in a crystal breaks the rotational symmetry of electron spins. The spin states in this case are projected onto a crystal's magnetic axis. We assume in the following that this symmetry breaking occurs in the two separate systems that form our tunneling junction. Depending on the orientation of the magnetic axes, two limiting cases have to be distinguished. The magnetic axis of sample and tip are either parallel or antiparallel. In the first case we have to sum up all electrons tunneling from spin-up states of the sample ($n_S^\uparrow$) to spin-up states of the tip ($n_T^\uparrow$), in the second case electrons tunneling from spin-up states of the sample to spin-down states of the tip and vice versa.
In the general case, where the two vectors enclose an arbitrary angle ϕ_M, we analyze the symmetry of the tunneling current with respect to different spin orientations. Within the perturbation approach the tunneling current is proportional to the square of the tunneling matrix element $M_{\mu\nu}$ [3]:

$$M_{\mu\nu} = -\frac{\hbar^2}{2m} \sum_\sigma \int d\mathbf{S} \left[\psi_{\mu\sigma}^* \nabla \chi_{\nu\sigma} - \chi_{\nu\sigma} \nabla \psi_{\mu\sigma}^* \right]. \tag{5.25}$$

The integral extends over the separation surface of sample and tip; the spin-polarized wavefunctions of the sample are given by $\psi_{\mu\sigma}$; wavefunctions of the tip are denoted by $\chi_{\nu\sigma}$; the summation extends over spin states. In density functional theory (DFT)[11, 12] the current, for a constant transition matrix element and within a perturbation, can be described by

$$I(\phi_M) = I_0(1 + P_S P_T \cos\phi_M), \tag{5.26}$$

where ϕ_M is the angle between $\mathbf{M}_S$ and $\mathbf{M}_T$. The equation originates from the description of measurements via the trace of a product of two density matrices, ρ_S and ρ_T. These matrices formalize the electron density of the two subsystems. The density of electrons of the sample system in the two states $\uparrow$ and $\downarrow$ is given by

$$n_S^{\uparrow} = \langle \uparrow |\rho_S| \uparrow \rangle = 1 + P_S,$$
$$n_S^{\downarrow} = \langle \downarrow |\rho_S| \downarrow \rangle = 1 - P_S, \tag{5.27}$$

where P_S is the polarization of the sample. The density matrix then is a 2×2 matrix:

$$\rho_S = \frac{1}{2}\begin{pmatrix} 1+P_S & 0 \\ 0 & 1-P_S \end{pmatrix}. \tag{5.28}$$

The density matrix for the tip is given by an identical form:

$$\rho_T = \frac{1}{2}\begin{pmatrix} 1+P_T & 0 \\ 0 & 1-P_T \end{pmatrix}. \tag{5.29}$$

The current is proportional to the trace of the product of these two matrices, $I \propto Tr\left[\rho_S \rho_T\right]$. In general, the directions in space for the spin-up and spin-down states of sample and tip are different. Therefore the density matrix of the tip has to be rotated by an angle ϕ_M with respect to the sample states. Rotating the density matrix of the tip by ϕ_M, e.g., around the x-axis, we get with the rotation operator $U_x(\phi_M)$,

$$\begin{aligned} U_x(\phi_M) &= \begin{pmatrix} 1\,0 \\ 0\,1 \end{pmatrix}\cos(\phi_M/2) + i\begin{pmatrix} 0\,1 \\ 1\,0 \end{pmatrix}\sin(\phi_M/2), \\ \rho_T(\phi) &= U_x^+(\phi_M)\rho_T U_x(\phi_M) \\ &= \frac{1}{2}\begin{pmatrix} 1+P_T\cos(\phi_M) & 0 \\ 0 & 1-P_T\cos(\phi_M) \end{pmatrix}. \end{aligned} \tag{5.30}$$

For the product $\rho_S \rho_T$, we get consequently

$$\begin{aligned} \rho_S\rho_T(\phi_M) &= \frac{1}{4}\begin{pmatrix} A + P_S P_T\cos\phi_M & 0 \\ 0 & B + P_S P_T\cos\phi_M \end{pmatrix}, \\ A &= 1 + P_S + P_T\cos\phi_M, \\ B &= 1 - P_S - P_T\cos\phi_M. \end{aligned} \tag{5.31}$$

Since the tunneling current is proportional to the trace of the product of the two density matrices, we obtain finally

$$I \propto \frac{1}{2}\left(1 + P_S P_T\cos\phi_M\right). \tag{5.32}$$

The constant of proportion is the paramagnetic tunneling current I_0. Equation (5.26) thus describes the tunneling current under the condition that we measure this current for spin-polarized states of sample and tip and if the spin states are projected onto two different directions in space in the two half-systems. The polarization $P_{S(T)}$ of sample and tip is then defined by

$$P_{S(T)} = \frac{n^{\uparrow}_{S(T)} - n^{\downarrow}_{S(T)}}{n^{\uparrow}_{S(T)} + n^{\downarrow}_{S(T)}}. \tag{5.33}$$

This follows directly from (5.27). Polarization is an integral quantity, e.g., in (5.26): the difference between the number of electrons in different spin states divided by the total number of electrons. If we omit the transition probability $|M_{\mu\nu}|^2$ across the tunneling junction and focus on the number of electrons in either spin-up or spin-down states on both sides, we can write for the paramagnetic current I_0 the following expression:

$$I_0 = \frac{1}{2}\left(n^{\uparrow}_S + n^{\downarrow}_S\right)\left(n^{\uparrow}_T + n^{\downarrow}_T\right). \tag{5.34}$$

The factor of 1/2 arises from the probability for tunneling into either spin-up or spin-down states. The sums can be decomposed into ferromagnetic and antiferromagnetic charge transitions, where the currents I_F (ferromagnetic) and I_A (antiferromagnetic) are given by

$$I_F = n^{\uparrow}_S n^{\uparrow}_T + n^{\downarrow}_S n^{\downarrow}_T, \qquad I_A = n^{\uparrow}_S n^{\downarrow}_T + n^{\downarrow}_S n^{\uparrow}_T. \tag{5.35}$$

The paramagnetic current is consequently

$$I_0 = \frac{1}{2}\left(I_F + I_A\right). \tag{5.36}$$

Then the product $P_S P_T$ can be expressed in terms of I_F and I_A according to

$$P_S P_T = \frac{\left(n^{\uparrow}_S - n^{\downarrow}_S\right)\left(n^{\uparrow}_T - n^{\downarrow}_T\right)}{\left(n^{\uparrow}_S + n^{\downarrow}_S\right)\left(n^{\uparrow}_T + n^{\downarrow}_T\right)} = \frac{I_F - I_A}{I_F + I_A}. \tag{5.37}$$

And the current through the magnetic tunneling junction is then uniquely expressed in terms of I_F, I_A, and ϕ_M:

$$I(\phi_M) = \frac{1}{2}(I_F + I_A) + \frac{1}{2}(I_F - I_A)\cos\phi_M. \tag{5.38}$$

$$\cos\phi_M = \frac{\mathbf{M}_S \cdot \mathbf{M}_T}{|\mathbf{M}_S| \cdot |\mathbf{M}_T|}.$$

For arbitrary tunneling matrix elements $M_{\mu\nu}$ the current can be computed numerically within Bardeen's formulation for the tunneling current [8]. Since the energy of the tunneling electrons is very low, and since the overlap of the sample and tip wavefunctions is computed far outside the core region of surface atoms, spin-orbit coupling can generally be neglected in the theoretical treatment. The ferromagnetic and antiferromagnetic current I_F and I_A are simply the transitions for eigenstates with the same spin (I_F) or opposite spin (I_A):

$$I_F = I(n_S^\uparrow \longrightarrow n_T^\uparrow) + I(n_S^\downarrow \longrightarrow n_T^\downarrow), \tag{5.39}$$

$$I_A = I(n_S^\uparrow \longrightarrow n_T^\downarrow) + I(n_S^\downarrow \longrightarrow n_T^\uparrow) \tag{5.40}$$

These values can be directly obtained from standard DFT methods and perturbation theory. Calculating the tunnel current for different angles ϕ_M requires us then only to compute the linear combination of ferromagnetic and antiferromagnetic currents multiplied by the appropriate coefficients. From these three dimensional current maps the constant current contours and the surface corrugations can be extracted in a straightforward manner. To conclude the theoretical analysis, we find that all necessary information for a treatment of spin-polarized transport within perturbation theory can be provided by two separate pieces of information: (i) the tunneling current I_F and I_A for transitions from spin states of the sample to the same spin states of the tip, (ii) the angle ϕ_M between the magnetic axes.

5.3 Landauer–Büttiker approach

The main advantage of the Landauer–Büttiker approach is its mathematical rigor and its inclusion of the different boundary conditions of the STM leads. In principle it should thus yield a more accurate description of the tunneling condition. In addition, the treatment includes interference effects between separate conductance channels. The original one dimensional derivation by Landauer, recaptured in a paper by Markus Büttiker in 1985 [4], is actually quite simple. The main assumptions are these:

1. The two leads consist of ideal metals, the dispersion of electron states is equal to the dispersion of a free electron gas, which means, in one dimension, that the number of electrons of a given k value is linear with k.
2. The electrons impinging on the vacuum barrier are either transmitted across the barrier or reflected back into the lead. The transition of electrons changes the occupation in the leads so that the highest occupied level of the source, μ_1, is higher than the Fermi level. Conversely, the lowest unoccupied level of the drain, μ_2, is lower than its Fermi level.
3. The true Fermi levels, or the chemical potentials of source (μ_A) and drain (μ_B), are defined by the number of occupied and unoccupied states. Both levels are characterized by the number of occupied states above being equal to the number of unoccupied states below.
4. The applied bias eV is the potential difference between the two true Fermi levels (see Figure 5.1 for the one-dimensional setup).

5. The actual changes in the occupation numbers occur over a small energy interval; the energy level of tunneling electrons is thus equal to the Fermi level.

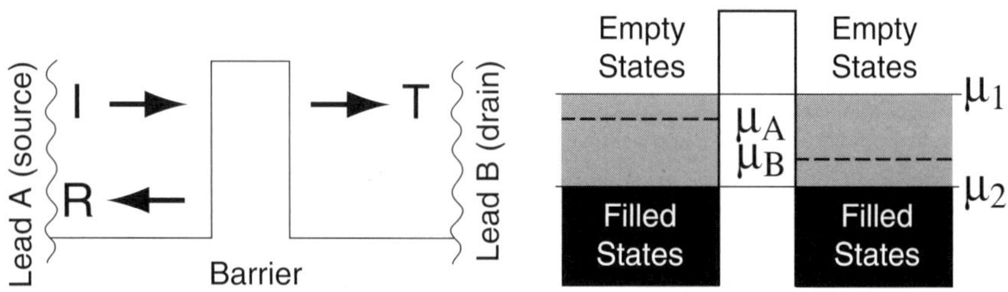

Fig. 5.1. The system of surface and tip is represented by two metallic leads and a vacuum barrier. The electrons entering the system at the left lead are partly transmitted and partly reflected by the vacuum barrier (left). Current transport across the barrier leads to a change of the highest occupied and the lowest unoccupied energy levels from μ_A, μ_B to μ_1, μ_2 (right).

The total current through the potential barrier will then be the difference between transmitted and reflected current contributions. The current emitted from the source is given by

$$I_S = ev\frac{\partial n}{\partial E}(\mu_1 - \mu_2)\,. \tag{5.41}$$

Here, e is the electron charge, v the electron velocity at the Fermi level, n the number of electrons, and E the energy. The number of electron states in a confined system is proportional to k with $2\pi n = k$; accounting for two spin orientations we get:

$$n = \frac{k}{\pi}, \qquad \frac{\partial n}{\partial k} = \frac{1}{\pi}. \tag{5.42}$$

Since the energy is given by $E = pv = \hbar k v$, we get for the density of states $\partial n/\partial E$,

$$k = \frac{E}{\hbar v}, \qquad n = \frac{E}{\hbar \pi v} \qquad \Rightarrow \qquad \frac{\partial n}{\partial E} = \frac{1}{\hbar \pi v}. \tag{5.43}$$

And since the transmission probability across the barrier is T, the net current through the tunneling junction is given by

$$I = \frac{2e}{h}(\mu_1 - \mu_2)\,T. \tag{5.44}$$

For the following it is important to note that the range of states below μ_2 will not contribute to the current, since all states are fully occupied. Neither will the range above μ_1, because all states are empty. The only relevant changes occur therefore in the range between these values. The energy levels μ_1 and μ_2 are calculated from the chemical potentials using the feature that the number of occupied and unoccupied states in each lead must be balanced with respect to the Fermi level. The total number N of carriers in each lead, for positive and negative velocities is given by

$$N = 2\frac{\partial n}{\partial E}(\mu_1 - \mu_2). \tag{5.45}$$

For the left lead, the number of occupied states above μ_A is due to impinging and reflected electrons; the number of unoccupied states below is the difference between the total number of carriers in this range and the number of occupied states. Thus

$$(1+R)\frac{\partial n}{\partial E}(\mu_1 - \mu_A) = [2-(1+R)]\frac{\partial n}{\partial E}(\mu_A - \mu_2). \tag{5.46}$$

Within the right lead, the number of occupied states above the Fermi level is due to transmissions across the barrier from the source, while the number of unoccupied states below μ_B is due to back transmissions into the source:

$$T\frac{\partial n}{\partial E}(\mu_1 - \mu_B) = (2-T)\frac{\partial n}{\partial E}(\mu_B - \mu_2). \tag{5.47}$$

From these relations the bias potential $eV = \mu_A - \mu_B$ is determined by a simple calculation, using the condition that $R + T = 1$, and we obtain

$$eV = \mu_a - \mu_B = R(\mu_1 - \mu_2). \tag{5.48}$$

Inserting into (5.44) this leads finally to Landauer's original result for the conductance between two metal leads separated by a vacuum barrier:

$$G(V) = \frac{I}{V} = \frac{2e^2}{h}\frac{T}{R} \approx \frac{2e^2}{h}T. \tag{5.49}$$

The result is strictly valid only at zero temperature. For a compact notation of this important equation within the framework of transition matrices, and under ambient thermal conditions, see the next section.

5.3.1 Scattering and perturbation method

Within a general framework of many-channel transitions the Landauer–Büttiker equation is generally rewritten in matrix notation. For a simple interface, comprising the two leads and a vacuum barrier, the conductance is then described by [4]

$$G(V) = \frac{2e^2}{h}\sum_i T_i = \frac{2e^2}{h}Tr(t^+(V)t(V)). \tag{5.50}$$

Note that the trace over the matrix product is in this case equivalent to the sum over all the conduction channels. The situation changes, however, if different pathways through an interface exist, e.g., the tunneling current through a molecule adsorbed on a surface. In this case the matrix product also contains off-diagonal elements, which describe the simultaneous transition of an electron through more than one conduction channel of the interface. The question how important these interference effects are for the actual tunneling image is still without a conclusive answer. It has been shown that interferences may play a major role in the images of benzene adsorbed on rhodium surfaces [13]. However, this seems not to have been established on a wider scale. Apart from these effects, the formalism is leading to the same results in tunneling simulations as perturbation methods.

5.4 Keldysh–Green's function approach

During the last few years, theoretical treatments of the tunneling process on the basis of a nonequilibrium Green's function formalism [5] have become increasingly popular [14, 15]. The most complete treatment of the problem considers the Hamiltonian of a system comprising two leads and a barrier region [6, 16, 17, 18]. The intricacies of the theory were treated in the previous chapter; here we merely state the result. The current through an interface including inelastic effects is given by (see previous chapter)

$$I = \frac{e}{h} \int_{\mu_A - eV}^{\mu_B + eV} dE\, Tr \left[\Sigma_A^{<}(E) G^{>}(E) - \Sigma_A^{>}(E) G^{<}(E) \right] . \qquad (5.51)$$

Here, the self-energies $\Sigma_A^{<}(E)$ describe the coupling to the infinite leads, while $G^{<}(E)$ is the nonequilibrium Green's function of the conductor interface. If we analyze the time scales involved in tunneling processes, then we get for normal tunneling conditions ($I \approx 1$ nA) an interval between single electron processes of about 10^{-10} s. Considering that this interval is one hundred to one thousand times longer than the typical time scale of lattice excitations, it seems safe to neglect interactions of electrons in the barrier. In this case (5.51) reduces essentially to the Landauer–Büttiker formula (see previous chapter):

$$\begin{aligned} I(V) = & \frac{2e}{h} \int_{-\infty}^{+\infty} dE \left[f(\mu_S, E) - f(\mu_T, E) \right] \\ & \times Tr \left[\Gamma_S(E) G^{\mathrm{R}}(E) \Gamma_T(E) G^{\mathrm{A}}(E) \right] , \end{aligned} \qquad (5.52)$$

where f denotes the Fermi distribution functions, $G^{\mathrm{R}}(E)$ and $G^{\mathrm{A}}(E)$ are the retarded and advanced Greens functions of the barrier, and $\Gamma_{S(T)}$ the surface and tip contacts. This formalism has become increasingly popular in particular for applications in molecular electronics [19], or the analysis of transport properties through interfaces [14, 15]. Due to the wide range of interactions included in the formalism, Keldysh's method is the most accurate today. Its main problem is the computational cost, which either has to be made up for by approximations in the description of the solid state systems, or by limiting the number of atoms in the interaction range. For this reason it has generally been implemented in tight-binding models. This limitation, and the ensuing lack of numerical precision were the main objections against its general use. However, it was shown recently that the method can also be implemented within a plane wave basis set [20]. Since state-of-the-art DFT methods rely on plane waves for an accurate description, this is a strong argument in favor of replacing perturbation methods by scattering methods.

5.5 Unified model for scattering and perturbation

5.5.1 Scattering and perturbation

From a theoretical point of view a tunneling electron, e.g., in a scanning tunneling microscopy measurement, is part of a system comprising two infinite metal leads and an interface, consisting of a vacuum barrier and, optionally, a molecule or a cluster of atoms with different properties from those of the infinite leads. The system can be said to be open, the number of charge carriers is not constant, and out of equilibrium, the applied potential and charge transport themselves introduce polarizations and excitations within the system. The theoretical description of such a system has advanced significantly over the last years; to date the most comprehensive description is based either on a self-consistent solution of the Lippman–Schwinger equation [21] or on the nonequilibrium Green's function approach [6, 19, 22, 23, 24]. Inelastic effects within, e.g., a molecule-surface interface can be included by considering multiple electron paths from the vacuum into the surface substrate [25]. Within the vacuum barrier itself, inelastic effects play an insignificant role. Here, as in most experiments in scanning tunneling microscopy, the problem can be reduced to a description of the tunneling current between two leads–the surface S and the tip T–thought to be in thermal equilibrium. The bias potential of the circuit is in this case described by a modification of the chemical potentials of surface and tip system, symbolized by μ_S and μ_T. This reduces the tunneling problem to the Landauer–Büttiker formulation [4, 19], or (see previous chapter)

$$I = \frac{2e}{h}\int_{-\infty}^{+\infty} dE\,[f(\mu_S, E) - f(\mu_T, E)] \times Tr\left[\Gamma_T(E)G^{\mathrm{R}}(E)\Gamma_S(E)G^{\mathrm{A}}(E)\right].$$

Here, f denotes the Fermi distribution function, $G^{\mathrm{R(A)}}(E)$ is the retarded (advanced) Green's function of the barrier, and Γ_S, Γ_T are the surface and tip contacts, respectively. They correspond to the difference of retarded and advanced self-energy terms of surface and tip; we define them by their relation to the spectral function $A_{S(T)}$ of the surface (tip) [19]:

$$A_{S(T)}(E) = i\left[G^{\mathrm{R}}_{S(T)}(E) - G^{\mathrm{A}}_{S(T)}(E)\right] = G^{\mathrm{R}}_{S(T)}(E)\Gamma_{S(T)}(E)G^{\mathrm{A}}_{S(T)}(E). \tag{5.53}$$

At present, these equations are evaluated within localized basis sets, and in a matrix representation. From a theoretical point of view this requires one either to represent the electronic properties of the two surfaces also in a localized representation [23, 24], or to transform the plane-wave basis set of most density functional methods to a local basis. The use of local basis sets compromises the numerical accuracy in the tunneling barrier, since the vacuum tails of the surface wavefunctions decay too rapidly: the constant-current contours in this case are too close to the surface. The following presentation shows that this limitation can be lifted by a clever application of the Dyson equation, and the tunneling current and interaction energy can then be calculated also within a plane-wave basis set and for a system with broken lateral symmetry like the general system in STM experiments [20]

5.5.2 Green's function of the vacuum barrier

In this section we present a formulation of the problem that is based on the Green's functions of the two surfaces, given in a real space representation based on the electronic eigenstates of the two systems. We show how the multiple scattering formalism described in (5.53) can be evaluated in real space, and how it relates to the perturbation expansion of the tunneling problem. We start with an eigenvector expansion of the surface and tip Green's functions, given by

$$G_S^{\mathrm{R(A)}}(\mathbf{r}_1, \mathbf{r}_2, E) = \sum_i \frac{\psi_i(\mathbf{r}_1)\psi_i^*(\mathbf{r}_2)}{E - E_i' + (-)i\eta}, \tag{5.54}$$

$$G_T^{\mathrm{R(A)}}(\mathbf{r}_1, \mathbf{r}_2, E) = \sum_j \frac{\chi_j(\mathbf{r}_1)\chi_j^*(\mathbf{r}_2)}{E - E_j' + (-)i\epsilon}. \tag{5.55}$$

Throughout this section the wavefunctions ψ and χ denote the Kohn–Sham states of surface and tip, respectively, resulting from a density functional calculation. The setup of the system is shown in Figure 5.2a. The energy eigenvalues of the Green's functions are shifted due to the applied bias voltage (see Figure 5.2 b), so that $E_i' = E_i - eV/2$, $E_j' = E_j + eV/2$. The spectral function A_S describes the charge density matrix, from (5.53) we obtain

$$A_S(\mathbf{r}_1, \mathbf{r}_2, E) = 2\eta \sum_i \frac{\psi_i(\mathbf{r}_1)\psi_i^*(\mathbf{r}_2)}{(E - E_i')^2 + \eta^2}. \tag{5.56}$$

The spectral function is related to Γ_S by (5.53). With the ansatz for Γ_S,

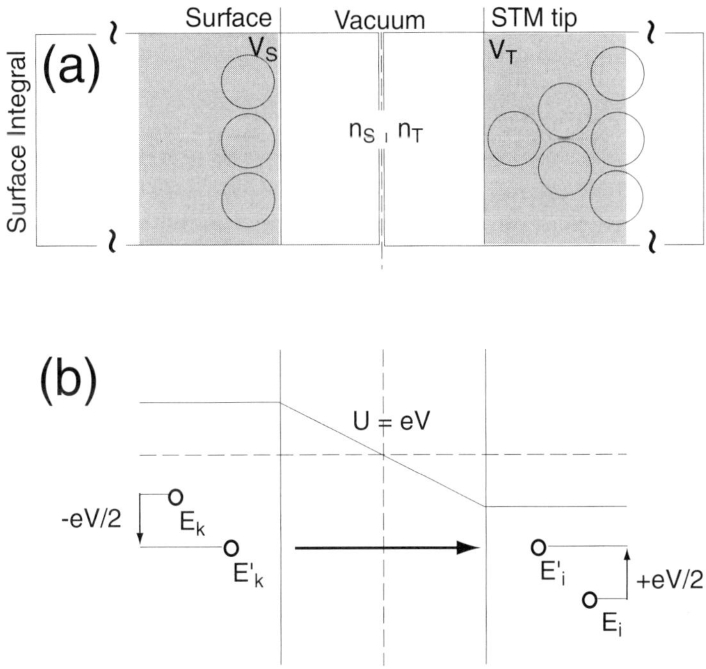

Fig. 5.2. (a) The system under consideration, and the surface integrals used in deriving the zero order current. (b) The effect of finite bias potentials: in this case the eigenvalues are shifted by $\pm eV/2$.

$$\Gamma_S(\mathbf{r}_3, \mathbf{r}_4, E) = C \sum_j \psi_j(\mathbf{r}_3)\psi_j^*(\mathbf{r}_4), \tag{5.57}$$

where C is a constant, we perform the double volume integration of (5.53). In this case the orthogonality of surface states reduces the expression to a compact form:

$$\begin{aligned} C \int d^3r_3 d^3r_4 G_S^R(\mathbf{r}_1, \mathbf{r}_3, E)\Gamma_S(\mathbf{r}_3, \mathbf{r}_4, E)G_S^A(\mathbf{r}_4, \mathbf{r}_2, E) \\ = C \sum_{ijk} \frac{\psi_i(\mathbf{r}_1)\psi_k^*(\mathbf{r}_2)\delta_{ij}\delta_{jk}}{(E - E_i' + i\eta)(E - E_k' - i\eta)}. \end{aligned} \tag{5.58}$$

Comparing the result with (5.56) we obtain for the contacts of surface and tip

$$\Gamma_S = 2\eta \sum_k \psi_k(\mathbf{r}_3)\psi_k^*(\mathbf{r}_4), \qquad \Gamma_T = 2\epsilon \sum_i \chi_i(\mathbf{r}_1)\chi_i^*(\mathbf{r}_2). \tag{5.59}$$

For the construction of the Green's function in the barrier we use the fact that the charge density is known from the separate calculation of surface and tip. In the limit of weak coupling, the total charge density of the interface is given by (see Figure 5.2)

$$n(\mathbf{r}, E) = \sum_i \psi_i(\mathbf{r})\psi_i^*(\mathbf{r})\delta(E - E_i') + \sum_j \chi_j(\mathbf{r})\chi_j^*(\mathbf{r})\delta(E - E_j'). \qquad (5.60)$$

This indicates that a zero-order approximation for the Green's function of the vacuum barrier can be constructed as sum of surface and tip Green's functions, or

$$G_{(0)}^{\mathrm{R(A)}}(\mathbf{r}_1, \mathbf{r}_2, E) = G_S^{\mathrm{R(A)}}(\mathbf{r}_1, \mathbf{r}_2, E) + G_T^{\mathrm{R(A)}}(\mathbf{r}_1, \mathbf{r}_2, E). \qquad (5.61)$$

The diagonal elements $\mathbf{r}_1 = \mathbf{r}_2$ of this Green's function are just equal to the total charge density. For the off-diagonal elements $\mathbf{r}_1 \neq \mathbf{r}_2$ we demonstrate by two separate estimates that this choice is justified. First, from the Schrödinger equation,

$$\left[-\frac{\hbar^2}{2m}\nabla^2 + V_S(\mathbf{r}_1) + V_T(\mathbf{r}_1)\right](G_S(\mathbf{r}_1, \mathbf{r}_2) + G_T(\mathbf{r}_1, \mathbf{r}_2)) = 0, \qquad (5.62)$$

it follows that the Green's function is exact if

$$V_S(\mathbf{r}_1)G_T(\mathbf{r}_1, \mathbf{r}_2) + V_T(\mathbf{r}_1)G_S(\mathbf{r}_1, \mathbf{r}_2) = 0. \qquad (5.63)$$

In the surface region, $V_T = 0$ and $G_T \approx 0$. In the tip region, $V_S = 0$ and $G_S \approx 0$. In the vacuum region both terms are products of functions centered at different sides of the vacuum barrier and decaying exponentially; they are consequently very small in comparison to other terms. Thus (5.63) approximately holds throughout the whole system. Second, let us write the well-known property of the Green's function

$$\frac{\partial G(z)}{\partial z} = -G^2(z). \qquad (5.64)$$

Substituting (5.61) into this formula results in the following condition: $G_S(z)G_T(z) = 0$. This is approximately satisfied in the whole system because in the surface region, $G_T \approx 0$; in the tip region, $G_S \approx 0$, and in the vacuum region, we obtain a product of two exponential functions centered at opposite ends of the system:

$$G_S G_T \propto \frac{e^{-\kappa_S|\mathbf{r}-\mathbf{r}'|}}{|\mathbf{r}-\mathbf{r}'|}\frac{e^{-\kappa_T|\mathbf{r}-\mathbf{r}''|}}{|\mathbf{r}-\mathbf{r}''|} \approx 0.$$

5.5.3 Zero-order current

Now all the necessary components for calculating the trace in the non-equilibrium formalism are given in terms of the real space surface and tip wavefunctions. We obtain the following expression for the trace:

$$
\begin{aligned}
Tr\left[\Gamma_T G^R_{(0)} \Gamma_S G^A_{(0)}\right] &= 4\eta\epsilon \sum_{ik} \\
&\times |A_{ik}|^2 \left[\frac{1}{(E-E'_k)^2+\eta^2} + \frac{1}{(E-E'_k+i\eta)(E-E'_i-i\epsilon)}\right. \\
&+ \left.\frac{1}{(E-E'_k-i\eta)(E-E'_i+i\epsilon)} + \frac{1}{(E-E'_i)^2+\epsilon^2}\right],
\end{aligned}
\tag{5.65}
$$

with the overlap integral A_{ik} given by

$$
A_{ik} = \int d^3 r \chi_i^*(\mathbf{r}) \psi_k(\mathbf{r}). \tag{5.66}
$$

The sum of fractions involving energies and ϵ, η, which results from the multiplication of Green's functions, can be written in a more compact way as

$$
\frac{(E-E'_k+E-E'_i)^2+(\eta+\epsilon)^2}{[(E-E'_k)^2+\eta^2][(E-E'_i)^2+\epsilon^2]}.
$$

In the limit $\eta, \epsilon \to +0$ the second term in the numerator will vanish, and since

$$
\lim_{\eta \to 0} \frac{\eta}{(E-E'_i)^2+\eta^2} = \pi\delta(E-E'_i), \tag{5.67}
$$

the transmission probability reduces to

$$
\sum_{ik} |A_{ik}|^2 4\pi^2 \delta(E-E'_k)\delta(E-E'_i)(E-E'_k+E-E'_i)^2. \tag{5.68}
$$

The calculation of the matrix elements A_{ik} involves an integration over infinite space, which cannot directly be performed. To convert the volume integrals into surface integrals we use the fact that the vacuum states of surface and tip are free electron solutions with characteristic decay constants, complying with the vacuum Schrödinger equation:

$$
\frac{\hbar^2}{2m}\left(\nabla^2+\kappa_i^2\right)\chi_i(\mathbf{r}) = 0 \Rightarrow \chi_i(\mathbf{r}) = -\frac{\nabla^2}{\kappa_i^2}\chi_i(\mathbf{r}), \tag{5.69}
$$

$$
\frac{\hbar^2}{2m}\left(\nabla^2+\kappa_k^2\right)\psi_k(\mathbf{r}) = 0 \Rightarrow \psi_k(\mathbf{r}) = -\frac{\nabla^2}{\kappa_k^2}\psi_k(\mathbf{r}). \tag{5.70}
$$

In addition, we make use of the following identities

$$
\begin{aligned}
\chi_i^*\nabla^2\psi_k &- \nabla(\chi_i^*\nabla\psi_k) - \nabla\chi_i^*\nabla\psi_k, \\
\psi_k\nabla^2\chi_i^* &= \nabla(\psi_k\nabla\chi_i^*) - \nabla\chi_i^*\nabla\psi_k.
\end{aligned}
$$

After some trivial manipulations, and making use of Gauss's theorem, this allows us to convert the volume integral into an integral over the separation surface (see Figure 5.2):

$$A_{ik} = \frac{1}{\kappa_i^2 - \kappa_k^2} \underbrace{\int dS\,[\chi_i^*(\mathbf{r})\nabla\psi_k(\mathbf{r}) - \psi_k(\mathbf{r})\nabla\chi_i^*(\mathbf{r})]}_{=M_{ik}}. \qquad (5.71)$$

The relation is valid only if $\kappa_i^2 \neq \kappa_k^2$. In practice that does not limit the generality of the approach, since surface and tip workfunctions are generally different. The surface integral is well known; apart from the universal constant $\hbar^2/2m$ it describes the tunneling matrix element in the perturbation approach [3, 26]. Integrating over the energy range, we obtain from (5.53),(5.68), and (5.71) the tunneling current in the zero-order approximation

$$I_{(0)} = \frac{4\pi e}{\hbar} \sum_{ik} [f(\mu_S, E_k') - f(\mu_T, E_i')] \left| \frac{(E_k' - E_i')M_{ik}}{\kappa_i^2 - \kappa_k^2} \right|^2 \delta(E_i' - E_k'). \qquad (5.72)$$

The decay constants are proportional to the eigenvalues shifted by the bias voltage of the tunneling junction:

$$E_i = \frac{\hbar^2\kappa_i^2}{2m} = E_i' - \frac{eV}{2}, \qquad E_k = \frac{\hbar^2\kappa_k^2}{2m} = E_k' + \frac{eV}{2}. \qquad (5.73)$$

Including the effect of finite bias voltages thus leads to the following result:

$$I_{(0)} = \frac{4\pi e}{\hbar} \sum_{ik} \left[f\left(\mu_S, E_k - \frac{eV}{2}\right) - f\left(\mu_T, E_i + \frac{eV}{2}\right)\right] \times \left| \left(-\frac{\hbar^2}{2m} - \frac{eV}{\kappa_i^2 - \kappa_k^2} \right) M_{ik} \right|^2 \delta(E_i - E_k + eV). \qquad (5.74)$$

It can be seen from this formulation that the obtained tunneling spectrum, or the dI/dV curves, will increase quadratically with the applied bias voltage. This is actually observed in spectroscopy experiments [27]. The second term in parentheses, giving the bias dependency in the zero-order scattering approach, is a correction to the standard Bardeen approach, which can be recovered in the limit of zero bias. In this case we confirm the result by Feuchtwang and Pendry et al. [28, 29] that the Bardeen method is just the zero-order approximation, in the limit of zero bias, to a full scattering treatment [3, 26]:

$$I_B = \frac{4\pi e}{\hbar} \sum_{ik} [f(\mu_S, E_k) - f(\mu_T, E_i)] \left| -\frac{\hbar^2}{2m} M_{ik} \right|^2 \delta(E_i - E_k). \qquad (5.75)$$

This result and its interpretation in terms of scattering theory is well accepted. Here, it shows once more that the choice for the zero- order Green's function of the interface is justified.

Corrections to the Tersoff–Hamann approach

The additional approximation in the Tersoff–Hamann approach concerns only the shape of the tip orbital, in particular the substitution of the matrix element M_{ik} by

$$-\frac{\hbar^2}{2m}M_{ik} \propto \frac{\hbar^2}{2m}\psi_i(\mathbf{R}), \tag{5.76}$$

where $\mathbf{R}$ is the position of the STM tip. Since this does not affect the rest of the derivation, the bias dependency will also affect the result of a derivation, which is based on an analytical form of the tip wavefunctions. This means, that the modified Tersoff–Hamann result, including the bias dependency, will be the following:

$$I_{TH} \propto \left| \left(-\frac{\hbar^2}{2m} - \frac{eV}{\kappa_T^2 - \kappa_k^2} \right) \psi_k \right|^2, \tag{5.77}$$

where κ_T is the decay length of the tip s-orbital.

5.5.4 First-order Green's function

The approach can be extended to higher orders. In the first order expansion of the Dyson series the Green's function is given by

$$G^{\mathrm{R}}_{(1)} = G^{\mathrm{R}}_{(0)} + G^{\mathrm{R}}_{(0)} V G^{\mathrm{R}}_{(0)}. \tag{5.78}$$

To calculate the first-order Green's function for systems out of equilibrium, the equation has to be solved self-consistently [21, 23]. Self-consistency can in principle also be achieved by basing the calculation on the Kohn–Sham states ψ and χ of charged surfaces. Under tunneling conditions, however, the leads are in thermal equilibrium and the systems only weakly coupled. V in this case is the potential $V_S + V_T$ within the vacuum barrier:

$$\begin{aligned} G^R_{(1)}(\mathbf{r}_1, \mathbf{r}_2) = \quad & G^R_{(0)}(\mathbf{r}_1, \mathbf{r}_2) \\ & + \int dr_3 G^R_{(0)}(\mathbf{r}_1, \mathbf{r}_3) \left[V_S(\mathbf{r}_3) + V_T(\mathbf{r}_3) \right] G^R_{(0)}(\mathbf{r}_3, \mathbf{r}_2). \end{aligned} \tag{5.79}$$

This leads to six additional first-order terms, described by

$$\begin{aligned} G_{(1)} = \quad & G_{(0)} + G_S V_T G_S + G_T V_S G_T \\ & + G_S V_T G_T + G_S V_S G_T + + G_T V_S G_S + G_T V_T G_S. \end{aligned} \tag{5.80}$$

Here, the first line corresponds to excitations on either side of the tunneling junction; the second line describes the effects due to transitions. In the following we focus on transitions; we note, however, that excitations can be included in the formulation by a suitable adaptation of many-body theory. Writing the first term of the second line explicitly, and with the shortcut $f^{\pm}_{ik} = (E - E'_i \pm i\eta)(E - E'_k \pm i\epsilon)$, the integration then has to be performed only for the halfspace in which the potential is not zero. The integrals can be rewritten as surface integrals with the help of the Schrodinger equation:

$$\left(-\frac{\hbar^2}{2m}\nabla^2 + V_S \right) \psi_i = E_i \psi_i, \qquad \left(-\frac{\hbar^2}{2m}\nabla^2 + V_T \right) \chi_i = E_i \chi_i. \tag{5.81}$$

The first term in the second line

$$G_S^{\rm R} V_T G_T^{\rm R} = \sum_{i,k} \frac{\psi_i(\mathbf{r}_1)\chi_k^*(\mathbf{r}_2)}{f_{ik}^+} \int d^3r \psi_i^*(\mathbf{r}) V_T(\mathbf{r}) \chi_k(\mathbf{r}), \qquad (5.82)$$

can then be calculated, and we obtain for the integral

$$\begin{aligned}
&\int_{\Omega_T} d^3r \psi_i^*(\mathbf{r}) V_T(\mathbf{r}) \chi_k(\mathbf{r}) \\
&= \int_{\Omega_T} d^3r \left[\psi_i^*(\mathbf{r}) E_k \chi_k(\mathbf{r}) + \psi_i^*(\mathbf{r}) \frac{\hbar^2}{2m} \nabla^2 \chi_k(\mathbf{r})\right] \\
&= \int_{\Omega_T} d^3r \left[\chi_k(\mathbf{r}) E_k \psi_i^*(\mathbf{r}) + \psi_i^*(\mathbf{r}) \frac{\hbar^2}{2m} \nabla^2 \chi_k(\mathbf{r})\right] \\
&= \int_{\Omega_T} d^3r \left[\chi_k(\mathbf{r}) E_i \psi_i^*(\mathbf{r}) + \psi_i^*(\mathbf{r}) \frac{\hbar^2}{2m} \nabla^2 \chi_k(\mathbf{r})\right] \\
&= \int_{\Omega_T} d^3r \left[-\chi_k(\mathbf{r}) \frac{\hbar^2}{2m} \nabla^2 \psi_i^*(\mathbf{r}) + \psi_i^*(\mathbf{r}) \frac{\hbar^2}{2m} \nabla^2 \chi_k(\mathbf{r})\right] \\
&= -\frac{\hbar^2}{2m} \int_S dS \left[\chi_k(\mathbf{r}) \nabla \psi_i^*(\mathbf{r}) - \psi_i^*(\mathbf{r}) \nabla \chi_k(\mathbf{r})\right] = -\frac{\hbar^2}{2m} M_{ki}^* . \qquad (5.83)
\end{aligned}$$

Since the perturbative treatment is completely symmetric with respect to surface and tip system, we equally find for the second term, by integration over the surface region Ω_S,

$$G_S^{\rm R} V_S G_T^{\rm R} = \sum_{i,k} \frac{\psi_i(\mathbf{r}_1)\chi_k^*(\mathbf{r}_2)}{f_{ik}^+} \int d^3r \psi_i^*(\mathbf{r}) V_S(\mathbf{r}) \chi_k(\mathbf{r}), \qquad (5.84)$$

$$\begin{aligned}
&\int_{\Omega_S} d^3r \chi_k(\mathbf{r}) V_S(\mathbf{r}) \psi_i^*(\mathbf{r}) \\
&= \int_{\Omega_S} d^3r \left[\chi_k(\mathbf{r}) E_i \psi_i^*(\mathbf{r}) + \chi_k(\mathbf{r}) \frac{\hbar^2}{2m} \nabla^2 \psi_i^*(\mathbf{r})\right] \\
&= \int_{\Omega_S} d^3r \left[\psi_i^*(\mathbf{r}) E_i \chi_k(\mathbf{r}) + \chi_k(\mathbf{r}) \frac{\hbar^2}{2m} \nabla^2 \psi_i^*(\mathbf{r})\right] \\
&= \int_{\Omega_S} d^3r \left[\psi_i^*(\mathbf{r}) E_k \chi_k(\mathbf{r}) + \chi_k(\mathbf{r}) \frac{\hbar^2}{2m} \nabla^2 \psi_i^*(\mathbf{r})\right] \\
&= \int_{\Omega_S} d^3r \left[-\psi_i^*(\mathbf{r}) \frac{\hbar^2}{2m} \nabla^2 \chi_k(\mathbf{r}) + \chi_k(\mathbf{r}) \frac{\hbar^2}{2m} \nabla^2 \psi_i^*(\mathbf{r})\right] \\
&= -\frac{\hbar^2}{2m} \int_S dS \left[\psi_i^*(\mathbf{r}) \nabla \chi_k(\mathbf{r}) - \chi_k(\mathbf{r}) \nabla \psi_i^*(\mathbf{r})\right] = -\frac{\hbar^2}{2m} M_{ki}^* . \qquad (5.85)
\end{aligned}$$

In the last line we took into account that the surfaces of the two integrations point in opposite directions. The first-order Green's function of the interface is consequently

$$G_{(1)}^{\mathrm{R(A)}} = G_{(0)}^{\mathrm{R(A)}} - \frac{\hbar^2}{m} \sum_{i,k} \frac{\psi_i(\mathbf{r}_1) M_{ki}^* \chi_k^*(\mathbf{r}_2) + \chi_i(\mathbf{r}_1) M_{ik} \psi_k^*(\mathbf{r}_2)}{f_{ik}^{+(-)}}.$$

It is evident that each subsequent iteration in the interface Green's function can also be formulated in terms of Bardeen matrix elements: in principle, the Green's function and thus the current can therefore be evaluated to any order.

5.5.5 Interaction energy

Finally, we calculate the interaction energy between the surface and the tip in the low-coupling limit. It has been shown recently by an analysis of first-order perturbation expressions for the tunneling current and the interaction energy that the two variables should be linear with respect to each other. From the first-order Green's function we may construct the density matrix $\hat{n} = i/2\pi(G^{\mathrm{A}} - G^{\mathrm{R}})$. The interaction energy is then [26]

$$E_{\mathrm{int}} = \frac{i}{2\pi} \int_{-\infty}^{+\infty} dE Tr \left[\left(G_{(1)}^{\mathrm{A}}(E) - G_{(1)}^{\mathrm{R}}(E) \right) (V_S + V_T) \right]. \tag{5.86}$$

The density matrix is calculated from the first-order term of the Green's function, since zero-order terms will only lead to a shift of eigenvalues in the presence of the opposite lead. The explicit form of the density matrix is

$$\hat{n}(\mathbf{r}_1, \mathbf{r}_2, E) = -\frac{i}{2\pi} \sum_{i,k} \left(\frac{1}{f_{ik}^-} - \frac{1}{f_{ik}^+} \right) \left[\psi_i(\mathbf{r}_1) M_{ki}^* \chi_k^*(\mathbf{r}_2) + \chi_i(\mathbf{r}_1) M_{ik} \psi_k^*(\mathbf{r}_2) \right]. \tag{5.87}$$

The trace $Tr[\hat{n}V]$ leads to four terms:

$$\begin{aligned} Tr[\hat{n}V] = &-\frac{i\hbar^2}{2\pi m} \sum_{i,k} M_{ki}^* \Bigg[\underbrace{\int d^3r \psi_i(\mathbf{r}) V_S \chi_k^*(\mathbf{r})}_{=-\hbar^2/2mM_{ki}} + \underbrace{\int d^3r \psi_i(\mathbf{r}) V_T \chi_k^*(\mathbf{r})}_{=-\hbar^2/2mM_{ki}} \Bigg] \\ &-\frac{i\hbar^2}{2\pi m} \sum_{i,k} M_{ik} \Bigg[\underbrace{\int d^3r \psi_k^*(\mathbf{r}) V_S \chi_i(\mathbf{r})}_{=-\hbar^2/2mM_{ik}^*} + \underbrace{\int d^3r \psi_k^*(\mathbf{r}) V_T \chi_i(\mathbf{r})}_{=-\hbar^2/2mM_{ik}^*} \Bigg] \\ &-\frac{i}{2\pi} \left(\frac{\hbar^2}{m} \right)^2 \left[|M_{ik}|^2 + |M_{ki}|^2 \right]. \end{aligned} \tag{5.88}$$

The energy terms $f_{ik}^{\pm}$ lead to the following result:

$$\begin{aligned}\frac{1}{f_{ik}^-} - \frac{1}{f_{ik}^+} &= \frac{(E - E_i' + i\eta)(E - E_k' + i\epsilon) - (E - E_i' - i\eta)(E - E_k' - i\epsilon)}{[(E - E_i')^2 + \eta^2][(E - E_k')^2 + \epsilon^2]} \\ &= 2i\frac{\eta}{[(E - E_i')^2 + \eta^2]}\frac{E - E_k'}{[(E - E_k')^2 + \epsilon^2]} \\ &\quad +2i\frac{\epsilon}{[(E - E_k')^2 + \epsilon^2]}\frac{E - E_i'}{[(E - E_i')^2 + \eta^2]}.\end{aligned} \tag{5.89}$$

In the limit $\eta, \epsilon \to 0+$ this gives

$$\lim_{\eta,\epsilon\to 0+} = 2i\pi\frac{\delta(E - E_i')}{E - E_k'} + 2i\pi\frac{\delta(E - E_k')}{E - E_i'}. \tag{5.90}$$

The energy integration now has to be performed over the infinite energy interval. The only terms to consider are

$$2i\pi \int_{-\infty}^{+\infty} dE \left[\frac{\delta(E - E_i')}{E - E_k'} + \frac{\delta(E - E_k')}{E - E_i'}\right]. \tag{5.91}$$

Here we suppose that, physically speaking, all transitions across the barrier will lead to an increase of bonding and thus interaction energy. We therefore count every component separately:

$$\begin{aligned}&2i\pi \int_{-\infty}^{+\infty} dE \left[\frac{\delta(E - E_i')}{E - E_k'} + \frac{\delta(E - E_k')}{E - E_i'}\right] \\ &= \frac{2i\pi}{|E_i' - E_k'|} + \frac{2i\pi}{|E_k' - E_i'|} = \frac{4i\pi}{|E_i' - E_k'|}\end{aligned} \tag{5.92}$$

The final result for the interaction energy to first order is therefore

$$E_{\text{int}} = -4\left(\frac{\hbar^2}{m}\right)^2 \sum_{i,k} \frac{|M_{ik}|^2}{|E_i - E_k + eV|} \tag{5.93}$$

The absolute value of the denominator is due to integrating the infinite energy interval in two steps, and taking each result separately as a contribution to the interaction energy. The calculation of the interaction energy involves only the computation of the tunneling matrix elements. As shown previously, the interaction energy will therefore be proportional to the tunneling current [26]. To summarize, tunneling currents and interaction energies can be calculated in real space within the nonequilibrium Green's function formalism based on the separate wavefunctions of surface and tip. The zero-order expansion is equal to the Bardeen approach for zero bias; the bias dependency has been explicitly included in this new formulation. Higher-order Green's functions can be described in terms of Bardeen matrix elements, which demonstrates that the Green's functions, and thus the tunneling currents, can be computed to any order.

5.6 Electron–phonon interactions

The calculation of electron–phonon interactions in a tunneling junction is a genuine many-body problem and has therefore been restricted to tight-binding models till very recently. This restriction has now been removed due to the work of Lorente, Persson, and Brandbyge [30, 23]. The application of the method within DFT surface-structure simulations makes it possible to treat electron–electron and electron–phonon interactions during the tunneling process. This work is at the cutting edge of theory development at the time of publication of this monograph. A solution for electron–phonon interactions has been developed, which we present in the next section. For electron–electron interactions, similar methods should reach their state of maturity within the near future. The theoretical method is based on ground state density functional theory for the description of the surface and the STM tip, and its extension via the perturbative approach of Keldysh into the nonequilibrium regime. So far, it has been shown to provide accurate descriptions of the change in transport properties of molecules adsorbed on metal surfaces, when combined with the standard Tersoff–Hamann approach for the tunneling problem.
In the following we present a method to implement the procedure also in Bardeen's model of tunneling. The theoretical model to include the electron-vibration coupling into the many-body Bardeen formulation goes back to the work of Zawadowski, Appelbaum and Brinkman, Pendry, Crampin, and Lorente [31, 32, 29, 33, 34], it allows one to evaluate the changes in the conductance across the vibrational threshold. Within this framework the tunneling current is described by the expression

$$
\begin{aligned}
I(V) = \frac{2e^2}{\hbar\pi}\left(\frac{\hbar^2}{2m}\right)^2 & \\
& \times \int_{\epsilon_F}^{\epsilon_F + eV} d\omega \; Tr\left(\{\overleftarrow{\nabla}_1 - \overrightarrow{\nabla}_1\}\Im G_T^{\mathrm{R}}(\mathbf{r}_1, \mathbf{r}_2, \omega)\right. \\
& \times \left.\{\overleftarrow{\nabla}_2 - \overrightarrow{\nabla}_2\}\Im G_S^{\mathrm{R}}(\mathbf{r}_2, \mathbf{r}_1, \omega)\right). \qquad (5.94)
\end{aligned}
$$

The trace (Tr) and the nabla operators describe a surface integration with respect to the flux of the nabla operators. The arrows indicate the direction in which the derivative operates. Here, the rule is that the expression in parantheses is cyclic: a nabla operator with an arrow to the left acts on the previous expression, in case there is no previous expression it acts on the last expression, and so forth. The main advantage of this notation is that it keeps the length of equations within reasonable limits. The Green's functions are many-body Green's functions. If they are replaced by their single-particle counterparts, we recover the usual Bardeen formulation [29] (see previous chapter):

$$
G_S^{\mathrm{A(R)}}(\mathbf{r}, \mathbf{r}', \omega) = \sum_\lambda \frac{\psi_\lambda(\mathbf{r})\psi_\lambda^*(\mathbf{r}')}{\omega - \epsilon_\lambda - (+)i\eta}, \qquad (5.95)
$$

$$G_T^{\mathrm{A(R)}}(\mathbf{r},\mathbf{r}',\omega) = \sum_m \frac{\psi_m(\mathbf{r})\psi_m^*(\mathbf{r}')}{\omega - \epsilon_m - (+)i\eta}. \tag{5.96}$$

Here, A(R) refers to the advanced (retarded) Green's function, and $S(T)$ to the surface (tip). The local vibration can be introduced via a perturbation Hamiltonian of the following form:

$$\hat{H}_1 = \sum_{\mu,\nu} U_{\mu,\nu}\hat{c}_\mu^\dagger \hat{c}_\nu \,\delta Q(\hat{b}^\dagger + \hat{b}), \tag{5.97}$$

where $\delta Q = \sqrt{\hbar/(2M\Omega)}$ with Ω the vibration frequency of the localized mode, and M the reduced mass associated with the vibration. In the harmonic approximation the potential $U_{\mu,\nu}$ can be replaced by the derivative of the effective one-electron potential:

$$\hat{U}_q(\mathbf{r}) = \left\langle \frac{\partial \hat{H}_1(\mathbf{r},Q)}{\partial Q} \right\rangle, \tag{5.98}$$

with the brackets averaging over the harmonic oscillator states. The perturbation approach to electron-vibration coupling is based on the assumption that the tunneling can be considered as a single-particle problem, while the many-body perturbation occurs on one side of the junction only, the surface region. The out-of-equilibrium condition within this interface can be included using the Keldysh formalism [35, 36] for the nonequilibrium Green's function $G_S^{\gtrless}(\mathbf{r},\mathbf{r}',\omega)$:

$$G_S^{\gtrless}(\mathbf{r},\mathbf{r}',\omega) = 2i(1-f_\lambda)\Im G_S^{\mathrm{R}}(\mathbf{r},\mathbf{r}',\omega), \tag{5.99}$$

where $f_\lambda = n_F(\epsilon_\lambda)$ is the Fermi distribution for an eigenvalue ϵ_λ. Within a one-electron basis this correlation function can be simplified to

$$G_S^{\gtrless}(\mathbf{r},\mathbf{r}',\omega) = -i2\pi \sum_\lambda (1-f_\lambda)\psi_\lambda(\mathbf{r})\psi_\lambda^*(\mathbf{r}')\delta(\omega-\epsilon_\lambda). \tag{5.100}$$

It is important to note that the correlation function $G_S^{\gtrless}$ contains an inelastic part, which is due to the excitation of a phonon mode, and an elastic part, which is due to the crossing of electron paths during transition (see [37]). The inelastic component of the correlation function $\delta G_S^{\gtrless} = \delta G_{\mathrm{ine}}^{>} + \delta G_{\mathrm{ela}}^{>}$ is described in terms of the self-energies $\Sigma_S^{\gtrless}$:

$$\delta G_{\mathrm{ine}}^{>}(\mathbf{r},\mathbf{r}',\omega) = \int\!\!\int d\mathbf{r}_1 d\mathbf{r}_2\, G^R(\mathbf{r},\mathbf{r}_1,\omega)\Sigma_S^{\gtrless}(\mathbf{r}_1,\mathbf{r}_2,\omega)G^A(\mathbf{r}_2,\mathbf{r}',\omega), \tag{5.101}$$

which in turn can be written as

$$\Sigma_S^{>}(\mathbf{r}_1, \mathbf{r}_2, \omega) = -i2\pi U_q(\mathbf{r}_1) U_q(\mathbf{r}_2) \sum_{\lambda} (1 - f_\lambda) \psi_\lambda(\mathbf{r}_1) \psi_\lambda^*(\mathbf{r}_2) \delta(\omega - \Omega - \epsilon_\lambda). \tag{5.102}$$

From these relations the change of the conductance $\delta(dI/dV)$ at a bias voltage corresponding to an existing phonon mode is fairly straightforward to evaluate, and the result will be

$$\begin{aligned}\delta\left(\frac{dI}{dV}(\omega)\right)_{\text{ine}} &= \frac{\pi}{2} \sum_{m,\lambda} (1 - f_\lambda) \left| \int d\mathbf{S} \cdot (\delta\psi_\lambda \nabla \psi_m^* - \psi_m \nabla \delta\psi_\lambda^*) \right|^2 \\ &\times \delta(\epsilon_m - \omega) \delta(\omega - \Omega - \epsilon_\lambda).\end{aligned} \tag{5.103}$$

In the quasistatic approximation (neglecting the frequency dependency by setting $\omega \approx \epsilon_F$, this amounts to

$$\Delta\left(\frac{dI}{dV}\right)_{ine} = \frac{\pi}{2} \sum_{k,\lambda} | \int d\mathbf{S} \cdot (\delta\psi_\lambda \nabla \chi_k^* - \chi_k \nabla \delta\psi_\lambda^*)|^2 \delta(E_F - E_k) \delta(E_F - E_\lambda). \tag{5.104}$$

In a local basis set the perturbed sample wavefunctions are given by

$$\delta\psi_\lambda(\mathbf{r}) = \sum_{\mu} \psi_\mu(\mathbf{r}) \frac{\langle \mu | U_q | \lambda \rangle}{\epsilon_\lambda - \epsilon_\mu + i\delta}. \tag{5.105}$$

The elastic component is somewhat more complicated, but an approximation, which takes into account the cancellation of the logarithmic divergence due to the elastic Green's function, leads to a formally identical result with the modified perturbed wavefunctions, which are just the imaginary parts of the expression (5.105):

$$\delta\psi_\lambda(\mathbf{r}) = \sqrt{2}\pi \sum_{\mu} \psi_\mu(\mathbf{r}) \langle \mu | U_q | \lambda \rangle \delta(\epsilon_\lambda - \epsilon_\mu). \tag{5.106}$$

The main feature of the elastic contribution is that it is purely negative. This is an important consequence of this theory; the change in conductance is a mixture of positive (inelastic) and negative (elastic) contributions.

Implementation in standard DFT codes

The theoretical model can be implemented in every standard DFT code. It involves in principle only the calculation of the change of the wavefunctions due to phonon excitations. These perturbations to the electronic ground state $\delta\psi_\lambda$ can then be used as the input for a fully first-principles simulation of the change of conductance in crossing the energy threshold of phonon states, which can be directly compared to experiments.

The distorted wavefunctions $\delta\psi_\lambda$ can be evaluated to first order by expressing the matrix elements in terms of wavefunction overlaps. Since $\langle\mu|U_q|\lambda\rangle = \langle\mu|\delta U_q|\lambda\rangle/\delta Q$, we use a formulation by Head-Gordon and Tully [38, 39] to obtain

$$\langle\mu|\delta U_q|\lambda\rangle = \begin{cases} 0, & \mathbf{k}_{\parallel}^{\mu} \neq \mathbf{k}_{\parallel}^{\lambda}, \\ \delta\epsilon_\lambda, & n^\mu = n^\lambda, \mathbf{k}_{\parallel}^{\mu} \neq \mathbf{k}_{\parallel}^{\lambda}, \\ (\epsilon_\lambda - \epsilon_\mu)\langle\mu|\delta\lambda\rangle, & n^\mu \neq n^\lambda, \mathbf{k}_{\parallel}^{\mu} = \mathbf{k}_{\parallel}^{\lambda}. \end{cases} \tag{5.107}$$

In this case, both the matrix elements $\langle\mu|v|\lambda\rangle$ and the perturbed wavefunctions $\delta\psi_\lambda(\mathbf{r})$ can be obtained from the electronic structure of groundstate DFT calculations. The variation of an eigenstate $|\delta\lambda\rangle$ is calculated from the central differences of two displaced configurations. On the technical side it has to be noted that the implementation depends on the DFT code used. In pseudopotential codes it will be buried quite deeply within the code, since the overlaps require rescaling the pseudostates obtained, for example, in pseudopotential codes with the overlap matrix $\hat{S}^{1/2}$.

To sum up, all necessary steps to calculate the nonequlibrium changes of the wavefunctions due to electron–phonon interactions can be incorporated in present state-of-the-art DFT methods, and these changed wavefunctions can then be used, as in previous simulations, as an input to efficient STM simulations, including the electronic structure of the STM tip.

5.7 Summary

In this chapter we have presented an overview over the most common methods used in tunneling problems, which are, in increasing order of complexity: the Tersoff–Hamann model, the Bardeen model, the Landauer–Büttiker model, and the Keldysh model. The treatment of the tunneling junction in these models is described by one of the following: restricted to the surface only (Tersoff–Hamann); includes both sides of the junction, without considering interference effects (Bardeen); is based on elastic tunneling conditions (Landauer–Büttiker); includes the full nonequilibrium formulation of the problem (Keldysh). Readers interested in a general formulation of transport theory are referred to the previous chapter, where the whole framework is treated in some detail.

References

1. J. Tersoff and D. R. Hamann. *Phys. Rev. Lett.*, 50:1998, 1985.
2. J. Tersoff and D. R. Hamann. *Phys. Rev. B*, 31:805, 1985.

3. J. Bardeen. *Phys. Rev. Lett.*, 6:57, 1961.
4. M. Büttiker, Y. Imry, R. Landauer, and S. Pinhas. *Phys. Rev. B*, 31:6207, 1985.
5. L. V. Keldysh. *Sov. Phys. JETP*, 20:1018, 1965.
6. Y. Meir and N. S. Wingreen. *Phys. Rev. Lett.*, 68:2512, 1992.
7. A. A. Lucas. *Europhys. News*, 21:63, 1990.
8. W.A. Hofer and J. Redinger. *Surf. Sci.*, 447:51, 2000.
9. C. J. Chen. *Introduction to Scanning Tunneling Microscopy*. Oxford University Press, Oxford, 1993.
10. J. C. Chen. *Phys. Rev. Lett.*, 65:448, 1990.
11. P. Hohenberg and W. Kohn. *Phys. Rev.*, 136:B864, 1964.
12. W. Kohn and L. J. Sham. *Phys. Rev.*, 140:A1133, 1965.
13. P. Sautet and C. Joachim. *Ultramicroscopy*, 42:115, 1992.
14. J. Taylor, H. Guo, and J. Wang. *Phys. Rev. B*, 63:245407, 2001.
15. K. Reuter, P. L. de Andres, F. J. Garcia-Vidal, and F. Flores. *Phys. Rev. B*, 63:205325, 2001.
16. T. E. Feuchtwang. *Phys. Rev. B*, 10:4135, 1974.
17. T. E. Feuchtwang. *Phys. Rev. B*, 12:3979, 1975.
18. T. E. Feuchtwang. *Phys. Rev. B*, 13:517, 1976.
19. S. Datta. *Transport in Mesoscopic Systems*. Cambridge University Press, Cambridge UK, 1995.
20. K. Palotas and W. A. Hofer. *J. Phys: Cond. Mat.*, 17:2705, 2005.
21. M. Di Ventra and N. D. Lang. *Phys. Rev. B*, 65:045402, 2002.
22. J. Taylor, H. Guo, , and J. Wang. *Phys. Rev. B*, 63:245407, 2001.
23. M. Brandbyge, J.-L. Mozos, P. Ordejon, J. Taylor, and K. Stokbro. Theory of single molecule vibrational spectroscopy and microscopy. *Phys. Rev. B*, 65:165401, 2002.
24. F. J. Garcia-Vidal, F. Flores, and S. G. Davidson. *Progr. Surf. Sci.*, 74:177, 2003.
25. N. Lorente and M. Persson. *Phys. Rev. Lett.*, 85:2997, 2000.
26. W. A. Hofer and A. J. Fisher. *Phys. Rev. Lett.*, 91:036803, 2003.
27. J. A. Stroscio, D. T. Pierce, A. Davies, R. J. Celotta, and M. Weinert. *Phys. Rev. Lett.*, 75:2960, 1995.
28. T. E. Feuchtwang. *Phys. Rev. B*, 13:517, 1976.
29. J. B. Pendry, A. B. Pretre, and B. C. H. Krutzen. *J. Phys. Condens. Mat.*, 3:4313, 1991.
30. N. Lorente and M. Persson. *Faraday Discuss.*, 117, 2000.
31. A. Zawadowski. *Phys. Rev.*, 163:163, 1967.
32. J. A. Appelbaum and W. F. Brinkman. *Phys. Rev.*, 186:464, 1969.
33. J. Li, W.-D. Schneider, R. Berndt, and B. Delley. *Phys. Rev. Lett.*, 80:2893, 1998.
34. N. Lorente. *Verh. DPG*, 2003.
35. G. D. Mahan. *Many-Particle Physics*. Plenum Press, New York, 1990.
36. C. Caroli, R. Combescot, P. Nozieres, and D. Saint-James. *J. Phys. C: Solid State Phys.*, 5:21, 1972.
37. J. Rammer and H. Smith. *Rev. Mod. Phys.*, 58, 1994.
38. M. Head-Gordon and J.C. Tully. *Phys. Rev. B*, 46:1853, 1992.
39. M. Head-Gordon and J.C. Tully. *J. Chem. Phys.*, 103:10137, 1999.

6

Bringing Theory to Experiment in SFM

In previous chapters we have outlined the general setup of SPM experiments, and considered those interactions and processes that are dominant in their performance. More specifically, in the last two chapters the theoretical method for simulation of STM was outlined, and in this chapter we focus on building the theoretical background of SFM simulations. We begin with a discussion of how the total map of the tip–surface interaction is constructed, and then describe the theoretical framework of modeling the dynamics of the cantilever in this interaction field. Finally, we use a test example to explore how different interaction components affect the simulated images.

6.1 Tip–surface interactions in SFM

In Chapter 3 we summarized all the forces important in modeling SFM, particularly the macroscopic forces. Here we discuss how to decide which forces to include in modeling a given experiment, and how to integrate them with the microscopic forces.

The best way to describe these forces is to fit directly to experimental force/frequency change curves [1, 2]. The basic experimental procedure employed to measure frequency change vs. distance curves is to retract the tip *in situ* immediately after taking the last image to produce the curve. We should note that lateral thermal drift of the sample, which is unavoidable at room temperature, does not allow precise tip positioning over a specific atomic site. Therefore, these data represent an average over a spread of lateral positions. However, since the force curves are analyzed only with respect to long-range forces, this uncertainty does not generally affect the conclusions. In order to understand the procedure of fitting theory to experiment, here we consider two sets of data taken over the CaF_2 (111) surface [3]. Although the experimental data are given as a frequency change rather than directly as force, we leave discussion of modeling the cantilever dynamics until the next section and treat them as nominally equivalent here.

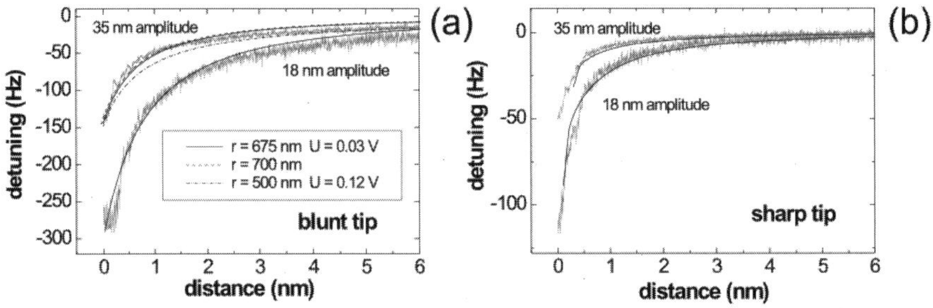

Fig. 6.1. Comparison of experimental and theoretical frequency change curves for (a) a blunt tip and (b) a sharp tip. Reprinted with permission [3].

Figure 6.1 shows two sets of experimental curves at two different oscillation amplitudes. The first tip (a) was prepared by bringing the sharp tip into contact with the surface to produce a much blunter tip, referred to as the "blunt tip". The second tip (b) is a standard unsputtered tip, which will be referred to as the "sharp tip". As discussed previously, in SFM imaging it is common practice to minimize long-range electrostatic interactions by applying a bias voltage, and this was done in both sets of experiments. For the two tips analyzed in detail here, the force curves were measured at oscillation amplitudes of 35 nm and 18 nm. The larger amplitude value was used for imaging measurements, while the smaller one is included for comparison. Clear differences in the curves can be seen for the different tips at both amplitudes: the tip–surface interaction is much more short-range for the "sharp" tip than for the "blunt" tip. The experimental curves are generally monotonic and smooth, but at short distances the slope for the sharp tip is much steeper than that for the blunt tip.

In order to fit these curves with theory, we have to make some assumptions about the tip–surface setup. Since the experiments were with a standard silicon tip and in UHV, we ignore any magnetic or capillary forces. This leaves van der Waals and electrostatic macroscopic forces as possible components of the interaction. We assume that the tip is of standard shape, and is well represented by a cone (see Chapter 3), and fix the value of the Hamaker constant at 1.0 eV. This corresponds to a characteristic value for the interaction between silicon and wide-gap insulator [4]. The remaining free parameters of the fit are only the tip radius and the bias voltage. The bias voltage applied in the theoretical model affects long-range electrostatic forces due to uncompensated tip and surface charge and image force interaction [2].

Best-fit results are shown as dashed lines in the graphs of Figure 6.1. The tip parameters for the fit are the same at both amplitudes, confirming that theory gives a consistent agreement. Distances given in the graphs of Figure 6.1 were obtained by shifting experimental data to align with theoretical curves assuming that closest approach of the tip to the surface occurs at 0.4 nm. We should stress that this is clearly a crude estimate. It is based on the results of theoretical modeling of the tip–surface interaction predicting that 0.4 nm is a typical tip–surface distance at which surface (tip) ions start to exhibit strong displacements from their equilibrium sites and may jump from tip to surface or vice versa [5, 6]. At shorter distances, perturbations in the tip–surface interaction due to tip contamination by surface ions may lead to instabilities in cantilever oscillations. Since these instabilities were not observed in these measurements, it was assumed that the tip–surface distance does not decrease below this critical value. Other possible methods for deciding the distance scale are to study the distance dependence of image features [3] or to use STM tunneling current as a reference [1].

For the sharp tip, the theoretical best fit was found for a tip radius of 100 nm and 0.00 V bias voltage. For the blunt tip the best fit was found for a tip radius of 675 nm and 0.03 V bias voltage. It should be noted that within the assumptions discussed above, these parameters are unique and a similar fit could not be found with an increased bias and reduced radius. This is due to the very different behavior of van der Waals and electrostatic forces as a function of distance [7]. In both cases, the macroscopic van der Waals force dominates the interaction and the long-range electrostatic force due to bias is insignificant. The latter is consistent with the fact that fitting was made to curves obtained under conditions where electrostatic forces have been minimized by the applied bias voltage. The overall agreement between theoretical and experimental curves at long range is better for the sharp tip than for the blunt tip. A tip radius above 500 nm appears unrealistically large even for a blunted tip. This is not surprising taking into account the idealized "cone terminated by a sphere" tip model used in the calculations. Of course, a real blunt tip can be expected to have an irregular shape with a number of nanoscale structures. These nanostructures will have overlapping ranges of interaction with the surface determining the overall shape of the force curve by superposition and causing deviation of force curves from what would be expected from an ideally spherical tip end. Also, the Hamaker constant is most certainly different from the 1 eV value used in our calculations due to partial oxidation of the silicon tip and contamination via the ambient, etc. Nonetheless, the obtained parameters are meaningful input values for theoretical calculations, since by using them for calculating the background forces, the absolute values for the frequency detuning of theory and experiment can be aligned with each other.

In case no experimental force curves are available, only images, then fitting becomes much more difficult. Effectively, every image represents a single point on a force curve and any fitting will be rather arbitrary. The only way to improve the fit is to use information about the experimental setup and environment, and to isolate only those interactions likely to be present.

However the theoretical macroscopic components are determined, the final stage just involves adding them to the microscopic forces to provide the total force map for that system. In general, for atomic resolution imaging, the macroscopic forces do not play a significant role in contrast formation, but modeling them as accurately as possible makes comparison with experiment much easier.

6.2 Modeling the tip

SFM tips are microfabricated from silicon in much the same way as computer chips and, as-produced, have a pyramidal shape. However, this only gives their structure on the micrometer scale, and there is no direct method for imaging the very end of the tip, the "nano-tip". Therefore additional information is needed to reconstruct tip structures. In particular, it is known that the tips are oxidized due to exposure to the atmosphere, and although the oxide layer can be removed by argon ion sputtering, they can be contaminated by residual water always present in a UHV chamber. Some recent atomically resolved images use untreated tips covered by an oxide layer [8] and specially prepared silicon tips cut from silicon wafers [9]. Metallic [10] and silicon tips covered by metal [11, 12] have also been used in SFM experiments.

In many SFM experiments atomically resolved images are obtained after tips were in contact with the surface and are most likely covered by the surface material. Tip crashes often happen spontaneously due to the strong tip–surface interaction, the presence of debris on the surface, and other artefacts. However, in many cases "gentle" contact is arranged intentionally, since it has been noted that this increases the chances of obtaining good atomic resolution. Tip contamination by the surface material has been explored in [5, 13] using classical molecular dynamics. A MgO cube tip was indented into the LiF surface and then retracted back from the surface. In another set of calculations the surface scanning has been simulated after indentation. In both cases stable clusters of the surface material were formed on the tip.

Still, this information gives only a very preliminary idea about the possible tip's chemical composition, and nothing about the geometric structure, stoichiometry, and charge of the nanotip. One solution to this difficult problem is to use idealized nanotips. This method has been used in recent *ab initio* studies of SFM on semiconductor surfaces [14, 15] and in atomistic simulations on ionic systems [16, 17]. It is a good basis for beginning the tip modeling process. Another direction of modeling is to try to find the most realistic tip model for a particular set of experiments.

The approaches using idealized tips are based on two main considerations. Firstly, they assume that the tip structure is too complex to be treated explicitly and is likely to change during experiments. Therefore one should try to reproduce only general qualitative features that can be responsible for image contrast in spite of all the complex issues discussed above. Secondly, to keep calculations practical, nanotips cannot be large and should include 10-30 atoms. For SFM a study of different possible nanotip models was performed to try to determine the closest match with the experimental behavior, e.g., on NaCl surfaces [18]. It was found that if the bottom of the tip was flat, i.e., no nanotip, then the interaction with the surface was averaged over several tip ions and no contrast was produced. When a nanotip is included, it must extend significantly beyond the main part of the tip to reproduce the interaction observed in experiment. Specifically, a nanotip of only a few atoms would not atomically resolve lower terraces of stepped surfaces, which contradicts experiments on the NaCl surface [18].

6.2.1 Silicon-based models

In SFM simulations, the most common perception in modeling on Si and other semiconductor surfaces has been that the main component of the tip–surface interaction responsible for image contrast on these surfaces is due to the interaction of a dangling Si bond at the end of the tip with the surface atoms. This dangling bond can be well described using relatively small 4- or 10-atom Si clusters saturated by H atoms [14] (see Figure 6.2(a)). Comparison of theoretical force curves calculated with this tip on the Si (111) surface demonstrate reasonable agreement with experiment (see Chapter 9), giving confidence that, at least for this particular system, it is a good model.

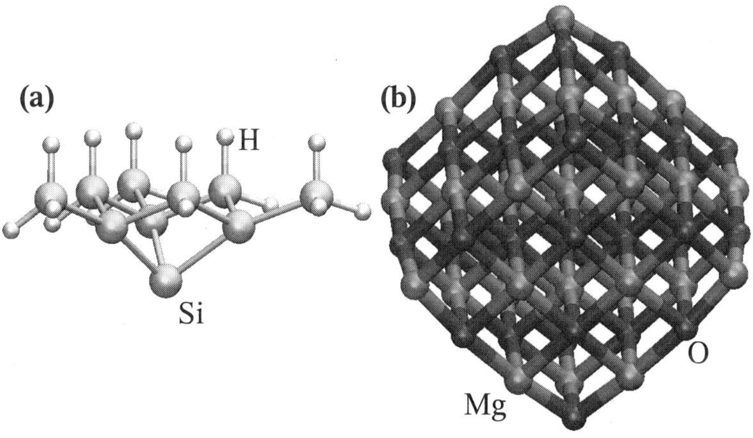

Fig. 6.2. Idealized nanotips used in simulations: (a) a dangling bond Si_10 tip and (b) a $Mg_{32}O_{32}$ cube tip.

On other surfaces, it is much more difficult to judge the accuracy of the model, since atom-specific force curves are not commonly obtained. However, extensive simulations do provide a background of general information, which is useful in understanding the role of different interactions in SFM. Studies of reactive semiconductor surfaces, such as Si (111) [14, 19], InP (110) [20, 21], and GaAs (110) [15, 22], demonstrated that the tip-surface interaction is dominated by the onset of covalent bonds between the tip dangling bond and atoms in the surface. Specifically, strongest interaction was seen with anions in the surface, where a greater source of bonding electrons can be found. This was shown to be a general phenomenon for all semiconducting and insulating surfaces by simulations on CaF_2, Al_2O_3, $CaCO_3$, MgO [23, 24] and TiO_2 [25, 26]. In each case the dangling bond silicon tip interacted most strongly with anions in the surface. The magnitude of the force also roughly scaled with the band gap of the materials, due to the increased electron density localization around atoms in more ionic materials. All these studies also showed the importance of atomic relaxations in both the tip and surface: displacements significantly influence the tip–surface force. Since the nature of the displacements depends strongly on the tip, the tip–surface interaction becomes even more sensitive to the model chosen.

Despite improvements in preparation of silicon tips [9, 27], there is yet little evidence that it is really possible to maintain a clean silicon tip during scanning. Hence, several studies have considered contaminated silicon models as more realistic tips. The obvious initial choice is to consider tips contaminated by material from the surface. For the GaAs and InP (110) surfaces, substitution of the apex silicon atom by a surface species signifcantly changed the interaction and predicted image contrast [21, 22]. For example, replacing the Si by Ga produces qualitatively similar interactions on the GaAs surface, but an As replacement reverses the contrast and Ga cations are now imaged as bright. A similar sensitivity was shown in studies of the CaF_2 (111) surface (see Chapter 9), where reasonable agreement with experiment was achieved using an oxygen contaminated silicon tip.

6.2.2 Ionic models

Another approach has been used to model tips for simulating SFM images of ionic surfaces. Here it has been assumed from the start that the tip is either oxidized by the ambient, or, more likely, covered by surface material. Hence, electrostatic forces between the tip and surface will be the most important, and mainly ionic tips have been considered [16, 17, 28]. The most common example is a MgO cube (see Figure 6.2(b)), which can orientated with either a Mg^{2+} or an O^{2-} ion at the apex, producing a strong positive or negative electrostatic potential gradient respectively (the potential from the oxygen apex is actually very similar to that for the oxygen-contaminated silicon tip [29]). This model was successfully used in interpreting atomic resolution images of the CaF_2 (111) surface (see Chapter 7).

Extensive studies of different NaCl-based tips [30] demonstrated the significant effect of the tip geometry, as well as chemistry, on the contrast pattern in DFM images. The strongest influence was seen for very soft tips, which show very large relaxations on approach to the surface, smearing the contrast pattern. Although experimental images of this kind have been observed, such soft tips are likely to be quite unstable under the influence of significant tip–surface interaction and prone to rearrangement to a stabler configuration.

Tips that presented an edge of several ions to the surface demonstrated contrast patterns that are qualitatively similar to more ideal tips, and the shift of maxima from atomic sites would not be detectable in most experiments. This gives some insight into the success of contacting the surface in producing tips that provide atomic resolution: it is not necessary to produce a perfect single probe, but rather any stable cluster with a sharp edge will serve.

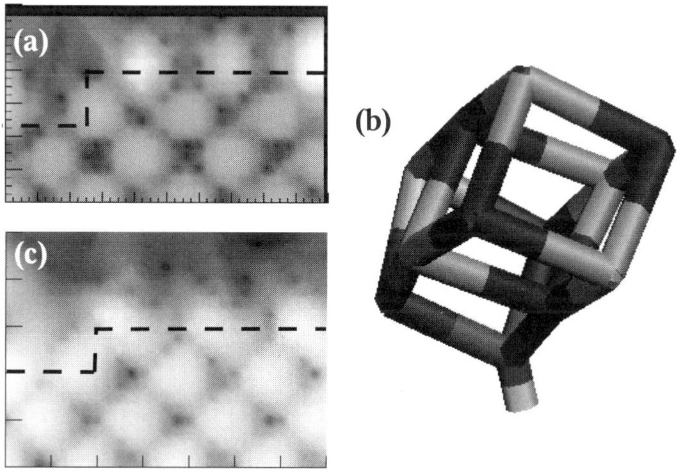

Fig. 6.3. (a) Simulated image of a NaCl step-edge using a symmetric Na-terminated NaCl cuboid tip. (b) NaCl cuboid tip contaminated by a hydroxyl group, and (c) an image produced using this tip. Reprinted with permission. Copyright 2004 IOP [30].

For NaCl tips contaminated by hydroxyl groups, when the dominant interaction is via a single OH group, the overall interaction is greatly reduced and displacements of the surface are almost zero. However, it is here that we see most clearly the effect of an asymmetric tip. The protrusion of the cuboid in one direction away from the OH at the apex (see Figure 6.3(b)) acts as a secondary, weaker, probe of the surface, and produces a corresponding distortion in the contrast pattern, and atoms are seen as almost triangular (see Figure 6.3(c)). If this secondary probe is closer than a surface lattice constant to the main probe, the combined interaction merely produces a symmetric pattern, with quantitative changes in interaction, but little qualitative changes from an ideal tip. A distinct secondary probe is required to produce asymmetries. Generally the different configurations of NaCl tips produce significantly different forces over identical terrace sites (see Figure 6.4), perhaps explaining some of the deviations of simpler tips from experimental results (see Chapter 9).

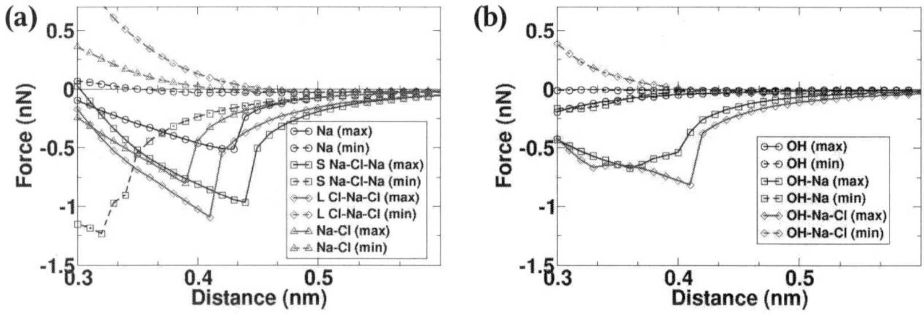

Fig. 6.4. Tip–surface maximum and minimum force curves over the island terrace with various configurations of the NaCl cuboid tip: (a) Na is a Na-terminated ideal tip, S Na-Cl-Na is a small three-atom edge tip, L Cl-Na-Cl is a large three-atom edge tip, and Na-Cl is a two-atom edge tip; (b) OH is an ideal OH tip, OH-Na is a two-atom edge tip, and OH-Na-Cl is a three-atom edge OH tip. Reprinted with permission from R. Oja and A. S. Foster, Nanotechnology 16:S7 2004. Copyright 2004 IOP [30].

An interesting aspect of the behavior of different tips is ion jumps from the surface to the tip. Figure 6.4 shows a comparison of force curves for different tips over terrace sites, and in all of the maximum curves, apart from the weakly interacting ideal OH tip, force jumps can be seen, for example, for the S Na-Cl-Na tip at about 0.45 nm in Figure 6.4(a). However, in each case the distance at which the jump occurs is different, which demonstrates that the different interactions produced by these different tips also result in different stability ranges for surface ions (no jumps of atoms from the tip to the surface were observed). If the tip is retracted after a jump, the resultant behavior depends strongly on the material of the tip and surface. In this case, where tip and surface are of the same ionic material a cation and anion pair is likely to be removed [31]. However, use of a different material in the tip, such as MgO, means that a chain of atoms will form, with alternating cations and anions being pulled from the surface until the chain snaps due to thermal motion [32].

6.3 Cantilever dynamics

In dynamic SFM, translating a calculated tip–surface interaction map into an image requires modeling the behavior of the oscillating cantilever in that interaction field. The general behavior of the cantilever can be described by the following equation of motion:

$$\frac{k}{\omega_0^2}\ddot{z} + \alpha\dot{z} + kz - F(z+h) = F_{exc}, \tag{6.1}$$

where F is the tip–surface force, F_{exc} describes the excitation of the oscillations, ω_0 is the oscillating frequency of the cantilever in the absence of any interaction with the surface ($\omega_0 = 2\pi f_0$), α is the damping coefficient, and h is the equilibrium height of the cantilever above the surface in the absence of interaction. In stable operation, the excitation will compensate exactly any dissipation of the oscillations, both intrinsic and due to the tip-surface interaction, so that the amplitude remains constant. Hence, the damping term and excitation term can be neglected. Further, if we assume that $F(z)$ does not depend on time, we can simplify equation (6.1) to the following conservative form:

$$\frac{k}{\omega_0^2}\ddot{z} + kz - F(z+h) = 0. \tag{6.2}$$

A general numerical solution of this equation is possible [33], but approximations exist under certain conditions. For *small oscillation amplitudes*, it is possible to use only the tip-surface force gradient at h to calculate the frequency change, giving the following equation of motion [34],

$$\frac{k}{\omega_0^2}\ddot{z} + \left(k - \left.\frac{\delta F(z)}{\delta z}\right|_{z=h}\right) z = 0, \tag{6.3}$$

which results in the often quoted relationship between the frequency change and the force gradient

$$\Delta f(h) = -\frac{f_0}{2k}\left.\frac{\delta F(z)}{\delta z}\right|_{z=h}. \tag{6.4}$$

For large amplitudes where (6.4) fails, the case in most experiments, it is also possible to approximate the cantilever motion as a perturbed harmonic oscillator [35, 36]. Then the frequency change can be calculated from [35]:

$$\Delta f(h) = -\frac{f_0}{k}A_0^2\left\langle F(z)\right\rangle. \tag{6.5}$$

If we assume that the force between the tip and sample can be expressed by a simple power law $F(z) = -Cz^{-n}$, where C is the force constant and n is the power order, we get the following expression for a full oscillation cycle:

$$\Delta f = \frac{f_0 C}{2\pi k A_0 d^n}\int_0^{2\pi}\frac{\cos x dx}{\left(1 + \left(\frac{A_0}{d}\right)(\cos x + 1)\right)^n}, \tag{6.6}$$

where d is the tip-surface closest approach and $x = f_0 t$. For large amplitudes, such that $A_0 \gg d$, a Taylor series expansion of the denominator of (6.6) around

$x_0 = \pi \, (x' = x - \pi, \cos \simeq (-1 + x_2'/2))$ and a substitution $\left(y = \sqrt{A_0/2dx'}\right)$ gives

$$\Delta f = -\frac{f_0 C}{\sqrt{2}\pi k A_0^{\frac{3}{2}} d^{n-\frac{1}{2}}} I_1, \tag{6.7}$$

where

$$I_1 = \int_{-\infty}^{\infty} \frac{dy}{(1+y^2)^n}. \tag{6.8}$$

Under these assumptions, since $\Delta f \propto f_0/kA_0^{\frac{3}{2}}$ for all inverse powers, it is possible to introduce a general normalized frequency shift that condenses all these parameters into a single value, γ_0:

$$\gamma_0 = \frac{\Delta f k A_0^{\frac{3}{2}}}{f_0}, \tag{6.9}$$

γ_0 is very useful for comparing parameter sets for different experiments. The range of γ_0 for which atomic resolution has been achieved is large, from -387 to -0.29 fNm$^{\frac{1}{2}}$, although most results are achieved between 0 and -30 fNm$^{\frac{1}{2}}$ [36]. For small amplitudes, it is possible to obtain atomic resolution in the repulsive part of the tip–surface interaction [37], and hence γ_0 would be positive. However, the amplitudes (0.25 nm) used are so small that the approximations used to calculate γ_0 break down.
An expression for the frequency change can also be derived by considering a Fourier expansion of the motion [6, 38, 39]. We can search for a solution of (6.2) in the form of a Fourier series:

$$z(t) = \sum_{n=0}^{\infty} a_n \cos(n\omega t), \tag{6.10}$$

where a_n are the Fourier coefficients. Substituting into (6.2) gives the following equation of motion:

$$\sum_{n=0}^{\infty} \left(1 - \left(n\frac{\omega}{\omega_0}\right)^2\right) a_n \cos(n\tau) + a_0 - \frac{F(z+h)}{k} = 0, \tag{6.11}$$

where ω is the oscillation frequency under the influence of the tip–surface interaction and we introduce dimensionless time $\tau = \omega t$. To find ω and $a_n, n = 0, 2, \cdots, \infty$ for a given oscillation amplitude $a_1 \, (= A_0)$, h, and $F(z)$, we multiply (6.11) by $\cos(j\tau)$ and integrate the result over the period of the main frequency $\tau = [0, 2\pi]$. This produces a system of nonlinear equations for $a_n, n = 0, 2, 3, \cdots, m$, which is approximate for finite m:

$$a_0 - \frac{1}{2\pi k} \int_0^{2\pi} F(z+h) d\tau = 0, \tag{6.12}$$

$$a_n - \frac{\int_0^{2\pi} F(z+h)\cos(n\tau)d\tau}{\pi k - \pi k n^2 + \frac{n^2}{a_1}\int_0^{2\pi} F(z+h)\cos(\tau)d\tau = 0}. \tag{6.13}$$

If we designate the left-hand side of (6.12) and (6.13) as $\phi_n(a_0, a_2, \cdots)$, we can rewrite this system of equations more compactly, i.e., $\phi_n(a_0, a_2, \cdots) = 0, n = 0, 2, 3, \cdots, m$, and solve it using a modified Newton method. As an initial step, we set all a_i except a_1 to zero. For each iteration thereafter the values of the amplitude increments (Δa_i) can be obtained by solving the set of equations

$$\frac{d\phi_n}{da_0}\Delta a_0 + \sum_{j=2}^{m} \frac{d\phi_n}{da_j}\Delta a_j = -\phi_n. \tag{6.14}$$

Unfortunately, using (6.14) as the foundation of the iterative procedure often leads to divergent results, and the Δa_i are used only to find a search direction. The absolute values of the increments are calculated by minimizing the residual function

$$\Phi(\lambda) = \sum_{i=0}^{m} \phi_i^2\left((a_i)_k\right) \tag{6.15}$$

with respect to the parameter λ, where

$$(a_i)_k = (a_i)_{k-1} + \lambda\Delta(a_i)_{k-1} \tag{6.16}$$

and k is the iteration number, and i is the index of the unknown coefficient. Finally, the frequency of the cantilever oscillations in the presence of the tip–surface interaction is given as

$$\omega = \omega_0\left(1 - \frac{1}{\pi k a_1}\int_0^{2\pi} F(z+h)\cos(\tau)\,d\tau\right)^{\frac{1}{2}}. \tag{6.17}$$

This is functionally equivalent to a general version of (6.6) [38]. Beyond these approaches, it is also possible to develop a full simulation of the cantilever dynamics, where the electronics of the SFM are included in the modeling. In this case, A_0, d, and f_0 are now time-dependent variables that proceed to their steady-state values modulated by the simulated electronics. Simulations of this kind have been performed [40, 41] providing much greater insight into the relationship between the cantilever's dynamics and the measured frequency change. However, perhaps the most important result [42] is that the approximations discussed above, and (6.17) in particular, are shown to be valid, at least in describing Δf. Hence, we will use the Fourier series method for generating simulated images in this and further chapters.

6.3.1 SFM at small amplitudes

Although the small- and large-amplitude approximations have been generally successful for interpreting many measured tip–surface interactions, a more general criterion for a given system is the comparison between the amplitude and the length scale of the interaction [43]. This is complicated in systems where both long-range forces (such as electrostatics) and short-range forces (such as chemical bonding) are present, since there are two very different length scales. An amplitude in between these length scales can actually make it more difficult to separate the chemical forces from the background [43]. This generally motivates the removal of long-range interactions or the use of oscillation amplitudes smaller than the shortest interaction length of interest. A further benefit of small-amplitude operation can be found when one is measuring the higher harmonics of the cantilever oscillations. The resolution of SFM is signficantly determined by the *localness* of the tip–surface interaction , and for small amplitudes this effectively means the gradient of the force. The higher gradients of the tip–surface force decay faster, providing a more local interaction, and potentially greater resolution. Dürig's analysis of the higher harmonics of cantilever oscillations [39] demonstrated that they coupled to the higher gradients of the tip–surface force. Hence, imaging via the higher harmonics should provide much greater resolution, especially at small amplitudes [44]. Figure 6.5 shows a successful application of this technique [44], where it was possible to provide a lateral resolution of 77 pm and directly resolve the atomic orbitals of a tungsten atom.

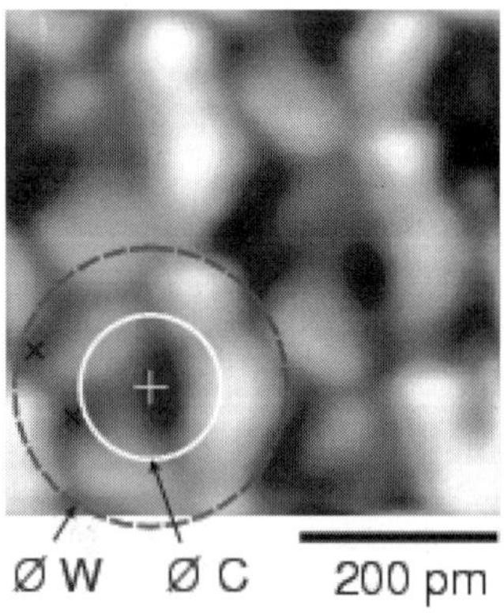

Fig. 6.5. Higher harmonic amplitude image of a tungsten tip imaged via a single carbon atom probe on the surface. The circles show the respective van der Waals radii of a W and a C atom. Reprinted with permission. Copyright 2004 AAAS [44].

6.3.2 Atomic-scale dissipation

As described in the previous section, to maintain constant amplitude, an excitation is applied to the cantilever based on the feedback signal: this can be characterized by an excitation amplitude, A_{exc}, which is usually also measured during an experiment. Atomic contrast in A_{exc} was first observed on Si(111) [45], and has since been seen on several different surfaces. In principle, using dissipation as the imaging signal offers certain advantages to the frequency shift. In particular, the signal is monotonic [46, 47], offering distance control comparable to STM. The signal itself appears much more sensitive to the nature of the tip, so that it is immediately apparent when a tip change has occurred. Figure 6.6 demonstrates how the damping contrast much more clearly highlights the tip change in the experiment than the topographic contrast. Other images [48] show inverted or no contrast in dissipation, while topographic images are clearly recorded.

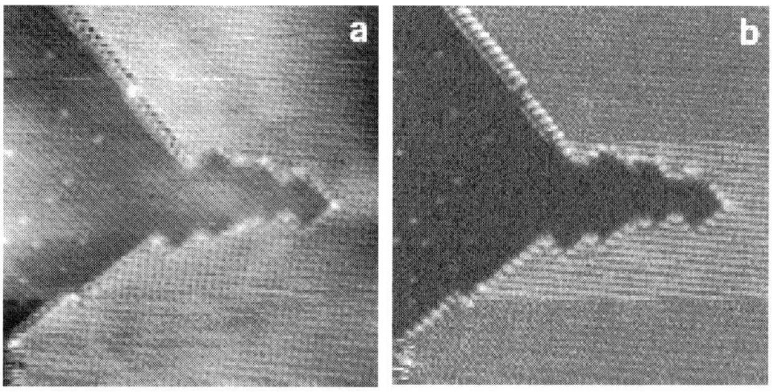

Fig. 6.6. (a) Topography and (b) A_{exc} images of a NaCl island on Cu(111). The tip changes after one-fourth of the scan, thereby changing the contrast in topography and increasing the contrast in A_{exc}. After two-thirds of the scan, the contrast from the lower part of the images is reproduced, indicating that the tip change was reversible. The image size is 324 nm^2 (Δf = -128 Hz, f_0 = 158 kHz, k = 26 N/m, A = 1.8 nm, Q = 24000, U_{bias} = 0 V) Reprinted with permission. Copyright 2000 by American Physical Society [18].

Despite the potential benefits, the initial lack of understanding behind the atomic-scale mechanism of dissipation means that images of A_{exc} remained objects of interest, but of little scientific worth. More recently, many simulations of the problem [38, 40, 46, 49, 50, 51, 52, 53, 54, 55, 56] have distilled out two likely mechanisms: the stochastic friction force mechanism and adhesion hysteresis.

In the *stochastic friction force* mechanism [46, 52, 53, 54, 55] energy is dissipated due to induced friction from the thermal fluctuations of atoms in the surface and the tip (similar to the behavior a massive Brownian particle immersed in a fluid of much lighter particles [55]). However, all estimates of the magnitude of energy dissipated by this mechanism are much smaller then those observed in experiments. A comprehensive treatment [55], including a realistic atomistic tip and surface, predict dissipation energies on the order of 10^{-8} to 10^{-9} eV per cycle, compared to 0.01 to 1 eV per cycle in experiments. This strongly suggests that the stochastic mechanism does not play a significant role in contrast in dissipation images.

The *adhesion hysteresis* mechanism is based on the tip following a different path through the tip–surface energy landscape on approach and retraction. This occurs when strong and reversible changes in the tip and/or surface structure are induced by the tip–surface interaction, resulting in a double potential well in the tip–surface potential energy surface. Including this non-conservative contribution to the total force means that it is now intrinsically time-dependent, and although the equation describing the cantilever motion is very similar to (6.17), its solution is more complex [56]. The equation describing the frequency of the cantilever (the only factor when one is considering exclusively conservative forces) must now be solved simulaneously with the time-dependent microscopic force and the state probability function describing in which potential energy state the system resides. Estimates of the contribution of the adhesion hysteresis to dissipation based on atomistic simulations [56] are very similar to experimental measurements, indicating that this mechanism is the most probable candidate for understanding dissipation imaging. Further simulations are needed to confirm this, but the dependence of the dissipation contrast on the mass of atoms would offer the possibility of identification of tip and surface species in the future.

6.4 Simulating images

Generally, simulating the cantilever dynamics provides a map of the topography (or Δf for constant height mode) over the surface unit cell for a given Δf (or height), and this can be immediately plotted (or interpolated) as the theoretical image. The nature of this image depends strongly on the ingredients in its production, and some important insights into the role of different parameters and interactions can be seen by studying image simulation. In this section we take a standard system, and show how the image of this system varies as we change the microscopic forces, introduce atomic relaxations, and change the scanning mode. We also look at how convolution between the tip and surface can change the appearance of imaged surface features.

6.4.1 Test system

In order to provide different kinds of surface site, the test surface for image simulations consists of a monolayer "strip" of NaCl on top of a NaCl surface. This is based on experimental atomic resolution images of a similar system [18]. The system is shown in Figure 6.7(a), and it is based on a NaCl island, but in the simulations this cell is periodic, so it becomes a continuous monolayer strip, rather than an island. The strip contains two step edges, and also a cation and anion kink site. For a tip, we use a NaCl cuboid (shown in Figure 6.7(b)), which can be oriented with either a Na or Cl at the apex.

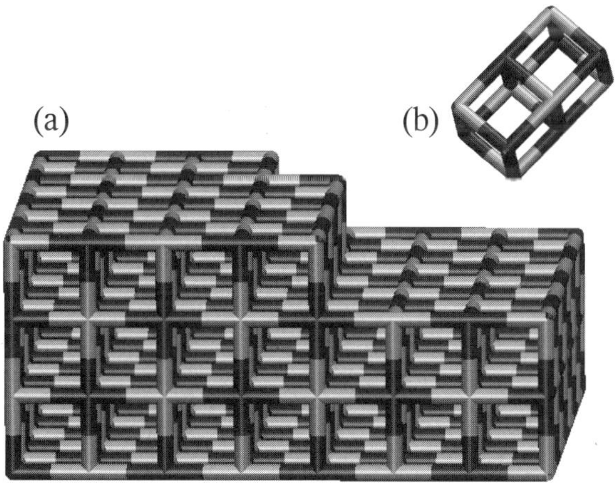

Fig. 6.7. (a) NaCl strip used as test surface. (b) NaCl cuboid tip. Na is represented by the dark color, and Cl as light.

Since we have both an ionic tip and ionic surface, using atomistic modelling is a reasonable approximation, which allows us much more freedom in the number of calculations we can perform. The surface is split into a grid of 29 × 21, giving 609 different surface points, and over each point the tip is moved through 50–80 different heights in the range 0.0–8.0 Å, for an overall total of about 40,000 positions. Parameters for the cantilever dynamics used in modeling oscillations are taken from experiments [18]: f_0 = 158,271 Hz; k = 26 N/m; A = 1.8 nm; Δf = -128 Hz; contrast = 0.15 Å. The Hamaker constant for the macroscopic van der Waals force is 6.45×10^{-20} J taken from data for the interaction of SiO_2 with NaCl [57]. Since we do not have a full force curve from experiment, we assume that electrostatic forces are compensated by applied bias (see Section 6.1), and the macroscopic forces are purely van der Waals. The radius of the tip in simulations is chosen to match experimental contrast at the experimental frequency change: this gives a value of 20 nm for the full interaction calculations, which is used in all the following sections unless otherwise stated. All images are oriented according to the surface orientation shown in Figure 6.8.

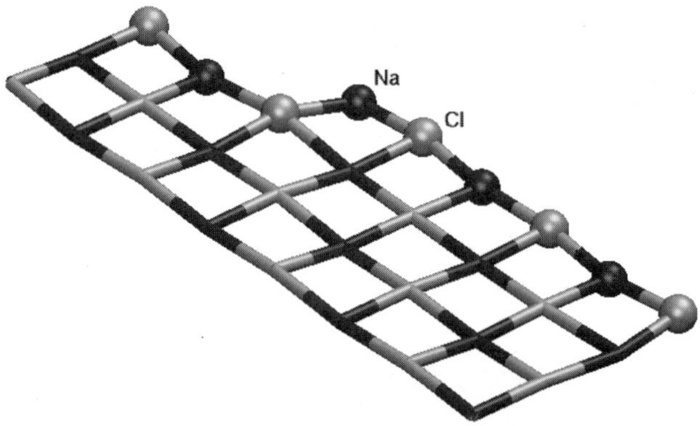

Fig. 6.8. Top layer of the NaCl strip.

6.4.2 Microscopic interactions

The first example we shall consider is the affect of changing the microscopic tip–surface interactions. This is a rather abstract concept, since in real modelling you always include all interactions. However, by separating out different parts of the interaction, we can see how each component influences the final simulated image and also emphasize why in general it is important always to include the *full* picture.

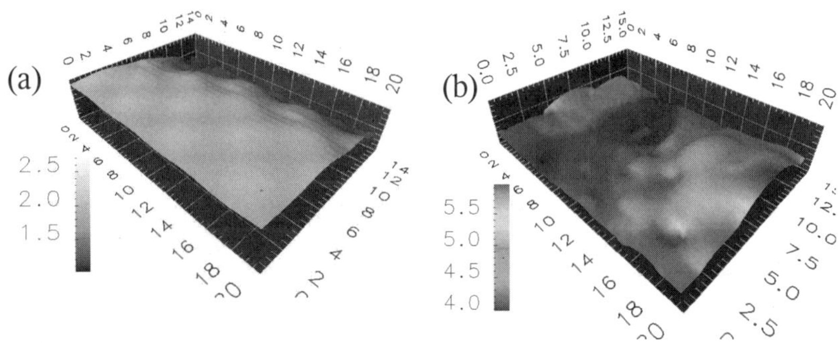

Fig. 6.9. (a) Simulated image of the NaCl test system with a Na-terminated tip using microscopic van der Waals only. All atoms are frozen. (b) Simulated image of the NaCl test system with a Na-terminated tip. All atoms are frozen. Images produced at frequency change of -140 Hz.

In the first example, we consider a system in which the only interactions between atoms are microscopic van der Waals (the Na and Cl ions are made neutral and nonpolarizable) and no atomic relaxation is included. The tip is orientated so that a Na atom is at the apex. We see clearly in Figure 6.9(a) atomic resolution on the terrace, with a contrast on the order of 0.1 Å at a tip–surface distance of about 2.5 Å. This demonstrates that the van der Waals interaction is chemically specific, and can provide a source of contrast in images. Generally, other forces dominate, but van der Waals dominated imaging has been seen on, for example, inert surfaces like xenon [58]. Note that as the tip passes the step edge, the interaction decreases as the number of atoms within a given radius from the tip apex is reduced. This is shown especially clearly in the top of the image, where the interaction is lowest as the tip passes the kink "vacancy" at the step edge.

Figure 6.9(b) shows a very different image, produced at the same frequency change as Figure 6.9(a), but now including electrostatic interactions. The atoms remain frozen at ideal positions, and polarization is not included. We see that there is something like atomic resolution on the right side of the image, but it is not clear. In fact, the whole image is dominated by two features: bright contrast on the right and dark on the left. From Figure 6.8 we can see that these contrast features correspond to Na and Cl kink sites. At kink sites, the low coordination of the ions means that their electrostatic potential is much less screened than normal terrace ions, and they produce a correspondingly increased interaction with the tip. For the Na^+-terminated tip, this results in strong attraction with the Cl^- kink site and strong repulsion over the Na^+ kink site. In comparison to the van der Waals image, the contrast is much larger, over 1.0 Å, and the required frequency change is obtained much farther from the surface, at about 4.5 Å, both reflecting the increased magnitude of the microscopic tip–surface interaction. Achieving an improvement in atomic resolution would require increasing the frequency change to move closer to the surface, thereby increasing the relative interaction of terrace ions.

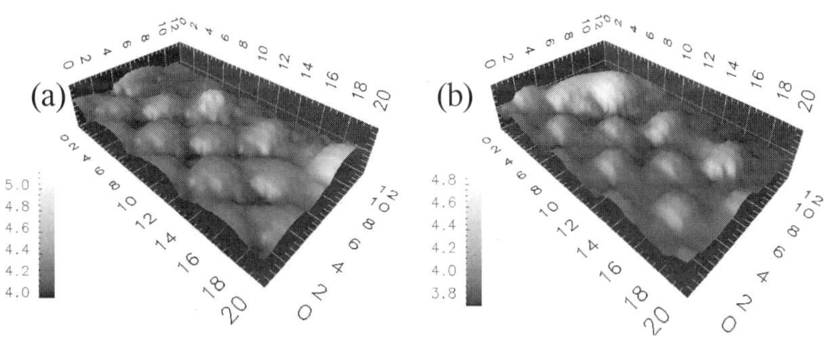

Fig. 6.10. (a) Simulated image of the NaCl test system with a Na-terminated tip. (b) Simulated image of the NaCl test system with a Cl-terminated tip. Images produced at frequency change of −140 Hz. Reprinted with permission. Copyright 2004 IOP [30].

In the images in Figure 6.10 we now include all interactions and full relaxations of the atoms. Both are taken at a frequency change of −140 Hz, resulting in a tip-surface distance of about 4.2 Å; this is slightly closer than for the frozen electrostatic model due to the effects of polarization and relaxation. Clear atomic resolution is seen on the terrace of the strip with a contrast of about 0.5 Å. The introduction of atomic relaxation has an effect over every site, with Cl^- ions displacing towards the Na^+ tip and Na^+ ions displacing away (and vice versa for the Cl^- tip). This is exaggerated for the low-coordinated ions at the step-edge and kink sites. Their displacements, especially towards the tip, are about double that of the terrace sites. These displacements produce an increase in the electrostatic potential over the ions, increasing the force. Combined with the "frozen atom" increase in electrostatic potential due to low coordination, this produces a significant increase in interactions at the step-edge and kink sites.

Figure 6.10(a) shows an image with a Na^+-terminated tip, resulting in Cl^- ions imaged as bright and the Cl^- kink site as the brightest feature. We can also see that a Cl^- ion next to the Na^+ kink site appears brighter than the terrace ions, emphasizing the large tip induced atomic displacements both at and near kink sites. The step-edge Cl^- ion in the middle of the image also appears slightly brighter than the terrace ions. In Figure 6.10(b) the Na^+ ions are now imaged as bright, with the Na^+ kink the brightest feature. Note that the Na kink and the step-edge Na to its left appear almost as one bright feature due to atomic displacements.

Distance dependence

In some ways it is misleading to show images at only a single frequency change (or height), since we can never claim to match the experimental setup exactly. In the best case experimental images exist at several different frequency changes, reducing the freedom of any fits in the calculations. Even if only a single experimental image exists, it is important to see over what range simulations match it (if any). Examples of this kind of comparison will be shown in Chapter 7, but here we consider how the images of the NaCl test system change as the surface is approached with the tip. Figure 6.11 shows in series of images spanning a frequency change of 60 – 160 Hz, and a tip–surface distance of 8.5 – 3.75 Å. At long range, contrast is only seen over the kink sites, as in the frozen atom image (see Figure 6.9(b)), but atomic resolution on the whole surface appears once the tip–surface distance is less than 5.0 Å. As the distance is further reduced, the relative tip-surface interaction over the terrace in comparison to the edge sites becomes smaller, and the contrast on the terrace increases. Only in the last two images does the increase in contrast over normal step-edge (nonkink) sites and the sites next to a kink become apparent. The development of contrast features as a function of tip–surface distance is very helpful in assigning ranges in experimental images.

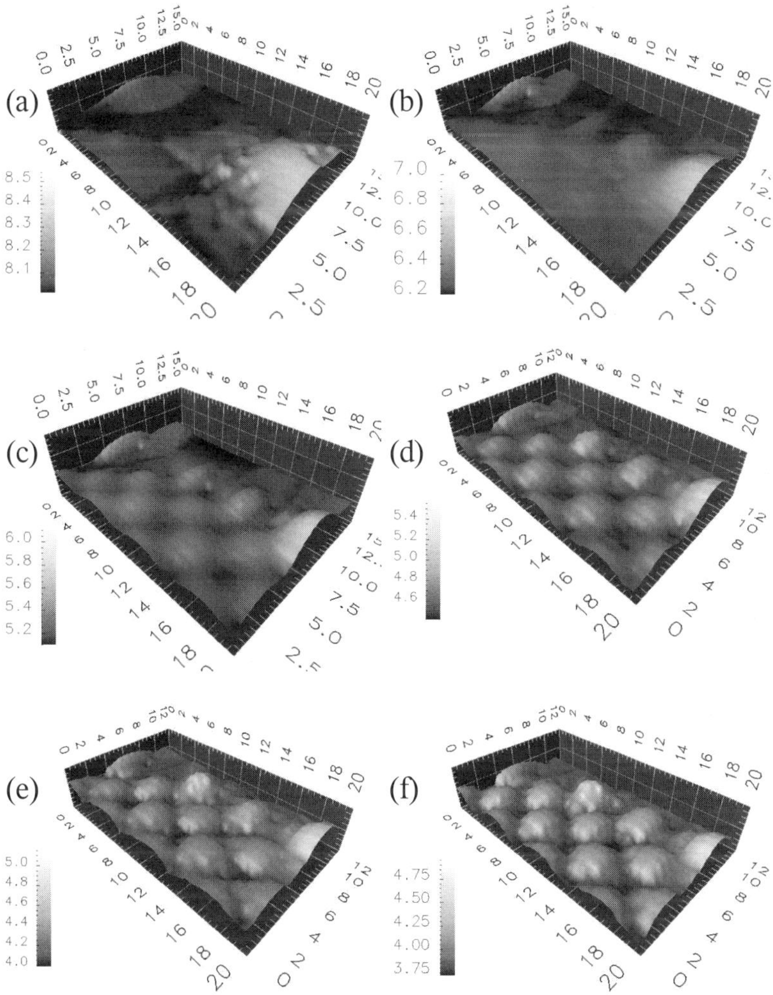

Fig. 6.11. Simulated images of the NaCl test system with a Na-terminated tip calculated at a frequency change of (a) −60 Hz, (b) −80 Hz, (c) −100 Hz, (d) −120 Hz, (e) −140 Hz, and (f) −160 Hz.

Imaging mode

In Chapter 2 we discussed the different possible modes of SFM imaging, specifically constant frequency change and constant height. Here we now consider how these different modes affect images of the same system. Experiments are generally performed only with one mode, so it is interesting to see whether there are any significant differences. Figure 6.12 shows a comparison of a constant-height frequency map at 4.5 Å, (a), with a topographic image at a constant frequency change of −140 Hz (about similar height). The differences are largely cosmetic; both show atomic resolution on the terrace, and increased contrast at the step-edge and kink sites. In the next section we will consider some cases in which a difference in images due to the scanning mode does occur.

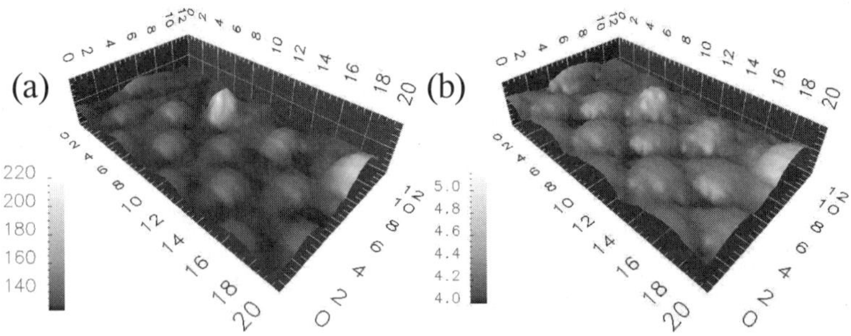

Fig. 6.12. Simulated images of the NaCl test system with a Na-terminated tip calculated at a (a) height of 4.5 Å and (b) a frequency of -140 Hz.

6.4.3 Tip convolution

One important issue, well known in contact SFM [59], but also relevant in dynamic SFM, is tip convolution. This occurs when an image shows not only surface features, but features of the tip; the image is a convolution of the tip and surface. In lower resolution contact SFM, with knowledge of the tip on the macroscopic scale, it is possible to deconvolve it from an image. However, in high- or atomic-resolution dynamic SFM, we do not have knowledge of the tip on a scale comparable to the resolution of the surface, so it is very difficult to deconvolve. An example of this convolution can be seen Figure 2.14(b), where the two dangling bonds of the tip result in two maxima over a single atomic site. Other changes in images can be seen when one is imaging surface features that are *sharper* than the tip itself, and the surface feature will actually image the tip. This is particularly relevant in imaging nanoclusters on surfaces, where a 1-nm-diameter cluster may be an order of magnitude smaller than the tip. Due to the focus on atomic resolution of generally flat surfaces, tip convolution has not yet been a major issue in dynamic SFM experiments. However, as more studies focus on adsorbates on surfaces and nanostructures, it will become increasingly important. Recent SFM images of gold clusters on the KBr surface [60] show contrast shadows around clusters that are very similar to the simulated images discussed below.
To study this, we consider three tip models imaging a cluster on the surface. The cluster is trapezoidal (a physical shape for metal clusters of this size on

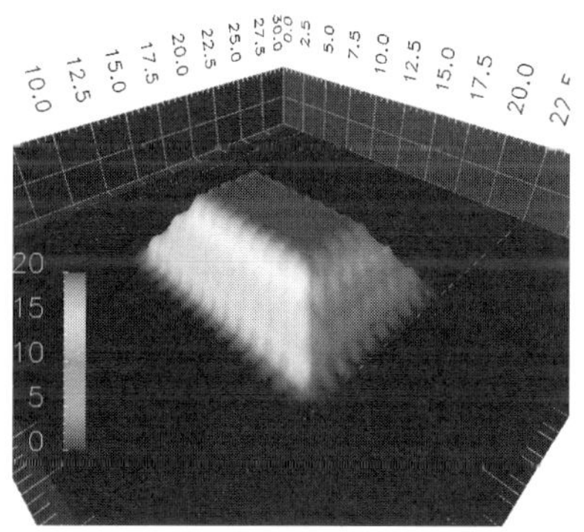

Fig. 6.13. Cluster used in simulating tip convolution. Note that in this and all topographic images in this section, the surface coordinates are marked in nm, but the the z-coordinate is in Å.

insulating surfaces [61, 62]) and about 8×12×2 nm in size (see Figure 6.13). The three tip models are shown in Figure 6.14: (a) a sharp tip with a width of about 5 nm at half the height shown, (b) a blunt tip with a width of 17 nm at half height, and (c) an asymmetric tip with the same width as the blunt tip, but split into a double tip. For this study we consider only the van der Waals force between the tip and cluster, and to calculate this for such arbitrary shapes, we build the tip and cluster from many thin cylinders and sum the interaction between them [63].

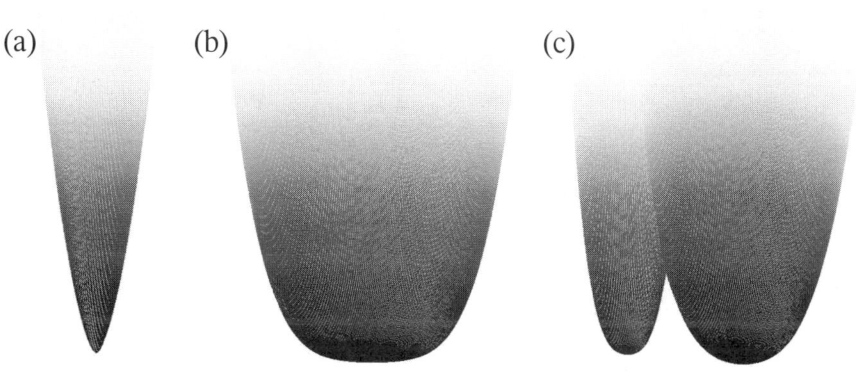

Fig. 6.14. The first 22 nm of the different tip models considered: (a) sharp tip, (b) blunt tip, (c) asymmetric tip.

Figures 6.15(a) and (b) show that the sharp tip gives a fairly representative image of the cluster in both imaging modes. The constant-height image provides a finer image of the cluster, but the difference is not very significant. Tip convolution is much more clearly demonstrated in the images with the blunt tip, shown in Figures 6.16(a) and (b). Although the outline of the cluster can be seen at the center of the image, convolution of the blunt tip with the cluster causes the contrast to be smeared out. Here the difference between imaging modes is more pronounced, with the constant-height image closer to the real cluster size, if not shape.

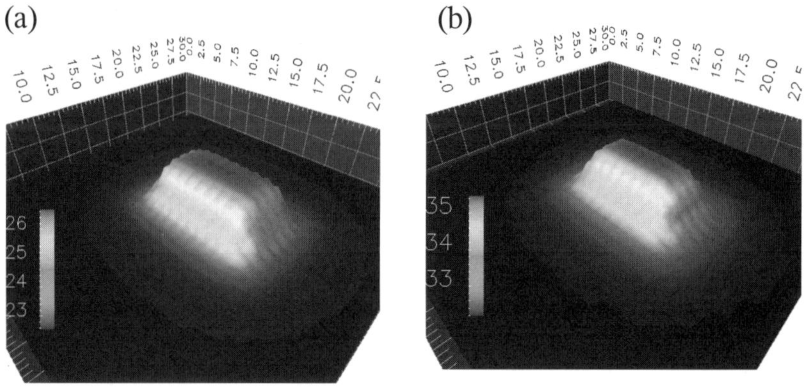

Fig. 6.15. Simulated images of the cluster with a sharp tip in (a) constant frequency change mode at −45 Hz, and (b) constant-height mode at 3.0 nm.

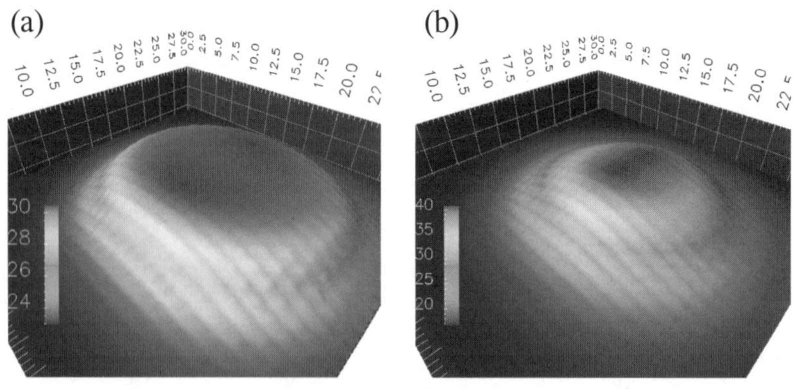

Fig. 6.16. Simulated images of the cluster with a blunt tip in (a) constant frequency change mode at −30 Hz, and (b) constant-height mode at 3.0 nm.

For the asymmetric tip, Figure 6.17(a) clearly shows the effect of the double apex. The outline of the cluster can be seen, with a similar, but extended, contrast to that of the sharp tip shown in Figure 6.15(a) at the center of the image. However, the lower part of the image shows that the cluster is effectively imaged again by the second apex, producing a weak-shadow contrast feature. This effect is present also in the constant-height mode image, Figure 6.17(b), but the frequency change images give a much better representation of the cluster.

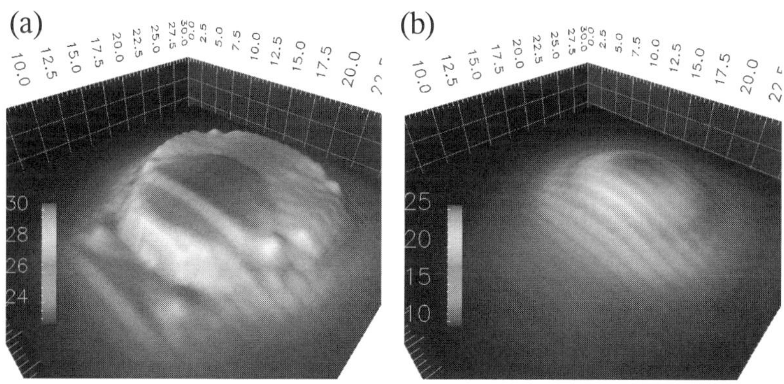

Fig. 6.17. Simulated images of the cluster with a asymmetric tip in (a) constant frequency change mode at −16 Hz, and (b) constant-height mode at 3.0 nm.

6.5 Summary

In this chapter we have shown how theorists actually proceed from a given SFM experimental result to arrive at a realistic simulation of the imaging process. It turned out that the key to successful modeling lies in the ability to successively refine the theoretical model, especially with regard to allowing flexibility in tip selection. This process is inherently iterative: it is usually not possible to arrive at a consistent model that agrees with experimental data without several iteration cycles to fine-tune the model. Contrary to what one might believe, theoretical modelling of SFM experiments is therefore no black box, at least not at the present stage. A general approach for real understanding in SFM simulations must include the following components:

- Justification for the interaction simulation method itself: empirical potentials can be useful, but must be carefully tested, and are usually inflexible.
- An attempt to model the real experimental tip if enough data exists, or at least several plausible models must be considered.

- For high-resolution imaging, tip and surface relaxations must be included since they have a significant influence on the interactions.
- The dynamics of the cantilever and experimental electronics must be treated at a level appropriate for the phenomenon being simulated.

References

1. M. Guggisberg, M. Bammerlin, Ch. Loppacher, O. Pfeiffer, A. Abdurixit, V. Barwich, R. Bennewitz, A. Baratoff, E. Meyer, and H.-J. Güntherodt. *Phys. Rev. B*, 61:11151, 2000.
2. A. S. Foster, L. N. Kantorovich, and A. L. Shluger. *Appl. Phys. A*, 72:S59, 2000.
3. C. Barth, A. S. Foster, M. Reichling, and A. L. Shluger. Contrast formation in atomic resolution scanning force microscopy on CaF_2 (111): Experiment and theory. *J. Phys.: Condens. Matter*, 13:2061, 2001.
4. R. H. French, R. M. Cannon, L. K. DeNoyer, and Y. M. Chiang. *Solid State Ionics*, 75:13, 1995.
5. A. I. Livshits and A. L. Shluger. Role of tip contamination in scanning force microscopy imaging of ionic surfaces. *Faraday Discuss.*, 106:425, 1997.
6. A. I. Livshits, A. L. Shluger, A. L. Rohl, and A. S. Foster. *Phys. Rev. B*, 59:2436, 1999.
7. A. S. Foster, A. L. Rohl, and A. L. Shluger. *Appl. Phys. A*, 72:S31, 2000.
8. K. I. Fukui, H. Onishi, and Y. Iwasawa. *Phys. Rev. Lett.*, 79:4202–4205, 1997.
9. F. J. Giessibl, S. Hembacher, H. Bielefeldt, and J. Mannhart. *Science*, 289:422, 2000.
10. R. Erlandsson, L. Olsson, and P. Mártensson. *Phys. Rev. B*, 54:R8309, 1996.
11. H. Hosoi, K. Sueoka, K. Hayakawa, and K. Mukasa. *Appl. Surf. Sci.*, 157, 2000.
12. W. Allers, S. Langkat, and R. Wiesendanger. *Appl. Phys. A*, 72:S27, 2001.
13. A. I. Livshits and A. L. Shluger. *Phys. Rev. B*, 56:12482, 1997.
14. R. Pérez, I. Stich, M. C. Payne, and K. Terakura. *Phys. Rev. B*, 58:10835, 1998.
15. S. H. Ke, T. Uda, R. Pérez, I. Stich, and K. Terakura. First-principles investigation of tip-surface interaction on a GaAs(110) surface: Implications for atomic force and scanning tunneling microscopies. *Phys. Rev. B*, 60:11631, 1999.
16. A. L. Shluger, A. I. Livshits, A. S. Foster, and C. R. A. Catlow. *J. Phys.: Condens. Matter*, 11:R295, 1999.
17. A. L. Shluger and A. L. Rohl. *Topics in Catalysis*, 3:221, 1996.
18. R. Bennewitz, A. S. Foster, L. N. Kantorovich, M. Bammerlin, Ch. Loppacher, S. Schär, M. Guggisberg, E. Meyer, and A. L. Shluger. *Phys. Rev. B*, 62:2074, 2000.
19. R. Pérez, M. C. Payne, I. Stich, and K. Terakura. *Phys. Rev. Lett.*, 78:678, 1997.
20. J. Tóbik, I. Stich, R. Pérez, and K. Terakura. Simulation of tip-surface interactions in atomic force microscopy of an InP(110) surface with a si tip. *Phys. Rev. B*, 60:11639, 1999.
21. J. Tóbik, I. Stich, and K. Terakura. *Phys. Rev. B*, 63:245324, 2001.
22. S. H. Ke, T. Uda, I. Stich, and K. Terakura. *Phys. Rev. B*, 63:245323, 2001.
23. A. S. Foster, A. Y Gal, J. M. Airaksinen, O. H. Pakarinen, Y. J. Lee, J. D. Gale, A. L. Shluger, and R. M. Nieminen. *Phys Rev. B*, 68:195420, 2003.

24. A. S. Foster, A. Y. Gal, J. D. Gale, Y. J. Lee, R. M. Nieminen, and A. L. Shluger. *Phys. Rev. Lett.*, 92:036101, 2004.
25. S. H. Ke, T. Uda, and K. Terakura. *Phys. Rev. B*, 65:125417, 2002.
26. A. S. Foster, O. H. Pakarinen, J. M. Airaksinen, J. D. Gale, and R. M. Nieminen. *Phys. Rev. B*, 68:195410, 2003.
27. T. Eguchi and Y. Hasegawa. *Phys. Rev. Lett.*, 89:266105, 2002.
28. A. Y. Gal and A. L. Shluger. *Nanotec.*, 15:S108, 2004.
29. P. V. Sushko, A. S. Foster, L. N. Kantorovich, and A. L. Shluger. *Appl. Surf. Sci.*, 144–145:608, 1999.
30. R. Oja and A. S. Foster. *Nanotechnology*, 16:S7, 2005.
31. A. L. Shluger, L. N. Kantorovich, A. I. Livshits, and M. J. Gillan. *Phys. Rev. B*, 56:15332, 1997.
32. T. Trevethan and L. Kantorovich. *Nanotechnology*, 16:S79, 2005.
33. H. Hölscher, U. D. Schwarz, and R. Weisendanger. *Appl. Surf. Sci.*, 140:344, 1999.
34. T. R. Albrecht, P. Grütter, D. Horne, and D. Rugar. *J. Appl. Phys.*, 69:668, 1991.
35. F. J. Giessibl. *Phys. Rev. B*, 56:16010, 1997.
36. F. J. Giessibl. *Rev. Mod. Phys.*, 75:949, 2003.
37. F. J. Giessibl, H. Bielefeldt, S. Hembacher, and J. Mannhart. *Ann. Phys. (Liepzig)*, 10:887, 2001.
38. R. García and R. Pérez. *Surf. Sci. Rep.*, 47:197, 2002.
39. U. Dürig. Interaction sensing in dynamic force microscopy. *New J. Phys.*, 2, 2000.
40. M. Gauthier, N. Sasaki, and M. Tsukada. *Phys. Rev. B*, 64:085409, 2001.
41. G. Couturier, R. Boisgard, L. Nony, and J. P. Aimé. *Rev. Sci. Instr.*, 74:2726, 2003.
42. M. Gauthier, R. Perez, T. Arai, M. Tomitori, and M. Tsukada. *Phys. Rev. Lett.*, 89:146104, 2002.
43. J. E. Sader and S. P. Jarvis. *Phys. Rev. B*, 70:012303, 2004.
44. S. Hembacher, F. J. Giessibl, and J. Mannhart. *Science*, 305:380, 2004.
45. R. Lüthi, E. Meyer, M. Bammerlin, A. Baratoff, , L. Howard, C. Gerber, and H.-J. Güntherodt. Atomic resolution in dynamic force microscopy across steps. *Surf. Rev. Lett.*, 4:1025, 1997.
46. M. Gauthier and M. Tsukada. Theory of noncontact dissipation force microscopy. *Phys. Rev. B*, 60:11716, 1999.
47. S. P. Jarvis, H. Yamada, K. Kobayashi, A. Toda, and H. Tokumoto. *Appl. Surf. Sci.*, 157:314, 2000.
48. S. Morita, R. Wiesendanger, and E. Meyer, editors. *Noncontact Atomic Force Microscopy*, chapter 20, page 395. Springer, Berlin, 2002.
49. S. Morita, R. Wiesendanger, and E. Meyer, editors. *Noncontact Atomic Force Microscopy*. Springer, Berlin, 2002.
50. N. Sasaki and M. Tsukada. *Appl. Surf. Sci.*, 140:339, 1999.
51. A. Abdurixit, T. Bonner, A. Baratoff, and E. Meyer. *Appl. Surf. Sci.*, 157:355, 2000.
52. L. N. Kantorovich. *Phys. Rev. Lett.*, 89:096105, 2002.
53. T. Trevethan and L. Kantorovich. *Nanotechnology*, 15:S34, 2004.
54. T. Trevethan and L. Kantorovich. *Nanotechnology*, 15:S44, 2004.
55. T. Trevethan and L. Kantorovich. *Phys. Rev. B*, 70:115411, 2004.

56. L. N. Kantorovich and T. Trevethan. *Phys. Rev. Lett*, 93:236102, 2004.
57. L. Bergström. *Adv. Coll. Int. Sci.*, 70:125, 1997.
58. H. Hölscher, W. Allers, U. D. Schwarz, A. Schwarz, and R. Wiesendanger. *Appl. Phys. A*, 72:S35, 2001.
59. P. Markiewicz and M. C. Goh. *Langmuir*, 10:5, 1994.
60. C. Barth and C. R. Henry. *Nanotechnology*, 15:1264, 2004.
61. S. Giorgio, C. Chapon, C. R. Henry, G. Nihoul, and J. M. Penisson. *Phil. Mag. A*, 64:87, 1991.
62. L. M. Molina and B. Hammer. *Phys. Rev. Lett.*, 90:206102, 2003.
63. K. Cooper, A. Gupta, and S. Beaudoin. *J. Coll. Int. Sci.*, 234:284, 2001.

7

Topographic images

So far, we have shown how interaction energies and tunneling currents can be obtained by suitable theoretical methods. This, however, is only the necessary basis for an actual simulation. Such a simulation involves not only a method, but also a suitable choice of physical parameters and features of e.g. the surface, and geometry and chemical composition of the SPM tip. In this chapter we shall investigate in detail, how an experimental result translates into the setup of a detailed simulation, and how these simulations are compared to experiments. We shall start, owing to the historical development of the instruments, with case studies of STM simulations, followed by studies of SFM simulations.

7.1 Setting up the systems

Unless a theoretical model is used to predict experimental results, it is common to start a simulation with a well defined set of experimental results. These results can involve a number of different experimental methods as well as measurements taken under different physical conditions. In fact, every unambiguous experimental result reduces the parameter space in a simulation. This is particularly important for STM experiments, as the result *always* involves a number of different physical processes, which can completely overshadow the characteristics of isolated surfaces. In the following, we shall present two main instructive examples of actual simulations. The first, where the electronic structure of the STM tip is the decisive variable, deals with STM simulations on oxygen covered Ru(0001) surfaces. The second, where tip surface interactions play a major role, are STM simulations of a close packed Al(111) surface.

7.1.1 Ru(0001)-O(2×2)

In many areas of research, e.g. catalysis, or complex transition metal oxides, it is important to identify metal and oxygen sites at surfaces in order to understand processes such as dissociation of molecules, or the role of impurities in the formation of striped phases [1] . Although STM topographs can be used to characterize the surfaces at the atomic level, they do not simply reflect the real position of surface atoms [2, 3, 4, 5]. If we restrict ourselves to adsorbed oxygen layers or oxide surfaces , experimental reports show [6, 7, 8, 9, 10, 11] that, depending on the system and the state of the tip, either oxygen or metal atoms appear as bright features in the STM images. Because the geometric and electronic structure of the surface, as well as the chemical state of the tip, play a role in determining the corrugation, contrast and shape of the image, it is necessary to perform *ab-initio* calculations to interpret properly the STM images.

Experimental images

The STM experiments were performed at room temperature in ultra-high vacuum. At low bias voltage and low tunnelling resistance the clean Ru(0001) surface is imaged as an hexagonal array of round protrusions separated by 2.7 Å, shown in Fig. 7.1(a). Scanning Tunnelling Spectroscopy experiments and simulations [12] indicate that the states dominating the current at standard distances are due to a surface resonance of p_z character, located close to the Fermi energy and spatially localized on top of the Ru atoms. Fig. 7.1(b) shows an STM image of a compact O adlayer with 2×2 periodicity with respect to the substrate recorded at standard gap resistance. The O atoms are visualized as circular depressions. The bright regions in the image correspond to (mobile) oxygen vacancies in the 2×2 superstructure, i.e. clean Ru patches. However, the shape of the image does not remain the same over the whole conductance range. As shown in Fig. 7.1(c), the depressions have a *circular* shape at low tunneling conductances, and a *triangular* shape in the high conductance regime. In this case experimental images were taken by forward and backward scans simultaneously; the tunneling conductance in the backward scan remained constant, conclusively proving that the change of shape cannot be due to a change of the STM tip in the experiments.

Simulating the electronic groundstate

At this stage, the main question posed by experimenters is the origin of the depressions. It is evident, that they must be related to the electronic structure of the oxygen covered surface. The first step in the theoretical simulations is therefore to calculate the electronic structure of the surface in its groundstate. This part of the simulation involves only standard DFT methods. In essence, oxygen atoms are put on top of the clean ruthenium (0001) surface, mimicked by a metal film of a few (usually five to seven) layers, and then relaxed until they are bound to the surface. Depending on the adsorption site the total energy of the system varies. The groundstate is defined as the state of lowest energy, which for a given oxygen coverage will be oxygen adsorbed at the three-fold hcp hollow positions, i.e. the hollow positions above Ru atoms in the subsurface layer. The oxygen atoms are located above the metal surface, at a distance of 1.16 Å. Apart from the total energy, this result is also suggested by LEED measurements and simulations [13].

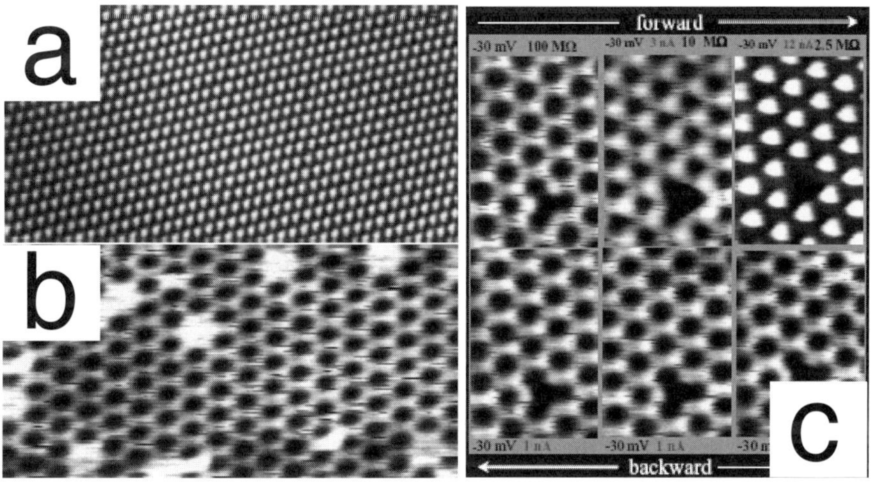

Fig. 7.1. (a) 10 nm by 5 nm STM image of a clean Ru(0001) surface. (b) 10 nm by 5 nm STM image of the O(2×2) superstructure on Ru(0001). The oxygen atoms appear as round holes in these conditions. theoretical frequency change. (c) Series of 2.2 nm by 3.2 nm dual-mode STM images. The upper panel shows images recorded at constant sample voltage decreasing gap resistances (from left to right, 100, 10 and 2.5 MΩ^{-1}). The reference images in the lower panel were all taken at a constant gap resistance of 30 MΩ^{-1}. F. Calleja et al., Phys. Rev. Lett. 92, 206101 (2004). Copyright (2004) American Physical Society, reprinted with permission.

The choice of a suitable STM tip model is partly motivated by the experimental situation and partly by the current values obtained in an experiment. It is, for example, possible to change the current at a given position above a metal surface in a simulation by up to one order of magnitude, depending on the apex structure of a clean tungsten tip . Furthermore, the actual corrugation in an experiment needs to be reflected by the tip structure, as high corrugation values make it necessary to simulate a tip apex by a cluster of at least two layers. It was also found in simulations that the resolution in STM images of close packed metal surfaces can only be obtained in simulations with an atomically sharp tip.

From a theoretical point of view the tip apex is part of a semi-infinite crystal surface. This feature, which has a substantial influence on the electronic tip structure, must also be reflected in simulated STM tip. For this reason it is generally not enough to mimic the tip by a cluster of only a few atoms. Such a tip would not have the correct bandstructure and composition of the surface electronic structure. But if the tip is mimicked by an infinite metal film, then the surface integral, which is part of every STM simulation procedure, cannot be evaluated numerically. Combining both perspectives, the experimental condition of atomically sharp tip, and the theoretical requirement of an apex structure reflecting the true STM tip as much as possible, suggests to use an infinite metal film with an apex of a few layers terminating in a single atom. This guarantees that the bandstructure of the tip metal reflects the actual properties of the semi-infinite crystal, and it also accounts for the atomically sharp apex used in most experiments. At the same time it makes the calculation of such a tip system numerically tractable, since the repeat unit in the DFT calculation of the STM tip then contains less than 50 atoms.

One could think of extending the size of the STM tip in order to create a smoother bandstructure and thus coming closer to experimental conditions. However, this comes at a high price. Every layer of an STM tip in either (110) or (111) orientation contains more than eight atoms. To approach the level of precision used in groundstate calculations of metal surfaces, the number of layers should be higher than six (nonmagnetic tip systems) or 11-13 (ferromagnetic tip systems) . While in principle feasible, given enough computing resources, this is not really necessary in simulations of topographic images, where the main contributions to the tunneling current come from only a few tip states at the very apex (on metals and under bias voltages of about 100 mV, we find usually less than ten states per k-point). Since the apex is well described up to the fourth or fifth nearest neighbor atoms, a tip model of only three layers and two pyramids on either side is generally sufficient. It should be mentioned, though, that such a tip model is not sufficient for tunneling spectroscopies (see the following chapters), where the limitation of the number of tip layers introduces artifacts due to the vertical boundary conditions, which are likely to alter a simulated spectrum in an unwanted fashion. Two of the tip models used in our topography simulations are shown in Fig. 7.2.

7.1.2 Al(111)

On close packed metal surfaces measured corrugations exceed the values obtained from constant density contours by up to one order of magnitude [14, 15, 16]. On Al(111) surfaces, in particular, the measured corrugation of about 70pm [15, 16] cannot be explained in a straightforward manner (see Fig. 7.3). This fact has been known since the 1980s, and the puzzle has been the focus of attention for more than fifteen years. It is easy to see why: if

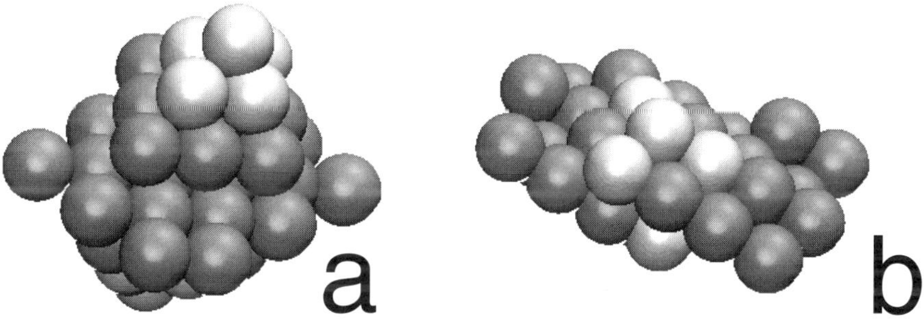

Fig. 7.2. STM tip models in topography simulations. The tip models have either (110) (a) or (111) (b) surface orientation. The apex pyramid (white atoms) is mounted on an infinite extended surface of three layers (dark atoms). All atoms in the images ar part of a single unit cell.

simulations of experiments in a simple case, like flat metal surfaces, leave up to 90% of the measured values unaccounted for, then interpretations of more subtle experiments are potentially imprecise by the same amount. It deprives theoretical work in this field of a sound scientific basis. This basis can only come from a detailed understanding of the physical processes involved in the imaging process.

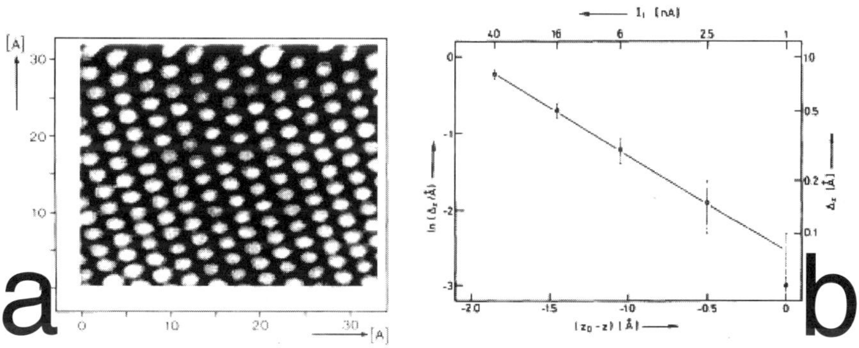

Fig. 7.3. (a) Topographic STM image on Al(111). (b) Corrugation amplitude in terms of tunneling current. J. Wintterlin, J. Wiechers, H. Brune, T. Gritsch, H. Höfer and R. J. Behm, Phys. Rev. Lett. 62, 59 (1989). Copyright (1989) American Physical Society, reprinted with permission.

Two separate models have been put forward to account for the deviations: it was either thought to be due to electronic effects, or due to the interactions between atoms at opposite sides of the tunneling junction. The first model has been favored in the work of Chen [17], where it was thought that states of d_{z^2} symmetry at the STM tip lead to an enhancement of corrugation. The assumption is backed to some extent by electronic structure calculations of Tsukada [18], who showed that a tungsten cluster displays a state of d_{z^2} symmetry at the Fermi level. The same line of reasoning was used in a paper by Jacobsen [19] in 1995. However, as Sacks reported recently, the obtained corrugation with these states is still one order of magnitude below experimental values [20]. Along a different line of research Doyen placed the emphasis on the tunneling process itself, accounting for increased corrugation by solving the scattering problem with a modified Dyson equation, obtaining corrugation enhancements of the right magnitude [21]. All methods based on an enhancement due to electronic structure are implicitly based on the assumption, that only very few electron transitions between surface and tip are responsible for the observed corrugation values. This assumption, however, is contradicted by explicit calculations within the Bardeen method [22], where typically a few hundred tunneling channels are obtained for bias voltages of 50 to 100 mV. Therefore, the theoretical understanding of the problem tended to ultimately favor interactions between surface and tip atoms. Here, the problem of dynamic processes in STM scans was until very recently treated by semi-empirical methods. Pair potentials were used by Soler [23] and by Clarke [24] to account for corrugation enhancements on graphite and copper surfaces, respectively. In this case it proved difficult to relate current values in the experiments, which are a measure for the distance between the two surfaces, to the forces and relaxations of atoms, since pair-potentials decay very rapidly beyond 300 pm: a detailed analysis of the interplay between interactions and tunneling currents remained elusive also with this method. The solution, for gold surfaces, was presented in 2001 [25]; it involved calculating the forces and relaxations of coupled systems, and to determine the effect on constant current contours within a first-principles approach. It was also shown that tunneling current and interaction energy are in fact proportional to each other [26]. This result, contradicting earlier assumptions by Chen [27], is confirmed by experimental data [28, 29]. However, it remained unknown, whether the same physical process applies to the case of aluminum surfaces, and equally, how the difference between the two surfaces can be accounted for. The solution to this problem required a first principles method for computing dynamic constant current contours in the simulation of STM experiments. Essentially, as shown in previous chapters, such a method can be based on the linearity between currents and interaction energies, and involves computing the corrections to the current obtained from the electronic structure of surface and tip.

Groundstate and elastic constants

The Al(111) surface is mimicked by a 13-layer film, the vacuum range above the surface has to be larger than about 1000 pm. This guarantees a smooth decay of the surface wavefunctions into the vacuum region. After initial relaxation of the outer layers, obtaining the groundstate positions of the ionic cores, the outermost atom is lifted by about 5 pm. In this case, the system is slightly out of equilibrium and the electronic relaxation shows that the reaction of the system is a retracting force on the surface atom. Since $F_{retract} = -k_{harm} \Delta z$, the harmonic constant of ionic motion can be determined, once the retracting force is known. This force is computed within most DFT codes using the Hellman-Feynman theorem. The surprising result of this calculation is that the aluminum surface is much more elastic than e.g. noble metal surfaces. This high elasticity is reflected in a small harmonic constant, which is only about a quarter of the harmonic constant of noble metals (see Table 7.1).

Metal surface	Cu(111)	Ag(111)	Au(111)	Al(111)
Harm. constant [eV/Å^2]	3.77	2.96	3.22	0.89

Table 7.1. Elastic constants of Al and noble metal surfaces. Aluminum is much more elastic than noble metals, reflected by a small value of the harmonic constant.

These values, computed by straightforward DFT from the surface alone, already provides a clear indication about the origin of high corrugation values on Al. If the enhancement of corrugation due to motion of the surface atoms is already about three for Au(111) [30], then it should be substantially higher for Al(111) . Harmonic constants also provide a measure for the necessity of including dynamic effects in an STM simulation. The apex atom of a tungsten tip, for example, possesses a harmonic constant well in excess of 10 eV/Å^2. A standard STM tip is therefore very rigid. The same statement also holds for oxygen atoms adsorbed on a metal surface. In this case the image obtained from the electronic surface and tip structure is sufficiently precise for a direct comparison with experimental values.

7.2 Calculating tunneling currents

Once the vacuum Kohn-Sham states of the surface and a model tip are calculated by standard DFT methods, the current between the two sides of the tunneling junction can be determined by computing the overlap of the wavefunctions for every position of the STM tip. Depending on the system such a calculation can be quite expensive. Two separate parameters of the current simulations are responsible for the required effort: (i) The lateral resolution

of the simulated image; and (ii) the Fourier expansion of the vacuum wavefunctions.
The lateral resolution is largely determined by the achievable resolution in an STM image. As a rule of thumb one gridpoint every 10 pm is sufficiently precise to match even the highest resolution images in state-of-the-art measurements. The number of two dimensional Fourier components is more difficult to determine, as it depends on the local resolution as well as the lateral unit cell. However, it is generally found that results converge very quickly once the expansion includes the reciprocal lattice vectors of at least the first and second Brillouin zone. From a practical point of view it should be noted that most DFT codes include only the irreducible wedge of the first Brillouin zone in the k-map of the reciprocal unit cell. Wavevectors in DFT calculations are consequently not complete and have to be expanded over the full Brillouin zone in a current calculation.

7.2.1 Ru(0001)-O(2×2)

Getting a first impression

It is usually straightforward, once the electronic structure of a surface is calculated, to obtain a charge density contour. Nearly every DFT program today contains a routine to this end, and the method is still widely used, despite evidence that it might only be safe in a distance range well above 0.4nm and under the condition that the feature size on the surface is well above the resolution limit in STM scans. To show the information gained as well as the limitations of the method, we have plotted three constant density contours above the oxygen covered surface. These plots are shown in Fig. 7.4.
It is interesting to note that the density contours give acceptable values for the corrugation, which is about 50 pm for normal tunnelling conditions. This seems to relate to the second feature, i.e. the slow decrease of the corrugation over distance. This in turn is related to the surface composition and the position of the oxygen atoms more than 0.1 nm above the surface rather than the exponential decay of vacuum wavefunctions. However, the resolution of the contour decreases substantially as the distance value approaches the range of actual measurements. This is quite understandable if one considers that we used a DFT code with a supercell geometry. Even if, as in our calculation, the vacuum range is larger than 2.5 nm, the vacuum decay of the wavefunctions and their representation in a two dimensional Fourier grid is still limited once the distance from the surface is higher than about 0.4 nm.
Apart from these methodical limitations of charge density contours, which could only be amended by a much larger cutoff in reciprocal space and a larger vacuum range - both of which make the calculation of the (2×2) unit cell already very expensive - there is also a disagreement between the experiments and the simulations in the shape of the contours at close distance. In the experiments (see Fig. 7.1) the shape of the contour is triangular, while it is

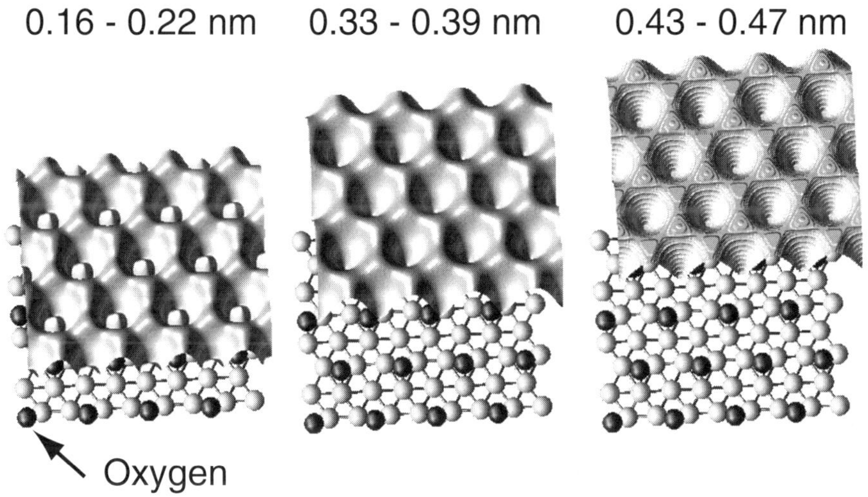

Fig. 7.4. Constant charge density contours on Ru(0001)-O(2×2). The bias voltage was assumed to be -30mV. Positions of Ru (green) and O (blue) atoms of the extended surface are shown. The three contours are in the very close distance range, generally considered too close for an actual scan. It can be seen that the resolution of the density contour decreases substantially in the range above 0.4nm (right figure).

circular at every distance range in the density contours. This indicates that even though the qualitative picture in the contours is roughly accurate (oxygen appears as a depression), the details of the picture and the actual quantities are missing.

Calculating constant current contours

For the following calculation of the constant current contours we used a tungsten (110) surface terminated by a single atom (see Fig. 7.2). Simulations with a (111) tip give slightly higher current values at a given distance, as the tip is less sharp, but do not change the overall picture. To compare with experiments the simulations were performed with two different gap resistances at a bias voltage of -30 mV. The results of the current calculation are shown in Fig. 7.5. Here, measured (left panels) and simulated (right panels) STM images at representative gap resistances are directly compared. When imaged with a W tip, the O atoms appear as depressions, while Ru is seen bright. As gap resistance decreases by one order of magnitude, the changing shape of the features associated with O and Ru is nicely reproduced by the simulations. Their shape is circular when the tip is relatively far away from the surface and triangular at closer tip-surface distances. This change is mostly due to the different geometry of Ru p_z orbitals, with rotationally symmetric lobes pointing outwards, and hybridized s/p_{xy} orbitals, with threefold rotational symmetry with respect to the adsorption site. At large distances p_z orbitals contribute most of the current (circular shape), while closer to the surface the contribution from s/p_{xy} orbitals increases (triangular shape). The inclusion of a realistic tip structure and the use of the Bardeen approach in the calculations is essential to obtain quantitative agreement between simulated and measured images in the studied range.

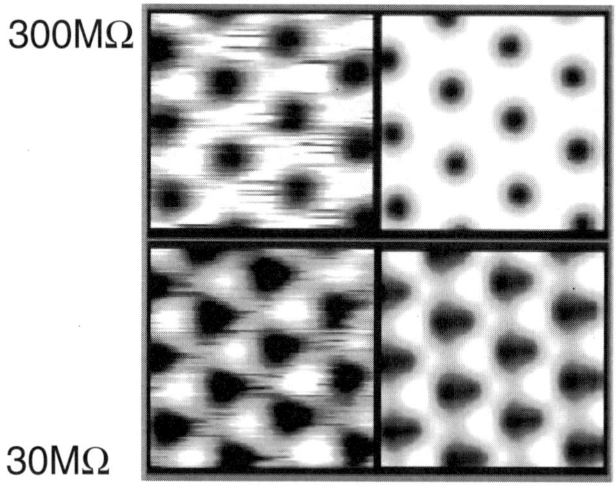

Fig. 7.5. Comparison of experimental (left panels) and simulated (right panels) STM images. In both cases the sample voltage was -30 mV. The simulations have been performed with tunneling currents of 0.03 nA (above) and 0.3 nA (below) and agree with the experimental ones for 300 MΩ (above) and 30 MΩ (below) gap resistances. The maximum corrugation is 50pm in all cases.

The simulated and measured images show near perfect agreement. In fact, the agreement seems too good to be true, considering that the calculation is based on a guess about the tip orientation and does not include interactions between surface and tip. On closer analysis, however, one finds that the main ingredient for an accurate description of the tunneling current are the shape and the eigenvalues of states at the tip apex atom. Whether this atom has three (111) or four nearest neighbors in the next layer, and whether it is higher elevated (110) does not influence the result in a decisive way. Also the surface electronic structure itself, with its rather high corrugation limits the changes due to the tip. This would be different, e.g. on flat metal surfaces. From a technical point of view it has to be considered that the surface of evaluation is only about 0.2-0.3 nm above the surface oxygen. This feature also makes the calculation in general more reliable than charge density contours, as the known problems with the vacuum decay of surface wavefunctions do not affect the numerical results to such an extent. And finally, the results presented here belong to the lower end of the measured conductance range. Agreement in the high conductance limit of about one $M\Omega^{-1}$ cannot be obtained. While the experiments still demonstrate triangular contours, the obtained conductance values lead to unphysically close distances. This is due to the limits of perturbation theory . In this range the only viable method is a fully self consistent calculation with both systems out of equilibrium.

Functionalizing the tip

It would be desirable, for an accurate comparison between experiments and simulations, to know the exact shape and chemical composition of the STM tip in experiments. This problem has been addressed by experimenters to some extent in the past years and selected measurements exist, where such a full control of the experimental situation was achieved. However, this is still only a minority of experiments. It seems all the more important, to use these selected experiments and to demonstrate the range of achievable images of the very same surface under different tip conditions. The wide range of experiments done on ruthenium surfaces include examples of such a functionalized STM tip. In essence, the oxygen adsorbed on the surface is not very stable and can rapidly diffuse along the metal rows. At the boundaries of a forming oxygen adlayer the oxygen is very mobile and can be picked up by the tip if the distance to the surface becomes very low. As the simulations show, the distance under scanning conditions is already quite small (about 0.4-0.5 nm in the calculated images), which facilitates the atomic transfer. On tungsten, oxygen is expected to adsorb at a hollow site. If this happens, then the foremost tungsten atom can be replaced by oxygen: the tip has been functionalized .

Simulating such a tip is straightforward. The only change in the tip composition is the replacement of the tungsten apex by oxygen. This has a substantial effect on the electronic tip structure as shown in Fig. 7.6. In the top panel we show a constant density contour above the tip for a clean tungsten tip (left), and for a tungsten tip covered by oxygen (right). It can clearly be seen that the maximum of charge density at the Fermi level then is no longer at the center of the tip, but at a rim around it. This is in line with expectations, as oxygen depletes metal surfaces of their charge at the Fermi level. It is essentially the same effect which makes oxygen appear as a depression on ruthenium.

If the surface is imaged with a functionalized tip, the maximum of a constant current contour is no longer found at the position of ruthenium atoms, as the main overlap between surface and tip wavefunctions is no longer located at the center of the STM tip. Instead, the current is highest, if the tip is on top of the oxygen atoms, as the higher charge density - and wavefunction amplitudes - at the rim will then provide maximum overlap with the wavefunctions above the ruthenium atoms. The contrast of the surface now appears to be reversed . This is shown in the bottom panels of Fig. 7.6.

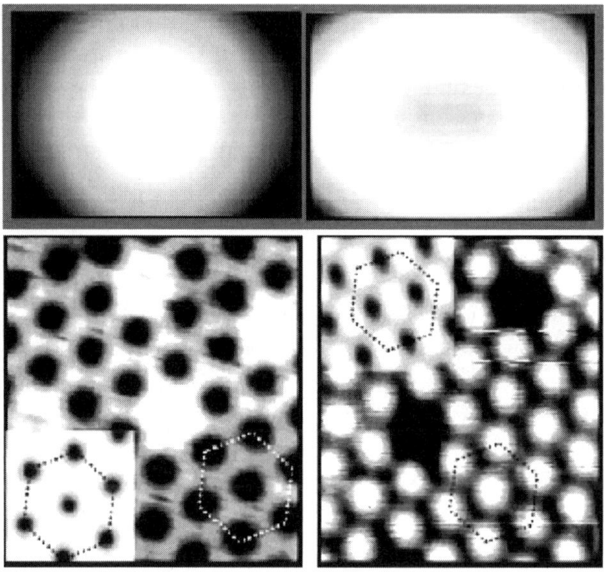

Fig. 7.6. (Top) Calculated density contour for a W- (left panel) and an O- (right panel) terminated tip in an area of 0.68 by 0.48 nm. The apex atom is located in the center of the panels. (Bottom) 4 nm by 5.2 nm STM image taken with a W tip at 0.1 V and 200 MΩ. The inset shows the image calculated under the same conditions (Left panel). 4 nm by 5.2 nm STM image recorded with an oxygen atom at the tip apex. The sample voltage was 0.35 V and the gap resistance 1.15 GΩ. The inset shows the image simulated with an O-terminated tip, calculated at 0.1 V and 1 GΩ and displaying the inversion of contrast (Right panel). Reprinted from [12], with permission.

It is a nice illustration of the fact that accurate theoretical descriptions often do not resolve an experimental issue, but make it more complicated. The theoretical answer to the seemingly simple experimental question: Is oxygen bright or dark in my experimental images? is not a definite answer, but another question: What tip did you use in the experiments? This might seem like an argument against the combination of experimental and theoretical work in a scientific endeavor, but is actually a good illustration of how increased precision in theoretical modelling inevitably leads to a more precise description of experimental conditions. And this, in turn, leads to the detection of effects which before where either not well understood or escaped notice because there existed no framework to classify them.

7.2.2 Al(111)

The second example, how theory is actually used to analyze and account for experimental results, is even more illuminating. It was already mentioned in the previous sections that it remained quite unclear for at least ten years, whether the very high corrugation on Al(111) was due to the effect of electronic surface structures, or whether they indicated some dynamical process which could not be accounted for theoretically. To convince ourselves that there really is a problem, we might first look, as in the previous example, at constant density contours above the aluminum surface. These contours are shown in Fig. 7.7

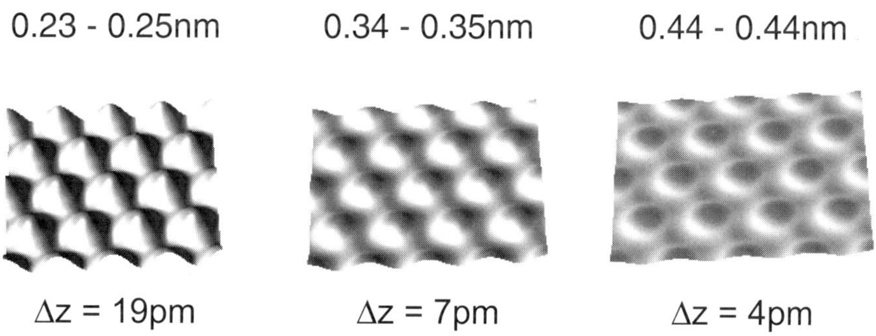

Fig. 7.7. Calculated constant density contours above Al(111) for a bias voltage of -50mV. The distance range from the surface is given on top of the images, the corrugation values at the bottom. For realistic distances larger than 0.3nm the corrugation is less than 10pm, in disagreement with experiments.

We obtain corrugation values which are unambiguously less than 10 pm for distances larger than 0.3 nm. We may conclude, from this calculation, that this simple model is definitely not suitable to account for experimental values, which are about one order of magnitude higher. As a first step towards improving the model we may consider the results with a realistic STM tip. The constant current contours calculated with a clean tungsten tip in (110) orientation are shown in Fig. 7.8.

This does not seem to improve the agreement between the simulations and the experimental values. One could repeat the simulations with different tips, e.g. tips contaminated by aluminum. In fact, such a simulation has actually been done and the result was very similar to the result obtained with a clean tungsten tip. So that one may conclude that the tip structure is not the important parameter in these experiment. This is also in line with expectations, as the high number of transitions on metal surfaces given a bias range in the range of thermal broadening (or about 80 mV), will yield a statistical distribution of wavefunction overlaps. Even if single states of the surface or tip lead to an enhanced corrugation due to their long vacuum tails, their contributions should not dominate the overall picture. Certainly not to such an extent that the corrugation of the surface is increased by one order of magnitude.

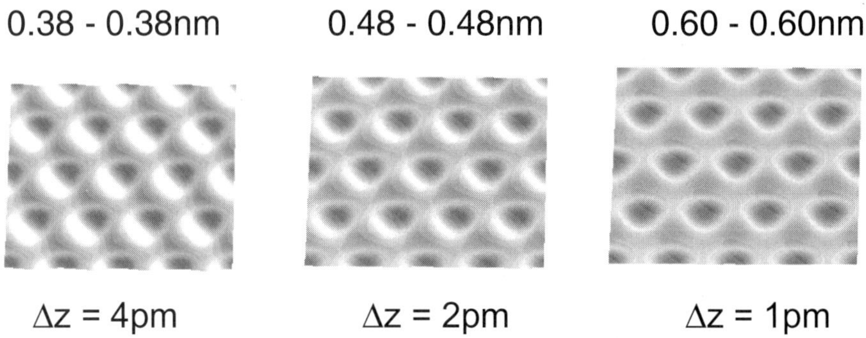

Fig. 7.8. Calculated constant current contours above Al(111) for a bias voltage of -50mV, using a tungsten tip in (110) orientation. The distance range from the surface is given on top of the images, the corrugation values at the bottom. The current values are 5 nA, 1 nA, and 0.1 nA, respectively. Compared to the constant density contours the corrugation amplitude does not change, the values are still one order of magnitude too low.

This situation, a complete disagreement between experiment and theory, is actually quite frequently encountered in theoretical research. To resolve the problem, it is usually necessary to proceed in two steps: (i) Analyze the assumptions which went into the model and, (ii) look for additional evidence, which might back a different set of assumptions. In case of Al(111) surfaces the electronic structure does not provide a clue as to what went wrong. The bandstructure is fairly well known, and quite typical for a metal surface. It also does not possess a surface state, which could become dominant in low bias experiments. In addition, it is non-magnetic. The only relevant information we have at this point is the very high elasticity of the surface. Given a certain interaction energy ΔE, which is proportional to the conductance [22] in the perturbation range, the relaxation of Al surface atoms will be about twice the value obtained on Cu or Au surfaces, since:

$$\frac{\Delta z_{Al}}{\Delta z_{Cu}} = \sqrt{\frac{k_{Cu}}{k_{Al}}} \tag{7.1}$$

Given that the k_{Al} is about one quarter of k_{Cu} (see Table 7.1), a current value of a few nA will lead to double the displacement of Al surface atoms compared to Cu surface atoms. We have of course simplified the problem a little, since the factor of proportion between current and interaction energy depends on the bandstructure and need not be equal for Al and Cu. We also assumed that relaxation of surface atoms occurs in the elastic range. Concerning the first point it turns out, in actual simulations, that the relation between tunneling current and interaction energy even leads to higher interaction energies at a specific current value; the relaxation of surface atoms at a given current value is therefore more than twice the value we obtain for Cu. Concerning the range of elasticity we know from simulations that the changes become irreversible at a defined energy threshold, which is about one eV . As long as interaction energies remain well below that limit we may safely apply the harmonic approximation to motion of surface atoms.

The theoretical model used in the dynamic simulations was introduced in section 3.3 of chapter 3. It is based on vertical displacement of surface atoms under the condition that the STM tip remains rigid. Apart from the relation between current and interaction energy, it is also based on a geometric correction, if the STM tip moves from the on-top to the hollow position of the surface. The main result of the theoretical model is contained in the following equations, which we repeat here for easier reading:

$$I'(z) = I(z) \exp \kappa \sqrt{\frac{P \alpha G}{k}} \tag{7.2}$$

The dynamic current $I'(z)$ is the current based on the electronic structure of surface and tip $I(z)$ corrected by the change due to the relaxation of surface atoms, which is described by the square root in the exponent and depends on the conductance, and the harmonic constant k.

$$P(d) = \cos\left(\frac{z}{\sqrt{z^2 + d^2}}\right) \cdot a \left[1 - b \tanh\left(\frac{4d}{d_0} - 2\right)\right] \tag{7.3}$$

$P(d)$ in this equation is the projection value, which changes as the STM tip moves from the on-top ($d = 0$) to the hollow position ($d = d_0$) of the surface, as interactions between surface and tip are limited to one atom only in the first case, and involve the three adjacent atoms in the second case. The parameters a and b depend on the symmetry of the surface, d_0 is the distance between the hollow site and the on-top site.

As shown in Fig. 7.9, this leads to a surface corrugation of up to 70 pm (-50 mV, Fig. 7.9 (left)). One obtains similar results for a clean tungsten tip, and a tip covered by aluminum, but only about half the corrugation value for the tip made of pure aluminum . The experimental corrugation amplitudes reported in Ref. [16] have remained a puzzle for more than fifteen years. Here, we find the solution of this puzzle: as for Al the surface atoms are less strongly bound to the surface than for noble metals, their outward relaxation under tunneling conditions is very large. Combined with the changes of forces, as the tip moves from the on-top to the hollow position, this gives rise to unexpected corrugation values.

The main increase of corrugation occurs between 1 $M\Omega$ and 10 $M\Omega$ tunneling resistance, defined as the ratio of bias potential and tunneling current. In this range the corrugation increases from 20 pm to 70 pm, the tunneling resistance decreases faster than exponential (Fig. 7.10). To analyze the stability in the limit of high corrugations we also computed the interaction energy in this situation. The value we obtain is less than 0.5 eV, corresponding to a an absolute displacement of Al atoms by 73 pm as the tip is in the on-top position.

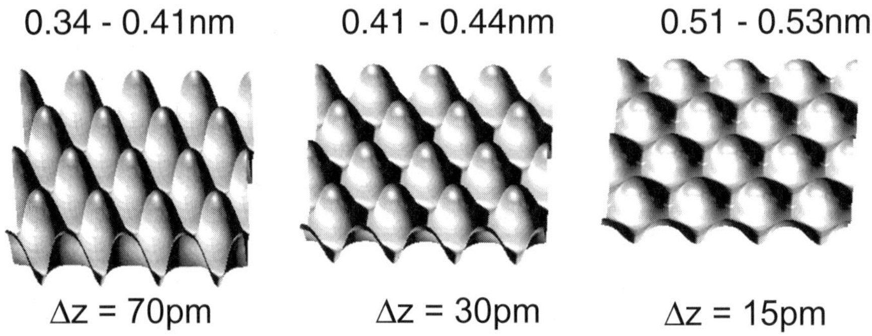

Fig. 7.9. Calculated *dynamic* constant current contours above Al(111) for a bias voltage of -50 mV, using a tungsten tip in (110) orientation. The distance range from the surface is given on top of the images, the corrugation values at the bottom. The current values are 45 nA, 10 nA, and 1 nA, respectively. Compared to the contours with dynamic adjustments the values are higher by about one order of magnitude.

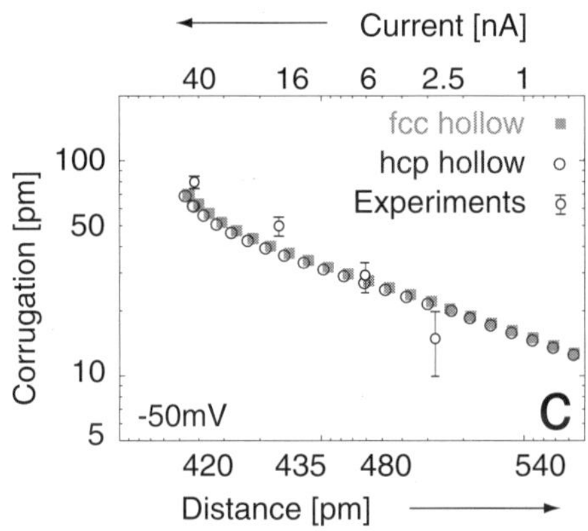

Fig. 7.10. Calculated corrugation amplitudes on Al(111). Interactions on this surface lead to very large enhancements and quite singular corrugations in excess of 70 pm. The apparent height of surface atoms increases mainly in the range from 10-1 MΩ tunneling resistance, its maximum value in the simulation is about 80 pm.

Comparing with our previous first principles calculations [30], these values are lower than the energy threshold of about 1 eV for the jump into contact and also substantially lower than the displacement of more than 130 pm related to it. It is thus safe to conclude that also this point of the simulation is well within the elastic range.

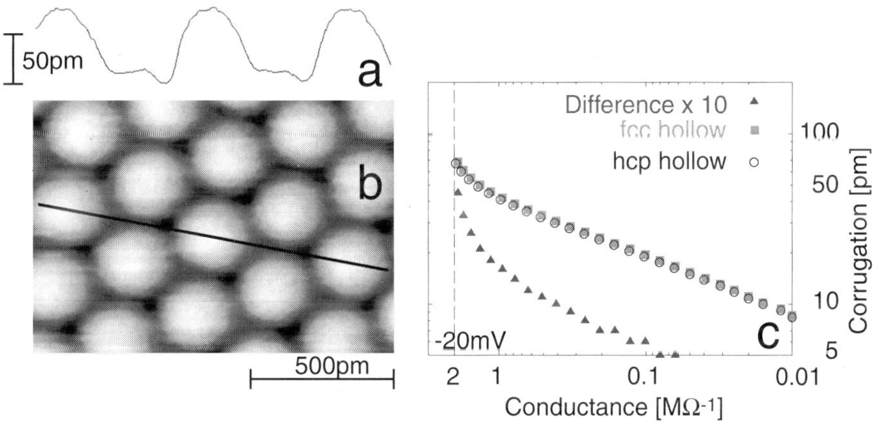

Fig. 7.11. Resolution of the position of subsurface atoms. The atoms of the subsurface layer are located at the hcp hollow site, which shows a slightly higher contour (about 5 pm) than the fcc hollow sites, where subsurface atoms are missing. (a) Experimental constant current profile along the $[11\bar{2}]$ direction of the Al(111) surface including hcp and fcc hollow sites. (b) Experimental image, with current profile indicated. (c) Simulated corrugation values and difference between hcp and fcc hollow sites.

Resolving the position of subsurface atoms

The high resolution and the excellent agreement between experiments and simulations allow to take the theoretical model one step further and determine the apparent height of two different hollow sites: on an fcc (111) surface like aluminum every other hollow sites is above an atom in the subsurface layer. In this case one could expect that the long range of wavefunctions into the vacuum might make it possible to resolve the position of subsurface atoms by their effect on the tunneling current. In Fig. 7.11 (left) we show the experimental image obtained by H. Brune in 1989, which clearly allows one to differentiate between the fcc hollow (no atom in the subsurface layer) and the hcp hollow (atom in the subsurface layer). Analyzing the experimental results one finds that the difference of about 5 pm is in the same range as the total corrugation based on a charge density contour. In this case there is simply no question of explaining the result from the electronic structure of the surface alone. The results of experiments and simulations are shown in Fig. 7.11. The experimental image taken at -20 mV is presented on the left, the simulated corrugation values and the difference between the fcc and the hcp hollow site on the right. Given that the experiment was performed at a conductance of 2 $M\Omega^{-1}$, we find excellent agreement between experiments and simulations.

7.2.3 Cr(001)

Chromium is a very hard metal which is used as surface coating in the industry. This is due to its durability under various chemical and thermal conditions. From this fact alone one could conclude that it will be most difficult to measure in an STM experiment, because the reactivity of a surface will be related to the decay of its surface charge: if the reactivity is very low, one expects that the interaction with molecules at the gas phase will also be small and that, consequently, the achievable tunneling current will be at the lower limit. Simulations of the surface confirm this preliminary understanding.

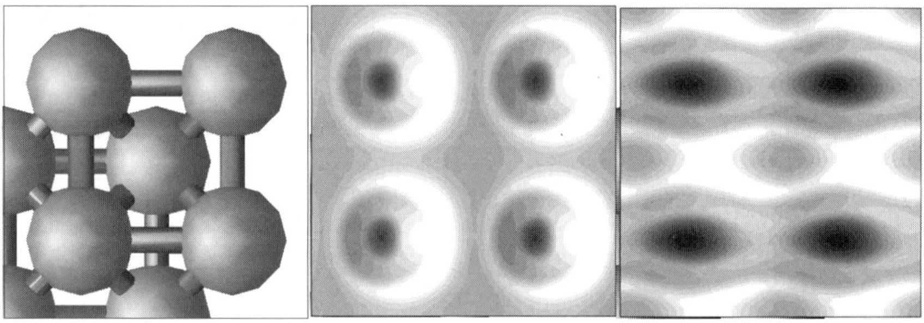

Fig. 7.12. Cr(001). Atomic positions (left), charge density contours at a distance of 3.3 Å(center) and constant current contours (right) with a tungsten tip model, at -50 mV and 1 nA. The protrusions appear at the hollow positions of the surface. The surface corrugation is very low (about 3 pm), the contrast of the current contour is equal to the contrast in the density contour.

The Cr(001) surface was simulated by standard DFT methods. To account for magnetic properties we used a projector augmented wave method. The individual layers show, as expected, anti-ferromagnetic ordering ; the magnetic moment of the surface layer is about 2.1 μ_B. This is well in line with simulations done a few years ago using a full potential code [31]. To calculate topographic images we simulated constant current contours for a bias voltage of -50 mV. The result of the simulation is shown in Fig. 7.12. The main problem, from an experimental perspective, is the low current (we obtain a maximum of only 2 nA in the simulation) and the high stiffness of the surface. Due to the low current, the tip has to approach the surface to the point of destruction in order to obtain substantial relaxation effects. The conclusion from the simulation seems to be clear: Cr(001) cannot be measured with atomic resolution, unless the tip is functionalized. Simulations with a functionalized tip are presented in the following sections. In case of Cr(001) we obtain in the charge density contours and the current contours the same contrast: the depressions in the density contours at distances above 3 Å correspond to depressions in the current contour: they indicate the atomic positions.

7.2.4 Fe(001)

If one considers the relation between a density contour and a current contour in simulations on Fe(001), it is clear that the situation might not always be as simple as for Cr, and considerable ambiguity may exist. The surface is notoriously difficult to resolve, and at present there is no clear understanding of how magnetic properties and lattice parameters are related. For a survey of recent work on ordered Fe(001) layers and their magnetic structure see for example [32]. Our aim at this point is to show that a certain level of complexity in the electronic structure, combined with very small corrugations, may lead to completely unpredictable results, if only the charge distribution above the surface is considered.

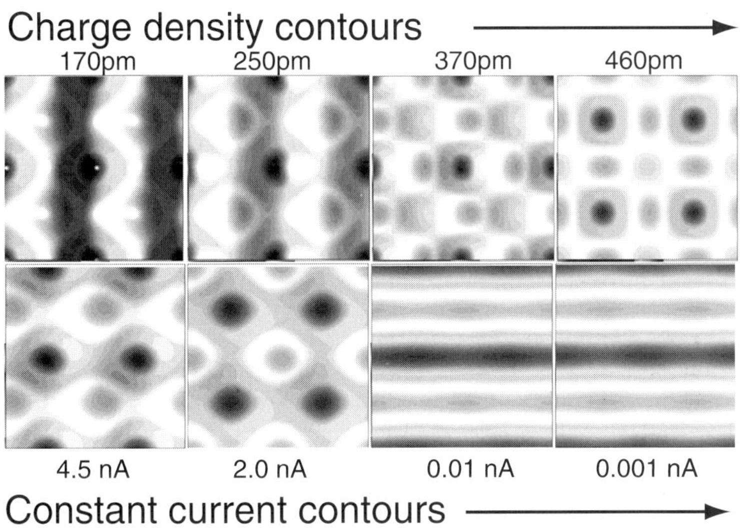

Fig. 7.13. Fe(001). Charge density contours at selected distances (top), and constant current contours from 1 pA to 4.5 nA (bottom). The charge density contour is only corrugated at very close distances below 300 pm (from 30 pm to 6 pm in the two frames), in this range the Fe atoms are revealed as protrusions. In the range above 300 pm the picture becomes somewhat difficult to interpret as the atomic positions are now minima, while the maxima of the density contours are at the bridge sites. Corresponding constant current contours show a negative corrugation only at one current value (2 nA), while the minima of the contour are at the bridge sites for higher currents (4.5 nA). Even at high currents the corrugation is well below 2 pm.

To this end we simulated a clean Fe(001) surface, relaxing the surface atoms in the process until the forces on surface and subsurface atoms were less than the usual threshold (0.01 eV/Å). Calculating a charge density contour above this surface we note that its actual shape depends very much on the distance from the surface. While the positions of the Fe atoms below a distance of 3 Å clearly shows up as a protrusion, it changes to a depression from this range to about 4.6 Å. Furthermore, the relative height of bridge and hollow sites in this range also changes, so that, from the viewpoint of a charge density contour, one even might obtain something akin to a rotated $(1/\sqrt{2} \times 1/\sqrt{2})$ unit cell. The corrugation of these electronic structures, however, is very low and does not reach 2 pm in the distance regime of STM operation. The constant current contours with a tungsten tip model are no less ambiguous. At very low distances the highest point of a contour at 4.5 nA is at the bridge sites of the Fe(001) surface. Decreasing the current to 2 nA this point is shifted to the hollow position between the Fe atoms. For currents lower than about 0.1 nA, and considering that the symmetry of the (110) tip will have an effect on the contours, it can be seen that the highest points are now corresponding to the positions of the Fe atoms. Since the corrugation in this distance range is already very low, this protrusion will probably not be detected, so that the surface appears essentially flat. The main point here is that there is no clear correlation between a density contour and a current contour. This fact should alert experimenters to the danger of overinterpreting STM images, in particular if the surface under consideration is hard and at the limit of STM resolution .

7.2.5 Metal alloys: PtRh(001)

Metal alloys are important for materials with a low thermal expansion and for applications in catalysis. In the first case the different expansion coefficients of different metals can close to cancel each other so that a material does not expand or contract over a large thermal range. In the second case the specific adsorption and dissociation properties of metals may even be improved if these metals form an alloy with a substrate matrix. Examples of the second category would be e.g. rhodium, or tin. From the viewpoint of STM experimenters the ability to differentiate the chemical species in STM scans adds additional information about chemical processes, since the position of adsorbates and dissociation products can then be analysed not only in terms of their position on the crystal matrix, but also with respect to advantageous positions at the boundaries of the alloy components .
Experimentally, differentiating between different metallic species on metal surfaces became possible in the early 1990's [33]. Theoretically, it was only established several years later, that the main parameters in experimental scans should be (i) the chemical composition of the STM tip; and (ii) the confinement of electron states of one species due to alloying [34]. The second feature leads to an enhancement of the density of states at the Fermi level very locally: this enhancement is then detected by STM. However, this also requires a suitable STM tip. Clean tungsten tips, as should be clear from the preceding sections, do not generally pick up the corrugation of the electronic surface structure. This is due to the convolution of states with non-radial symmetry,

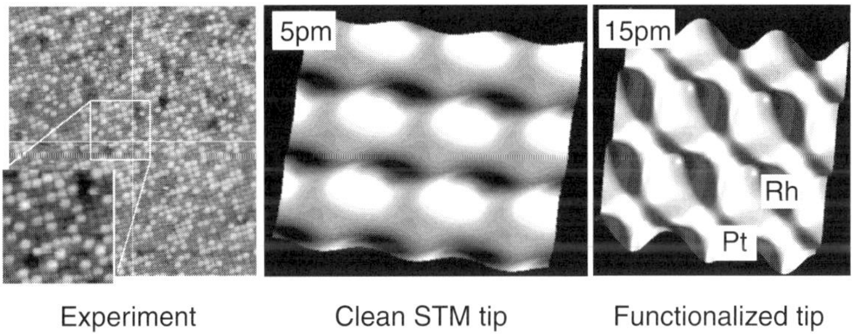

Fig. 7.14. PtRh(001). Current density contours at -100 mV/0.5 nA for a clean tungsten tip (left), and a tip contaminated by a rhodium atom (right). The chemical contrast between Rh and Pt (see labels in the right frame) is 15 pm with a functionalized tip and only 5 pm with a clean tip. Moreover, the Pt positions cannot be resolved with a clean tungsten tip. P. T. Wouda, B. E. Nieuwenhuys, M. Schmid and P. Varga, Surf. Sci. 359, 17 (1996). Copyright (1996) by Elsevier, reprinted with permission.

which may overlap with surface states even if the tip is at a different lateral position, as well as the high number of states contributing under typical tunneling conditions (in the simulations this number is generally higher than about 50, even at very low bias voltages).
In Fig. 7.14 we show the experimental scan, the simulated scan with a clean tungsten tip, and the scan simulated with an Rh contaminated STM tip. Only in this case does the corrugation of the simulated contour agree with the epxerimental value. This seems to point to a method frequently employed by experimenters to increase the contrast on a surface: the tip is crashed into the surface and picks up atoms of the surface material in the process. The reason this seems to work quite frequently is probably an increase of the tip states near the Fermi level, as the electron states of the additional material on the STM tip are confined to a relatively small space in case there is no extensive hybridization with tungsten states. As additional calculations reveal, the chemical nature of the contaminant is important. A Pt contamination of the tip leads, like in case of the clean tip, to a much smaller contrast and consequently lower resolution of the chemical surface structure .

7.2.6 Magnetic surfaces: Mn/W(110)

It is quite fashionable to explain the interest in magnetism , especially on the atomic scale, with the giant investments taken in the computer industry to produce reliable and small scale storage devices [35]. Quite apart from this application viewpoint, there is also a scientific interest in the way magnetic properties change with a change of the physical environment. So far, experimental research in this area has been hampered by the low resolution of existing methods (in the range of about 50 nm [36]). The combination of experiments with a spin-polarized STM or SFM, and refined theoretical models greatly enhance the possibility and quality of data at this extreme limit of resolution.

The STM tip in the experiments consisted of a tungsten polycrystal (paramagnetic tip) coated with ten to twenty layers of iron (ferromagnetic tip) . In addition, contamination of the tip by atoms of the sample surface (manganese) cannot be excluded, especially in view of the high tunneling currents of about 40 nA at very low bias voltages of 3 mV [37]. For this reason a number of separate tips have to be included in the analysis [38]. The most important ones are: a clean Fe tip, mimicking the polycrystal W wire coated with Fe; a Fe tip contaminated by a single Mn atom (low contamination of the tip); a Fe tip contaminated with a surface layer and a single Mn atom (high contamination of the tip). On the technical side we note that the free standing film consisted of five layers with (100) ordering and two additional layers for the apex. The STM tip models are displayed in Fig. 7.15.

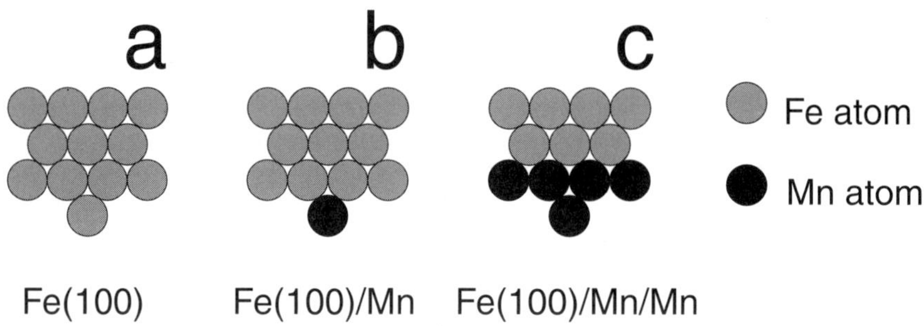

Fig. 7.15. STM tip models for spin-resolved measurements. The tip model consists of a five layer Fe(100) film with (a) a single Fe apex atom, (b) a single Mn apex atom, or (c) a Mn layer and a single Mn apex atom. These models mimick a clean ferromagnetic tip or a tip contaminated by surface atoms.

The angle ϕ_M between the magnetization vectors of surface and tip is in general unknown in the experiments. Therefore images for all possible angles have to be simulated for a comparison with experimental images. But this also means that a unique map from angles ϕ_M to corrugation amplitudes can be used to determine the actual angle from the apparent height of the atoms on the surface. We omit displaying the simulated image for the paramagnetic tip model, it is published in [39].

In practice two separate simulations were performed for every single tip model and the antiferromagnetic Mn overlayer: one simulation for ferromagnetic ordering in the tunneling transitions ($I_F(x, y, z)$), and one for antiferromagnetic ordering ($I_A(x, y, z)$). The two separate current maps were then compiled into a single image by defining an angle ϕ_M from the outset. In Fig. 7.16 we show the images with three different tips at a median distance of 450 pm (lower limit of stability).

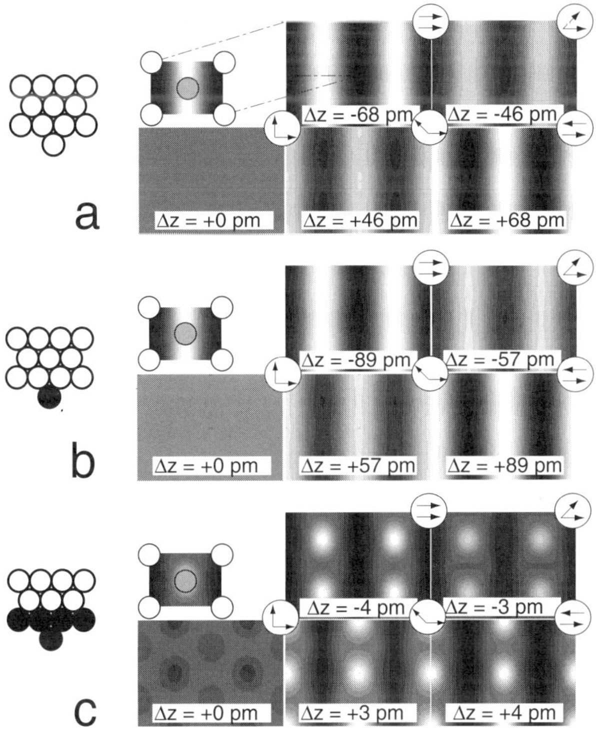

Fig. 7.16. Simulated STM images of W(110)Mn for three different STM tip models and a range of angles ϕ_M between the magnetic axis of sample and tip. The simulations with a clean tip (a) and a slightly contaminated tip (b) reveal a surface corrugation well in excess of experimental values, while the highly contaminated tip provides the best agreement with experiments (c). W.A. Hofer and A.J. Fisher, JMMM 267, 139 (2003). Copyright (2003) Elsevier, reprinted with permission.

The most remarkable result is the strong dependence of the apparent height of single atoms on the contamination of the STM tip. This is most obvious for the transition from a tip with low contamination (one Mn atom on a Fe(100) surface) to a tip with high contamination (one Mn atom and a surface layer on the surface). The position of individual atoms is only resolved in the simulated images with a highly contaminated surface. The decrease of the apparent height between the atoms has also been observed in the experiments. This is not the case for the clean Fe-tip and the tip with low Mn-contamination. In those cases the relative variation of the constant current contour is too low to be observable. The results prove once again, that tip contamination plays a crucial role in the quantitative results obtained in STM experiments. The second point of interest, especially for experimentalists, is the high magnetic contrast of the surface. Considering that close packed metal surfaces like Mn or Fe are notoriously difficult to image, a coating of the tip by magnetic layers may improve the contrast by more than one order of magnitude. It is also evident that the magnetic contrast vanishes, if the magnetic axes of the two surfaces are perpendicular. This entails a strong dependency of the magnetic contrast on ϕ_M, which in turn can be used to study the effect of impurities on the atomic scale and in real space.

7.3 Silicon (001)

The surface of Si(100) reconstructs in dimer rows along the (011) direction, the Si-Si dimer bond is 2.2 Å long, adjacent dimers are 3.8 Å apart [40]. The dimer reconstruction was subject of intense dispute around 1990, since photoemission spectra suggested a buckled dimer [41], while STM images clearly revealed a flat dimer structure [42]. The riddle has been solved by a combination of experimental and theoretical techniques. Experimentally, it was realized, that a tilted dimer in fast flip-flop motion would appear flat in STM images due to the low time-resolution of the STM. At temperatures below 90K the motion of dimers is frozen, individual dimers under these conditions appear tilted, as Wolkow showed in 1992 [43]. The same feature is observed if the buckling is pinned down by surface defects. The additional information, gained by STM simulations under zero temperature conditions, compared to charge density contour plots (see Fig. 7.17) is the exact distance range under experimental conditions (see Fig. 7.18). We also note that the agreement between the shape of the current contours in STM experiments and simulations is improved significantly.

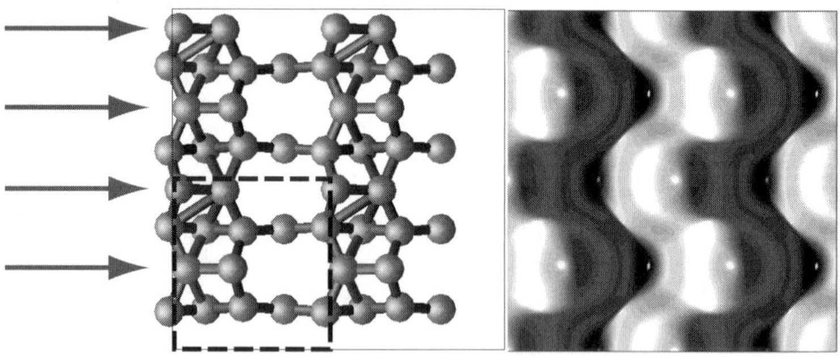

Fig. 7.17. Simulation of the buckled Si(001) surface. Adjacent dimers in one row are buckled in the opposite direction (left). In this case we simulated a 2×2 unit cell, which leads to the same buckling in adjacent rows. The charge density contours show that only one of the Si atoms is actually visible (right).

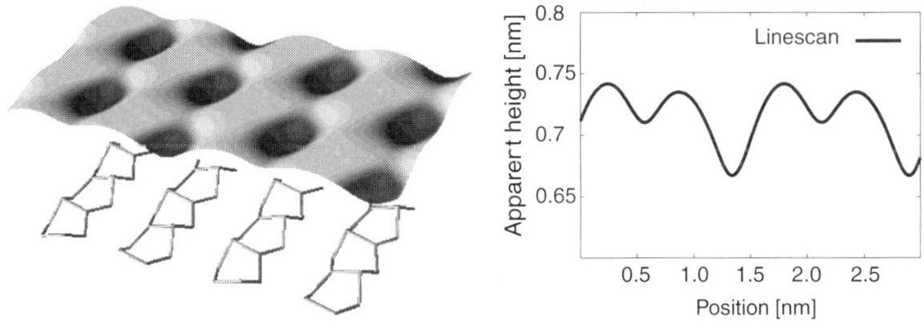

Fig. 7.18. Constant current contour plot for a bias voltage of -2 V and a current value of 50 pA. Adjacent Si-dimers are buckled in the opposite direction, the zig-zag pattern as well as the apparent height of about 0.6 to 0.8 Å is confirmed by experimental data [43].

It seems that the question of dynamic buckling is still to some extent discussed in the literature, even though the variable temperature experiments seemed to have proven beyond doubt that the flat dimer structure in room temperature experiments is a dynamic effect. Given the large distance between tip and surface, the assumption of current induced buckling [44] or buckling due to tip-surface interactions [24] lack experimental and theoretical confirmation.

7.3.1 Saturation of Si(001) by hydrogen

A silicon surface is very reactive. This is due to the dangling bond of the Si-dimer atoms , which reaches far into the vacuum and thus provides an adsorption site for atoms and molecules in the gas phase. The extent of the dangling bond can be estimated, if the silicon surface is saturated by hydrogen. In this case the electron charge in the vacuum range is substantially reduced. Saturating the whole surface with hydrogen by deposition from the gas phase leaves a basically inert surface. However, if a single hydrogen atom is removed from the surface by an STM tip, then the surface contains only one specific adsorption site for molecules. This fact can be used to position a molecule very accurately on the surface, furthermore, if a chemical reaction is induced, which removes another hyrogen atom from the surface during adsorption, then a self directed growth process with, in principle, a well defined growth direction can be initiated. Just why a dangling bond is so reactive can be seen from simulations of charge density and constant current contours. As the current at a specific location is proportional to the interaction energy (see previous chapters), we can study the effect by analysing simulated density and current contours.

To this end we simulated a 4×6 Si(001) unit cell, where all but one of the silicon atoms were saturated by hydrogen. The size of the cell is necessary to avoid an overlap between neighboring dangling bonds. The setup of the

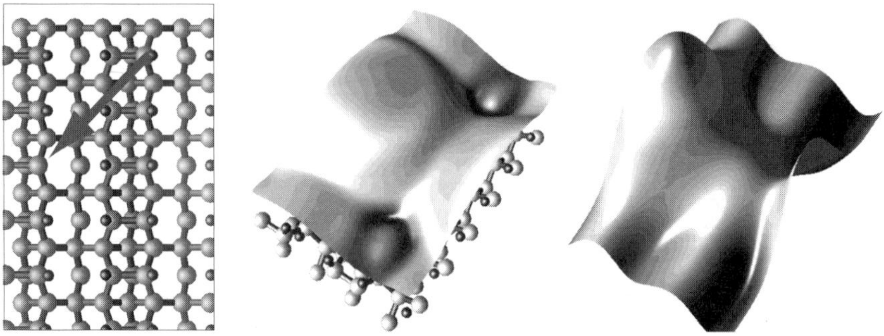

Fig. 7.19. Silicon (001) surface saturated by hydrogen but for a single location. This location, the dangling bond, is marked by a red arrow(left). A constant charge density contour shows the local extent of the dangling bond, which covers an area of about 1nm×1nm (center). A constant charge density contour has a different shape than the current contour, the apparent height of the dangling bond at a bias voltage of -2 V and a current value of 50pA is about 1.5 Å(right).

unit cell is shown in Fig. 7.19. It can be seen that the additional hydrogen atom, due to its change of the surface charge distribution removes the buckling of the surface, which consists now of flat dimers saturated by hydrogen. The charge density contour (center) contains only the charge of the dangling bond, which is situated in the bandgap of Si(001), somewhat below the middle (it is 0.6 eV above the valence band and 0.8 eV below the conductance band in simulations with standard DFT codes). The constant current contours also contain to some extent the contributions from the valence band of the surface. But as the vacuum of the saturated surface contains only very little charge, compared to the clean surface, these contributions should be minor. However, we observe a change of shape of the dangling bond: the peak becomes narrower and higher than the peak in the density contours. We attribute this effect to a genuine tip effect: as the overlap is a maximum, if the tip is centered at the position of the dangling bond, the slope of the protrusion must necessarily increase once the STM tip is included in the simulation. The apparent height of the dangling bond under normal tunneling conditions (-2 V/50 pA) is about 1.5 Å.

7.4 Adsorbates on Si(001)

Since the surface of Si(001) is so reactive it has been used as a template for studying adsorption processes . The additional advantage of silicon is that the covalent bonds are very localised and that diffusion barriers for the propagation of molecules from one adsorption site to another are forbiddingly high. It is therefore possible to study most processes under ambient conditions, while the large apparent height of the silicon dimers makes it possible to determine the location, the conformation, and the exact bonding site with great precision.

7.4.1 Acetylene C_2H_2 on Si(001)

The simplest hydrocarbon molecule is acetylene CH≡HC, which in vacuum possesses a triple carbon carbon bond. If this molecule attaches to a clean silicon surface, it has essentially two options: it can either adsorb on tip of a silicon dimer, the C-C bond in this case reduces to a double bond; or it can attach to two adjacent dimers, if the C-C bond is reduced to a single bond. There was some controversy a few years ago, about the preferred adsorption site. Different methods seemed to reach a different conclusion concerning the actual adsorption geometry under different thermal conditions (for an outline of the discussion, see [45]). There were essentially two diverging opinions: (i) There are only two adsorption sites, one on top of an Si dimer (called a cyclo-additon reaction), and one midway between two dimers. (ii) There are three adsorption sites, one on top of the dimer; one midway between two dimers, even though the orientation of the molecule - the C-C bond either parallel or perpendicular to the dimer rows - was under discussion; and a third one, which showed the same depression as the second one, but in addition an asymmetric feature (these three sites are shown in frame (A) of Fig. 7.20).
The main question, which arose from STM experiments, was the nature of the difference between the two adsorption sites, covering the area of two silicon dimers. In the experimental images, these two sites are clearly distinguished (feature II and III, see Fig. 7.20(A)). Since the C-Si bonds of organic molecules on silicon are very localized, the electronic structure remains quite unperturbed at short distances from the adsorption site. This makes it possible to use a relatively small unit cell. But in addition, the silicon lattice is very elastic. If, therefore, a molecule induces strain in the silicon lattice, the strain will shift Si atoms out of their groundstate position. The energy differences, arising from lattice strain, can be quite substantial and reach in specific cases values of about 0.5 eV. Together with the slight differences from exchange-correlation potentials and energy cutoff and k-space sampling, this makes for a large variety of adsorption energy values found in the literature (for a compilation, see [45])
Here, we are mainly concerned with topographic images and the comparison between experiments and theory. It can be seen that the simulated image (C) 1, of Fig. 7.20 agrees well with the experimental image. A thorough analysis also revealed that the apparent depression of about 0.3 to 0.4 Å is in line with experimental findings. The rotated configuration (C) 2 does not seem to appear in experiments, presumably because the adsorption energy in this case is lower by 0.1 eV. Concerning the adsorption sites with two-dimer footprints, the simulated images show a deeper depression than in the first case, but due to the small size of the unit cell, the question whether there is a difference

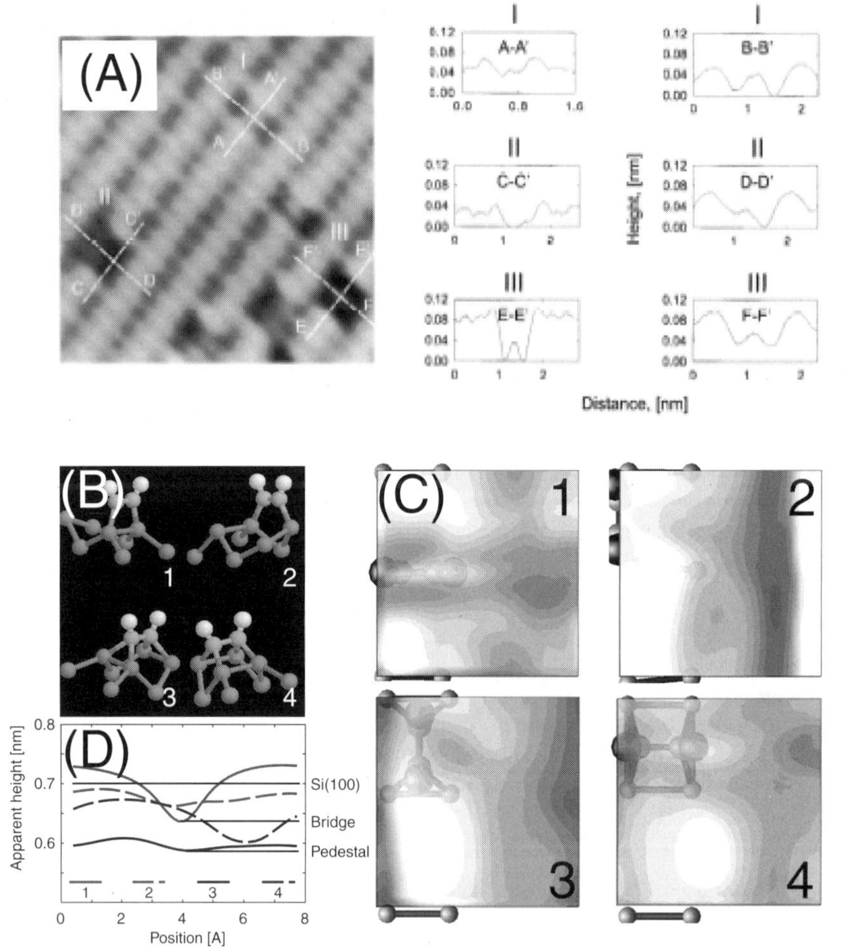

Fig. 7.20. Adsorption of acetylene on Si(001). The experimental STM scans show three different adsorption configurations, labelled I-III (A). They are due to the possibility of rehybridization of the carbon bond to either a double bond (configurations (1) and (2) in frame (B)), or to a single bond (configurations (3) and (4)) in frame (B)). The resulting STM images (frame (C)) in the simulation agree quite well with three of the configurations found in the experiments ((A), features I, II, and III). S. Mezhenny, I. Lyubinetsky, W. J. Choyke, R. A. Wolkow and J. T. Yates, Chem. Phys. Lett. 344 7 (2001). Copyright (2001) Elsevier, reprinted with permisssion.

between configurations 3 and 4, which would in one case show up as a slight asymmetry in the calculated constant current contour had to be left open [46]. Recently, the question was taken up by another group, which used a slight modification of the Tersoff-Hamann approach to calculate the STM contours, obtaining quite similar images to the ones presented here [47]. The authors interpreted features II and III of the experimental scans in a quite different manner: feature II supposedly arises from an end-bridge configuration (see (C)2 of Fig. 7.20), while feature III is thought to be due to two acetylene molecules adsorbed at the same dimer. From an energetic point of view, the interpretation is tempting, since it removes the problem of the large difference in adsorption energies between the bonding configurations (about 1 eV [45]), which makes it quite unlikely that the two species could exist in the same thermal environment for the interval it takes to perform an experimental scan. From the viewpoint of STM experiments, it is far less convincing, since the depression in features II and III of the experimental scans is much larger (about 0.8 compared to 0.4 Å) than that of feature I. It has to concluded, that at present the experimental features cannot be uniquely assigned to specific adsorption geometries.

7.4.2 Benzene C_6H_6 on Si(001)

While acetylene is the smallest organic molecule, benzene is the smallest molecule with a ring-like structure: its carbon ring is the building block of many organic molecules used in chemical synthesis. The carbon ring also provides a ready signature in STM images, because the delocalized π-electrons above and below the carbon nuclei provide the main overlaps with STM tip wavefunctions. These features have made the study of benzene quite attractive, and a large number of experimental and theoretical papers describe the adsorption of benzene on many metal and semiconductor surfaces (a survey from the experimental point of view can be found in [48, 49]).
Acetylene, as shown above, can attach to one or two dimers of the silicon surface. This feature is linked to the rehybridization of the carbon-carbon bond. In benzene, each carbon atom is attached to two neighboring atoms and a hydrogen atom. This leaves only one electron per atom, which is either delocalized, or can form a double bond, or it bonds to the dangling bond of a silicon surface. Due to the geometry of the molecule, which has a diameter of about 5 Å, it cannot attach to two adjacent silicon dimer rows. As the diameter across the carbon ring is about 3 Å, it will induce considerable strain into the silicon lattice, if it adsorbs in a configuration, where its central axis is parallel to a surface dimer. This makes it clear that the adsorption sites and their energetics are somewhat limited by the shape of the molecule itself. Consequently, one observes only three adsorption sites: (A) The 'butterfly', where the molecule straddles a single dimer; (B) the 'Tight Bridge', where it attaches to two adjacent dimers and the part of the ring, which remains unbonded, is

tilted upwards; and (C) a rotated 'Tight Bridge' (see Fig. 7.21(left)). Energetically, the 'Tight Bridge' site is favored by about 0.3 eV [40], which means that the 'Butterfly' will be transformed into a 'Tight bridge' quite rapidly. It should thus be the exception, rather than the rule, that both features can be observed in the same experiment. Benzene shows up as a protrusion in STM experiments. This is contrary to the result for acetylene. The reason is the size of the molecule. While single atoms like oxygen, or small molecules like acetylene deplete the surface charge of the contributions due to either dangling bonds (silicon) or surface charge (metals), they do not possess enough delocalized charge to lead to a substantial overlap with tip wavefunctions. The main effect is thus the reduction of charge. Benzene, however, possesses a ring of delocalized π electrons, which overlap with tip states; this ring is, moreover, substiantially elevated compared to the substrate surface.
STM simulations reveal one interesting difference to STM experiments: while on metals it is usually found that the simulated corrugation values are at the lower end of the experimental results, they are substantially higher than measured values on this surface (see Fig. 7.21(right)). To date, the reason for this difference is not quite clear. Disregarding the potential effect of a too small unit cell, which is evident from the constant current contour of the tight bridge, it seems the most likely origin of this deviation is either the interface between molecule and silicon substrate, or due to neglecting the bias dependency in these calculations (see the modifications of the Bardeen equation, if it is derived from the Keldysh formalism, at the end of Chapter 5).

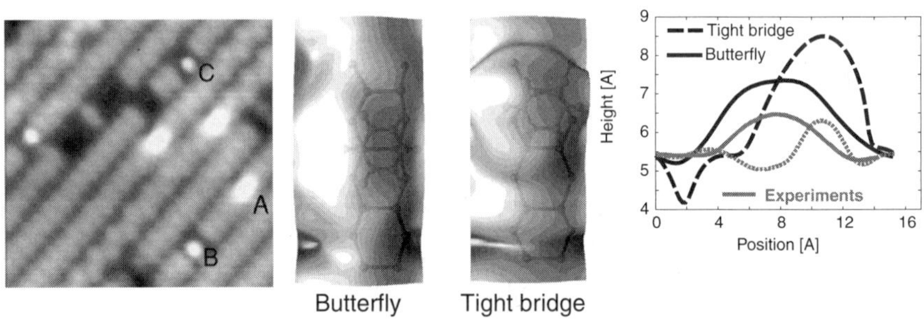

Fig. 7.21. Benzene C_6H_6 on Si(001). One observes three distinct adsorption sites for the molecule in experimental scans (left): (A) butterfly, (B) tight bridge, (C) rotated tight bridge. Two of the configurations have been simulated (center), the ensuing line scans agree qualitatively with the STM images, the actual corrugation values are, however, too high (right).

7.4.3 Maleic anhydride $C_4O_3H_2$ on Si(001)

As a final example, let us consider the adsorption of a highly polar molecule on silicon. Here, it was observed in STM experiments, that maleic anhydride adsorbs predominantly in the troughs between silicon dimer rows [50, 51]. In this case the energy component resulting from the strain of the silicon lattice plays a major role in the preferred adsorption site. As the lattice strain depends strongly on the coverage, the ensuing distributions of above trough and above dimer adsorption sites can change substantially with a variation of coverage.

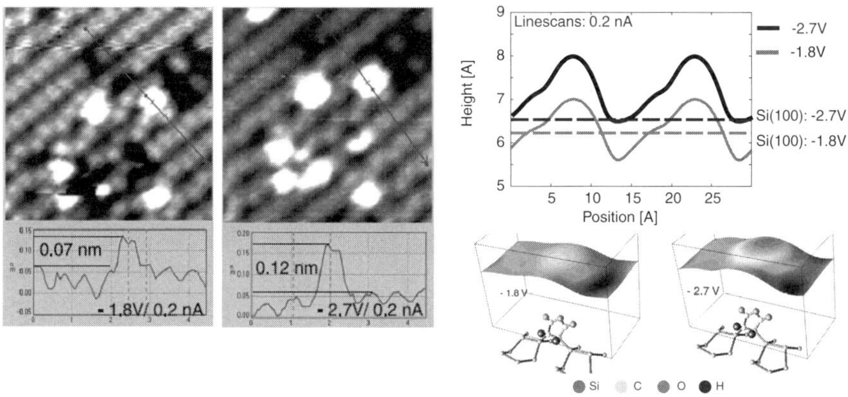

Fig. 7.22. Maleic anhydride on Si(001). Experimental images of scans at -1.8 V and -2.7 V, repsectively (left). Current contours for the experimental values (right, bottom), and linescans across two unit cells (right, top). The increase in this range is due to only one molecular state. W.A. Hofer, A.J. Fisher, T. Bitzer, T. Rada and N.V. Richardson, Chem. Phys. Lett. 355, 347 (2002). Copyright (2002) Elsevier, reprinted with permission.

The STM images show a protrusion by 0.07 nm (-1.8 V) and 0.12 nm (-2.7 V) at the position of the molecule (see Fig. 7.22(left)), which is well reproduced in the simulations (Fig. 7.22(right)). Please note that the linescans in the figure are for two adjacent unit cells. In this case the interesting feature in the scan is the increase of the molecular height by 0.05 nm if the bias voltage is increased by 0.7 V. The silicon surface itself will possess states in this energy range, so that the total contribution of the surface will be slightly enhanced and the current contour is about 0.03 nm higher (Fig. 7.22(right)). But the molecule itself increases its height by nearly double this amount. As a detailed analysis of the electronic structure of the molecule shows, this large incrase is due to only a single molecular state. In effect, passing the threshold of -2.5 V an additional state comes into play, which lights up the molecule's position. Passing the threshold, one therefore tunes into a single molecular state [50].

7.5 Titanium dioxide (110)

The importance of titanium dioxide (TiO_2) in a wide variety of applications, from photocatalysis to biomedical implants [52, 53, 54], has led to a considerable research effort to understand its properties. In most of these applications, it is TiO_2's surface properties that determine its behavior and hence the surface has developed into a benchmark transition metal oxide surface for studying many different processes [55]. The most stable (110) surface is characterized by rows of oxygen atoms bridging titanium ions (see Figure 7.23). The basic physical and electronic structure of the surface has been well studied both experimentally [56, 57] and theoretically [58, 59, 60, 61, 62], and now many investigations focus on defective surfaces, especially oxygen vacancies [63, 64], adsorption [65, 66, 67, 68], or even adsorption onto defective surfaces [69, 70].

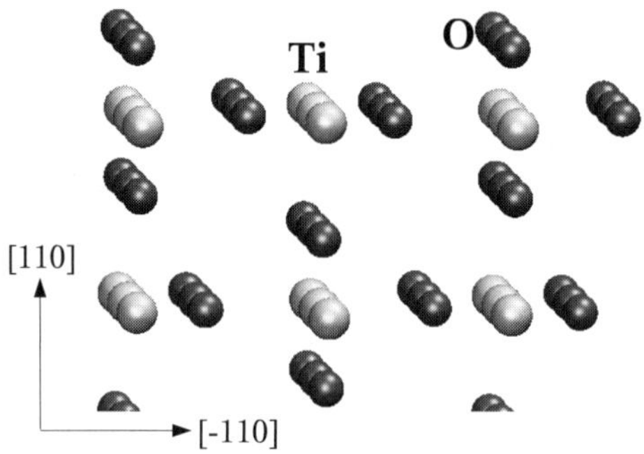

Fig. 7.23. Atomic structure of the TiO_2 (110) surface.

The tool of choice for studying such local processes on surfaces is SPM. Although in principle an insulator, TiO_2's small band gap (3 eV for the stoichiometric surface) means that it is accessible to both STM and SFM. Atomic resolution has been achieved on the (110) surface in both STM [56] and STM [71]. For STM, the source of contrast in images was identified through extensive cooperation between theory and experiment, identifying Ti atoms (see Figure 7.23) as the tunneling sites [56]. The first atomically resolved SFM images of TiO_2 [71] were simply interpreted based on the concept that the force between tip and sample was largely element independent, and therefore the protruding oxygen rows should appear as bright as they are closer to the tip; the reverse of STM images. Figure 7.24(a) shows an example SFM image of the surface.

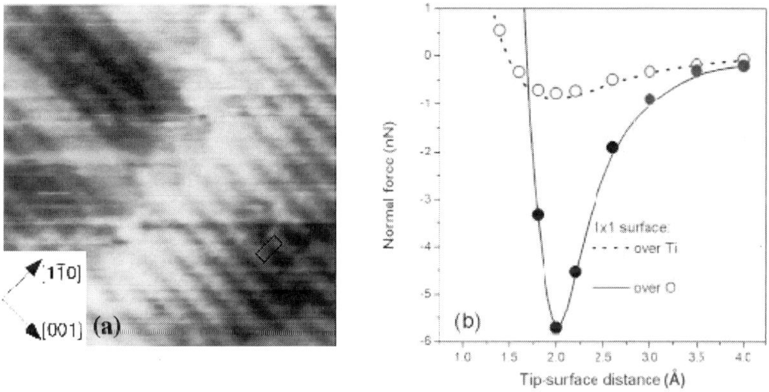

Fig. 7.24. (a) Experimental SFM image of the TiO_2 (110) surface ($A_0 = 15$ nm, $f_0 = 260$–290 kHz, $k = 26$–32 N/m, $\Delta f = -80$ Hz). Reprinted with permission. Copyright 1997 by the American Physical Society [71]. (b) Calculated force curves over the in-plane Ti and bridging O sites in the surface. Reprinted with permission. Copyright 2002 by the American Physical Society [72].

7.5.1 Simulations of ideal and defective surfaces

As mentioned in Chapter 6, initial studies of the TiO_2 surface used a silicon probe [72]; specifically a hydrogen terminated one atom Si tip. They found that the force over the bridging oxygen sites was much larger than over the Ti sites (see Figure 7.24(b)), in agreement with experimental speculations. However, the absence of any quantitative comparison between theory and experiment, and the very approximate tip model, means that the story remains incomplete. Again, despite the fact that the tips in these experiments are normally sputtered, it is difficult to believe that the tip apex is always pure silicon and a comparison of different tip models would be very useful.

This was attempted in a more extensive work considering three different tip models interacting with both the ideal and vacancy-defective surfaces [73]. The authors of that work found that the experimental contrast pattern could be reproduced with a larger silicon tip, and both an O- and Mg-terminated MgO tip (see Chapter 6). For the O-terminated tip, the contrast was reversed with respect to the other tip models, and the titanium ions were imaged as bright (see Figure 7.25(a)). Although the work confirmed the belief that oxygen vacancies would be seen as dark on bright oxygen rows (see Figure 7.25(b)), interpretation of atomically resolved images was again shown to be critically sensitive to the nature of the tip.

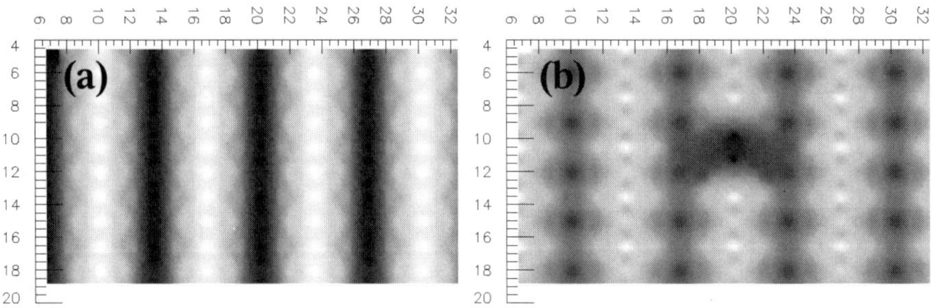

Fig. 7.25. Simulated images of the TiO_2 (110) surface: (a) an O-terminated MgO tip imaging the ideal surface; (b) a silicon tip imaging a vacancy-defective surface. Reproduced with permission. Copyright 2003 by the American Physical Society [73].

7.5.2 Acid adsorption on the TiO_2 (110) surface

Due to their particular relevance to catalysis, many studies have investigated the properties of adsorbed carboxylic (RCOOH) acid layers on the TiO_2(110) surface. The simplest member of this acid group, formate (HCOOH), has been studied extensively [74, 75, 76, 77, 78]. It undergoes a dissociative reaction upon adsorption into a carboxylate ion and a proton ($RCOOH \Rightarrow RCOO^- + H^+$). Some experimental [79, 80] studies on acetate (CH_3COOH) adsorption have also been performed, and the results suggest that the molecule also dissociates at the surface. Recent theoretical work supports this for both acetate and trifluoroacetate (CF_3COOH) [81], and confirms that the molecules adsorb to the surface in a symmetric bridging structure bonding to in-plane Ti atoms (see Figure 7.26).

This body of work has also motivated a study into whether acid molecules can be used to "mark" the TiO_2 sublattices, and hence yield an interpretation. A combined STM/SFM investigation of a formate (HCOOH) monolayer on the TiO_2 (110) surface [76] used the clear understanding of STM images to interpret the SFM images. Although the quality of images was not high, the experiments gave reasonable evidence that the bridging oxygen rows (see Figure 7.23) were imaged as bright in SFM. This prompted an extensive D-SFM study [82, 83, 84, 85, 86, 87] of both acetate and trifluoroacetate layers on the surface (see Figure 7.27). This remains the only fully systematic study of adsorption in atomically resolved D-SFM, and is also an important general study of imaging organic layers with this emerging technique.

Figure 7.27(b) shows an experimental image of a mixed monolayer containing both acetate and trifluoroacetate on the surface. The experiments assumed dissociative adsorption for both acids, and further, that since the molecules are of similar size, the adsorbed molecules would be equivalent in height. Hence, one might naively expect that the mixed layer would appear the same as images of a uniform acetate layer [76]. This was not the case, since the images

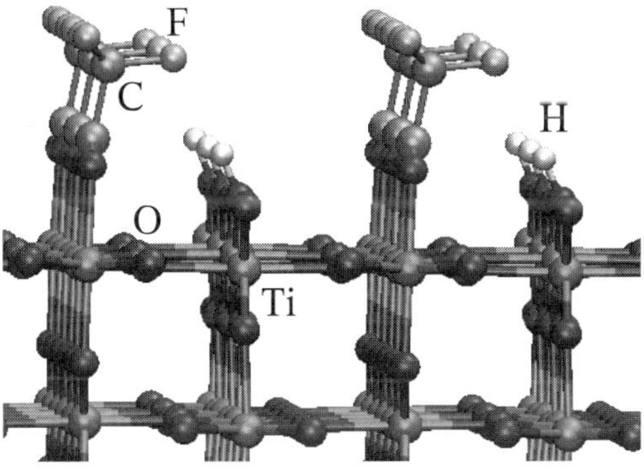

Fig. 7.26. Calculated atomic structure of trifluoroacetate adsorbed onto the TiO_2 (110) surface.

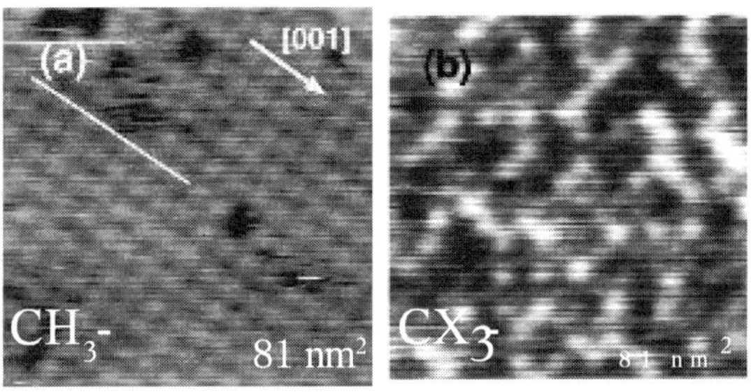

Fig. 7.27. Experimental SFM images of the TiO_2 (110) surface covered by (a) an adsorbed acetate monolayer and (b) an mixed monolayer of acetate and trifluoroacetate (A_0 = 3.4 nm, f_0 = 310 kHz, k = 14 N/m, Δf = -75 Hz). Reproduced with permission. Copyright 2001 by the American Physical Society [82].

effectively contained two magnitudes of bright contrast spots, the brightest matching the dose of acetate and the less bright the dose of trifluoroacetate, since in the absence of strong covalent bonds between the tip and surface, the force is strongly dependent on the interaction between the tip and surface electrostatic potential. Hence, the authors [82] speculated that the difference in contrast over molecules is due to the difference in dipole moment of the two species. Calculations of a similar system [81] showed that the dipoles of acetate and trifluoroacetate adsorbed on TiO_2 are in fact opposite, and a tip–surface interaction dominated by the dipoles could explain the experimental images. However, experimenters assumed a silicon tip in their analysis [82], but further calculations [88] demonstrated that a silicon tip will actually be repelled by the inert molecular layer and have a stronger interaction between the molecules, i.e., the molecules would be seen as dark in images. Despite the consistency of the dipole model, without information about the nature of the tip, it is still impossible to confirm that the experiments were really imaging the molecules. For formate, where individual molecules were imaged [76], we can speculate that the silicon tip was contaminated either by the ambient or the surface, and the interaction with the molecules was dominant. For the other acid species, and for imaging inert organic systems in general, it is clear that greater control of the tip is required to remove any ambiguity in interpretting images.

7.6 Calcium difluoride (111)

Due to the need for a conducting surface in STM studies, SFM really dominates in SPM experiments on insulating surfaces. Unfortunately, as emphasized throughout this book, the lack of information about the microscopic nature of the tip has meant that initial attempts at understanding experimental images were not very successful (see, for example, [89]). The first experiments on insulating surfaces, e.g., [90], focused on ionic alkali halide surfaces with geometrically identical positive and negative sublattices; this made it impossible to identify the species imaged as bright in either experiment or theory. The first real breakthrough in interpretation on insulating surfaces was achieved on the calcium difluoride (CaF_2) (111) surface, and as such, it represents an excellent example of a combined theoretical and experimental SFM study.

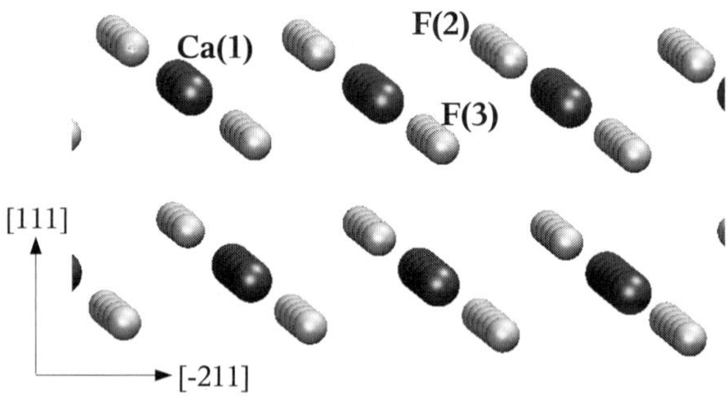

Fig. 7.28. Atomic structure of the CaF_2 (111) surface.

CaF_2 is a classic insulating material with a band gap of about 12 eV, characterized by a Ca^{2+} and F^- ionic lattice. However, its stable (111) surface (shown in Figure 7.28) contains three sublattices, which are not geometrically identical: a protruding F layer terminating the surface [91] (F(2) in Figure 7.28); a middle Ca layer (Ca(1)); and a lower F layer (F(3)). This asymmetry offered the possibility of producing different contrast patterns depending on the electrostatic potential of the tip, thereby allowing much easier identification of imaged species [92, 93, 94].

Simulating scanning

In experiments on CaF_2, real tips were unsputtered, covered by oxide from the ambient, and were contacted to the surface before imaging. Hence, the tip is very likely to be terminated by some form of ionic cluster, and the magnesium oxide model tip (see Chapter 6) was chosen to simulate scanning. Since both the surface and tip are ionic, the microscopic forces were calculated using the periodic static atomistic simulation technique and the MARVIN2 code. The parameters for the surface interactions were generated to match experimental bulk structural, elastic, and dielectric constants, and they gave good agreement with *ab initio* surface relaxations [93]. Parameters for the interactions between the MgO tip and the CaF_2 surface were taken from [95, 96]. The bottom two-thirds of the nanotip and the top six layers of the CaF_2 surface were relaxed explicitly.
Since we do not know in advance the nature of the electrostatic potential of the tip in a given experiment, scanning was simulated using the MgO tip orientated with an oxygen and a magnesium at the lower apex, producing a net negative and net positive electrostatic potential respectively. In both cases, a full surface map of the force field over the CaF_2 surface unit cell was calculated, beginning at a tip–surface height of 2 nm and approaching almost 0 nm with respect to the position of the Ca sublattice. In the final stages, the macroscopic background forces were fitted to match the experimental force curves and were added to the microscopic forces to give the total force. The oscillations of the cantilever were then modeled under the influence of this total force field. More details on this approach are described in Chapter 6. Note that to match the experimental method, simulated images were calculated in "constant height" mode, so that an image is a plot of the change in frequency across the surface at a constant height. Experimental parameters were used in the simulation where relevant: a cantilever amplitude of 23 nm, an eigenfrequency of 84 kHz, a spring constant of 6 N/m, and a mean frequency change of -155 Hz.

Standard Images

Many of the experimentally observed images of the CaF_2 (111) surface exhibit disklike and triangular contrast patterns. The interpretation of these two characteristic patterns is discussed in this section. They have been seen in several experiments using different tips and thus can be considered as "standard" images. The other contrast patterns seen on this surface are associated with more short-range scanning and tip–changes, and are discussed, and explained, in the next section. We will first discuss the properties of simulated images at a single constant height and compare them directly with experiment. Note, however, that the interaction ranges strongly depend on the true microscopic tip in an experiment, and any references to distance can be considered only as rough estimates. The images discussed in this Section were produced for a setup in which the shortest distance between the oscillating tip and the surface was 0.35 nm. At this height the average simulated contrast matches the experimental average contrast, which is a good measure of comparable interaction strength.

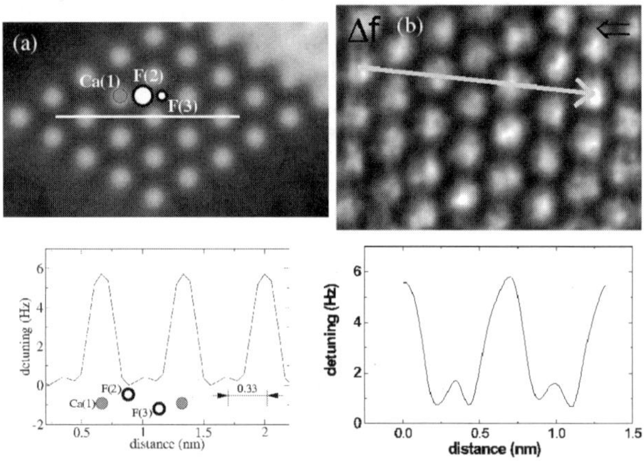

Fig. 7.29. (a) Simulated image and scanline produced using a tip with a negative electrostatic potential scanning at 0.35 nm. Atomic labels refer to similar conventions used in Figure 7.28. (b) Example experimental image and scanline demonstrating "disklike" contrast. The white lines in the images are along the $[\bar{2}11]$ direction and indicate the position of the scanlines. Reproduced with permission. Copyright 2002 by the American Physical Society [94].

Figure 7.29(a) shows a simulated image and scanline produced with a negative electrostatic potential tip at 0.35 nm. The image demonstrates a clear circular or "disklike" contrast with strongest brightness centered on the position of the Ca ions in the surface. The scanline shows that contrast is dominated by a large peak over the Ca ion, with a much smaller peak between the high and low fluorine ions. The smaller peak is due to a minimum in repulsion between the tip and F^- ions however, this peak is so small in comparison to the main peak over Ca that it has no effect on the contrast pattern.
The domination of Ca in the negative potential tip contrast pattern has two components: (i) The positive surface potential over the Ca^{2+} sites has a strong attractive interaction with the negative potential from the tip. Figure 7.30(a) shows clearly the domination of the attractive interaction over the Ca^{2+} ions. (ii) As the tip approaches the surface, the Ca^{2+} ions displace toward it and the F^- ions are pushed into the surface. Figure 7.30(b) shows that at 0.350 nm

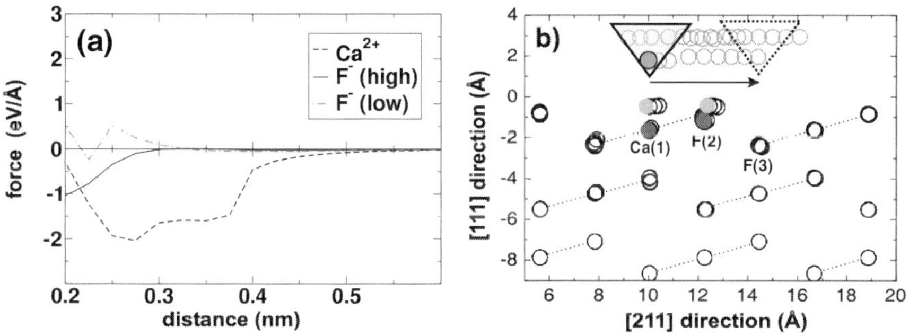

Fig. 7.30. Theoretical data from simulations with a negative electrostatic potential tip: (a) chemical force curves over the Ca, high and low F atomic sites; (b) full trajectories of atoms in a plane as the tip follows the [$\bar{2}$11] scanline at a height of 0.350 nm. Labels Ca(1), F(2), and F(3) are as in Figure 7.29. The atoms shaded light gray in the surface indicate initial positions of the most relevant atoms when the tip is over Ca(1) (leftmost tip position in figure), whereas atoms shaded dark gray are final positions when the tip is over F(3) (rightmost position). Note that trajectories of only the bottom four atoms (1 O^{2-} and 3 Mg^{2+} ions) of the tip have been included. Reproduced with permission. Copyright 2002 by the American Physical Society [94].

over the Ca(1) site, the Ca^{2+} ion displaces by 0.118 nm outwards also forcing the high F^- ion outward. However, as the tip moves toward the the F(2) site, the Ca^{2+} ion drops back to the surface and the high F^- ion is actually pushed in by 0.027 nm. The low F^- ions (F(3)) are not displaced significantly from their equilibrium positions at this scanning height. Displacement of ions from the surface greatly increases the range of the local surface electrostatic potential [97] and increases tip–surface interaction.

If we now compare simulated results with experimental results in Figure 7.29(b), we immediately see a clear qualitative agreement. The experimental image shows disklike contrast, and the scanline has a very similar form to that in the simulation. However, we can extend the comparison to a quantitative level; the simulation predicts (based on surface geometry) that the smaller peak should appear at 0.33 nm from the main peak over the Ca sublattice. If we take over 70 experimental scanlines from images that show disklike contrast, we find that the average position of the small peak is 0.32 ±0.05 nm, in excellent agreement with theory.

Figure 7.31(a) gives a simulated image and scanline when a tip is used with a net positive electrostatic potential from the apex. The contrast pattern is now clearly triangular, with the center of brightness over the high F^- ions, but also with an extension of the contrast toward the position of the low F ions in three equivalent directions forming the triangle. The simulated scanline shows a large peak over the high F^- position dominating the contrast, but we also see a shoulder to this main peak over the position of the low F^- ion. Since this

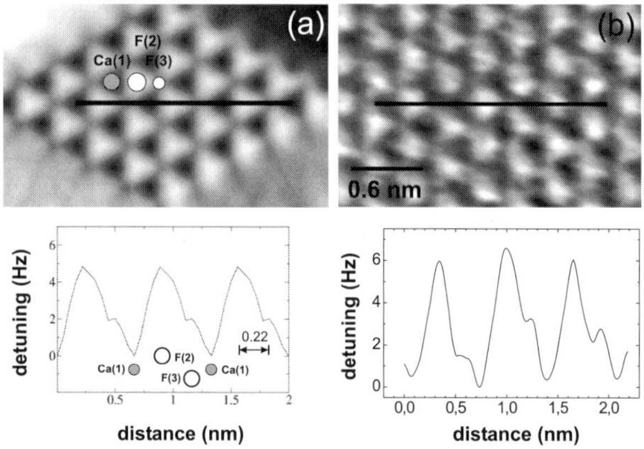

Fig. 7.31. (a) Simulated image and scanline produced using a tip with a positive electrostatic potential scanning at 0.35 nm. Atomic labels refer to similar conventions used in Figure 7.28. (b) Example experimental image and scanline demonstrating triangular contrast. The white lines in the images are along the $[\bar{2}11]$ direction and indicate the position of the scanlines. Reproduced with permission. Copyright 2001 by the American Physical Society [92].

shoulder is a significant fraction of the height of the main peak, it can be seen in images and is responsible for the triangular contrast pattern. The triangular contrast pattern has three components: (i) The negative surface potential over F^- sites gives a strong attractive interaction with the positive potential tip (see Figure 7.32(a)). This interaction is comparable to the interaction of the negative tip over the Ca^{2+} ions, since although the F^- ions have half the charge of the Ca^{2+}, the high fluorine protrudes 0.08 nm farther from the surface (see Figure 7.32(b)). (ii) The ions in both the F^- layers displace toward the tip as it approaches, and the Ca^{2+} ion is pushed inward. (iii) When the tip is over the low F^- ion, there is also some interaction of the tip with the next row of high F^- ions (see Figure 7.32(d)). Hence the shoulder has some component from this interaction, as well as the direct interaction with the low F^- sublattice. The role of displacements is discussed in detail in the next section.

Comparing with the experimental results in Figure 7.31(b), once more we see that there is a good qualitative agreement between experiment and theory. The experimental image shows triangular contrast, and the scanline shows large peaks with shoulders. Quantitatively we find that the average position of the shoulder with respect to the main peak in over 75 scanlines is 0.24 ± 0.04 nm, which compares very well with the theoretical prediction of 0.22 nm.

The described semiquantitative agreement of the theoretical images obtained with model tips with the experimental images supports the model of ionic tip, which may have two signs of the electrostatic potential probing the surface.

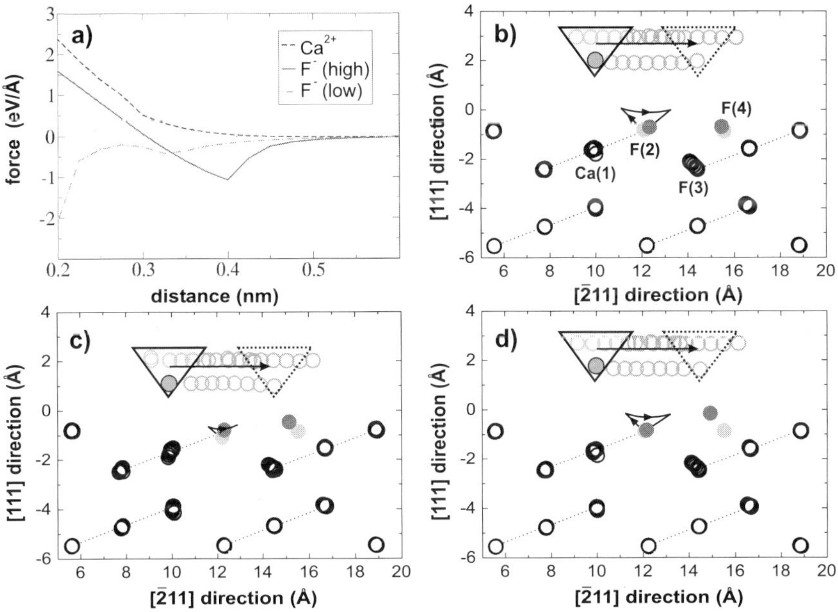

Fig. 7.32. Theoretical data from simulations with a positive electrostatic potential tip: (a) chemical force curves over the Ca, high and low F atomic sites; full trajectories of atoms in a plane as the tip follows the [$\bar{2}$11] scanline. Tip–surface distance is (b) 0.350 nm; (c) 0.250 nm; (d) 0.325 nm. Labels Ca(1), F(2) and F(3) are as in Figure 7.29. The atoms shaded light gray indicate initial positions of the most relevant atoms when the tip is over Ca(1) (leftmost tip position in figure), whereas atoms shaded dark gray are final positions when the tip is over F(3) (rightmost position). F(4) is a high fluorine atom out of the plane, but its trajectory has been projected onto the same plane as the other atoms for clarity. Note that trajectories of only the bottom four atoms (1 Mg^{2+} and 3 O^{2-} ions) of the tip have been included. Reproduced with permission. Copyright 2002 by the American Physical Society [94].

Although the MgO tip is clearly an idealized model, it seems to capture correctly both the possibility of different types of tip contamination by the surface and ambient ions and the strength of the short-range chemical interaction.

Distance dependence of images

Since we have demonstrated both qualitative and quantitative agreement between experiment and theory at a certain height, it is interesting to see whether the comparison is still favorable when a range of heights is considered. Figs. 7.33(a–c) show experimental images at increasing average frequency change, i.e., reduced tip–surface separation. Figures 7.33(a, b) clearly show the triangular contrast discussed above. Persistence of this pattern in experimental images obtained under different conditions shows that the triangular contrast pattern is not a unique feature seen only at a specific height, but is rather a distinct pattern related to the potential of the tip. However, in the image at closest approach we see that the contrast pattern has changed considerably. Figure 7.33(c) shows a "honeycomb" pattern, with the F sites now completely linked in bright contrast.

To understand whether this very distinct change in contrast could be explained within the same model as discussed in the previous section, further extensive modeling was performed. Figs. 7.33(d–g) show the development of contrast in simulated images as the scanning height (i.e., the closest distance between the turning point of tip oscillations and the surface) is reduced. Figure 7.33(d) demonstrates that even at very large distances the triangular contrast pattern is consistent, even if it is unlikely that experiments could measure such small chemical forces. As the tip approaches (Figs. 7.33(e, f)), theory predicts that the triangular pattern becomes even more vivid, as seen in the experimental images. Finally, at 0.275 and 0.25 nm separations in Figure 7.33(g, h), the simulated images develop the honeycomb contrast pattern seen clearly in experimental image Figure 7.33(c).

Further agreement can be seen by comparing the change in experimental and theoretical scanlines. Figures 7.34(a, b) show that at long range both experiment and theory demonstrate the large peak/small shoulder scanlines characteristic for the triangular contrast pattern. However, as the tip approaches closer, the magnitude of the shoulder increases until for scanlines from the honeycomb images it is clear that the shoulder is at least equal to the original main peak.

A more thorough understanding of this agreement in contrast development requires studying in detail the changes in forces and atomic displacements as the tip approaches the surface. Figure 7.32(a) gives the chemical force over the relevant sites in the CaF_2 surface as a function of distance for a tip with positive termination. For distances larger than 0.400 nm, the curves are as one would expect them, i.e., we find repulsion over the positive Ca^{2+} ion and attraction above the F^- ions. From the ionic interaction, we find some attraction over the low F^- ion and stronger attraction over the high F^- ion. Moving closer than 0.400 nm, we observe that the attraction over the high F^- ions reduces, and it increases over the low F^- site, until around 0.320 nm the greatest attraction is now over the low F^- site. This behavior can be understood by looking at atomic displacements as the tip approaches. At 0.350 nm there is strong displacement of the high F^- ion (F(2)) toward the tip (see Figure 7.32(b)), producing a very strong attractive interaction. However, as the tip moves closer, this F^- ion is driven back into the surface and the force is reduced. Frame (c) shows that the high F^- ion has been pushed effectively back into its original lattice position at a tip–surface separation of 0.250 nm. However, when the tip is over the low F(3) site, we see very little movement of the closest high F^- ion, but in fact, aided by the proximity of the Ca^{2+} ion, there is a much smaller barrier for displacement of the high F^- ion from

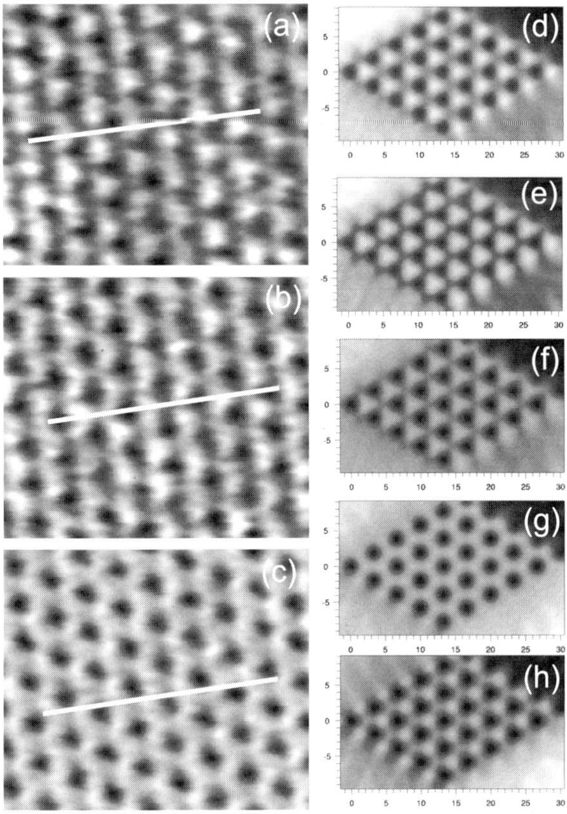

Fig. 7.33. (a–c) Experimental images taken as the tip approaches the surface [93]. (d–g) Simulated images at 0.500, 0.375, 0.325, 0.275, and 0.250 nm using a positive potential tip. Reproduced with permission. Copyright 2002 by the American Physical Society [94].

next nearest row (F(4)). Frame (d) shows how the next-nearest high F^- ion displaces very strongly to the tip at 0.325 nm when it is over the F(3) site. In summary, at a distance of 0.5 nm the interaction with the high F^- atom dominates, and we see only relatively small shoulders in scanlines over the low F^- sites. As the tip approaches, the nearest high F^- ion is pushed into the surface, reducing its dominance, and the interaction with low F^- and the next nearest high F^- ion increases the relative size of the shoulder. This corresponds to increasing vividness of the triangular contrast pattern in images. Finally, the contribution from the high F^- is balanced by the contribution from the low F^-/next-high F^- ion, and the main peaks and shoulders are equivalent in scanlines, and it is this equivalence that produces the characteristic honeycomb contrast pattern.

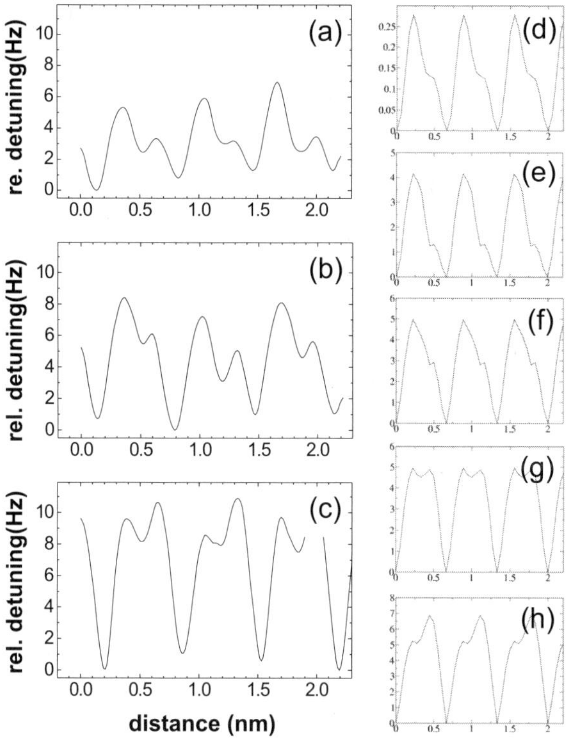

Fig. 7.34. Comparison of (a–c) characteristic experimental and (d–h) simulated scanlines taken from the images in Figure 7.33 as the tip approaches the surface. Reproduced with permission. Copyright 2002 by the American Physical Society [94].

These results alter the previous perception [97, 93] that, due to adhesion of the surface ions to the tip, SFM tips should be prone to rapid changes during short-range scanning. It was expected that large displacements of the surface and tip ions may lead to instabilities of the SFM operation and to tip crashes. Although this effect certainly has been observed in many experiments, it does not necessarily always result from large displacements and instabilities of the surface ions. As these results demonstrate, such effects can be reversible and may not disrupt imaging. The remarkable agreement between theory and experiment serves as an indirect, but powerful, indication that the displacements of the tip and surface ions (which cannot be imaged directly) play an extremely important role in contrast formation.

7.7 Summary

In reflecting on the simulations and the gradual resolution of the puzzle, it becomes clear that only a very limited part of the whole physical situation is actually accessible in the experiments. The change of the position of the SPM tip is a result of measurements of constant-current/height contours. But the change of the position of surface atoms under given experimental conditions cannot be determined. This makes simulations the only source of information on both the stability of a system under specific conditions determined from interaction energies, the elastic limit of a surface and tip system, and the relation between true surface properties (properties of its ground state) and virtual properties that are due to the measurement itself. It might seem that the last distinction is far-fetched. But one only has to consider that atomic positions on a surface can be determined by a number of different methods, e.g., electron diffraction, photon diffraction, and electron tunneling, to understand that different experimental methods might lead to different results. And in this case, the possibility in theory to switch on or off a particular effect makes it quite adaptable to a whole range of experimental data. This becomes even more important in the case of SFM.

References

1. E.W. Plummer. *Surf. Sci.*, 500:1, 2002.
2. N. D. Lang. *Phys. Rev. B*, 34:5947, 1986.
3. J. Jacobsen, B. Hammer, K. W. Jacobsen, and J. K. Norskov. *Phys. Rev. B*, 52:14954, 1995.
4. P. Sautet, J. Dunphy, D. F. Ogletree, and M. Salmeron. *Surf. Sci.*, 295:347, 1993.
5. F. Biscarini, C. Bustamante, and V. M. Kenkre. *Phys. Rev. B*, 51:11089, 1995.
6. L. Ruan. *Phys. Rev. Lett.*, 70:4079, 1993.
7. C.A. Ventrice. *Phys. Rev. B*, 49:5773, 1994.
8. U. Diebold. *Phys. Rev. Lett.*, 77:1322, 1996.
9. C.I. Carlisle. *Phys. Rev. Lett.*, 84:3899, 2000.
10. H. Over. *Science*, 287:1474, 2000.
11. Sh. K. Shaikhutdinov. *Phys. Rev. Lett.*, 91:076102, 2000.
12. F. Calleja, A. Arnau, J. J. Hinarejos, A. L. Vazquez de Parga, W. A. Hofer, P. M. Echenique, and R. Miranda. *Phys. Rev. Lett.*, 92:206101, 2004.
13. M. Lindroos, H. Pfnür, G. Held, and D. Menzel. *Surf. Sci.*, 222:451, 1989.
14. V. M. Hallmark, S. Chiang, J. F. Raboli, J. D. Swalen, and R. J. Wilson. *Phys. Rev. Lett.*, 59:2879, 1987.
15. J. V. Barth, H. Brune, G. Ertl, and R. J. Behm. *Phys. Rev. B*, 42:9307, 1990.
16. J. Wintterlin, J. Wiechers, H. Brune, T. Gritsch, H. Höfer, and R. J. Behm. *Phys. Rev. Lett.*, 62:59, 1989.
17. C. J. Chen. *Phys. Rev. Lett.*, 65:448, 1990.
18. M. Tsukada, K. Kobayashi, N. Isshiki, and H. Kageshima. *Surf. Sci. Rep.*, 13:265, 1991.

19. J. Jacobsen, B. Hammer, K. W. Jacobsen, and J. K. Norskov. *Phys. Rev. B*, 52:14954, 1995.
20. W. Sacks. *Phys. Rev. B*, 61:7656, 2000.
21. G. Doyen, D. Drakova, and M. Scheffler. *Phys. Rev. B*, 47:9778, 1993.
22. W. A. Hofer and A. J. Fisher. *Phys. Rev. Lett.*, 91:036803, 2003.
23. J. M. Soler, A. M. Baro, N. Garcia, and H. Rohrer. *Phys. Rev. Lett.*, 57:444, 1986.
24. A. R. H. Clarke, J. B. Pethica, J. A. Nieminen, F. Besenbacher, E. Laegsgaard, and I. Stensgaard. *Phys. Rev. Lett.*, 76:1276, 1996.
25. W. A. Hofer, R. A. Wolkow A. J. Fisher, and P. Grütter. *Phys. Rev. Lett.*, 87:236104, 2001.
26. W. A. Hofer and A. J. Fisher. *Phys. Rev. Lett.*, 91:036803, 2003.
27. C. J. Chen. *Introduction to Scanning Tunneling Microscopy.* Oxford University Press, Oxford, 1993.
28. A. Schirmeisen, G. Cross, A. Stalder, P. Grütter, and U. Dürig. *New J. Phys.*, 1:29, 2000.
29. S. Hembacher, F. J. Giessibl, J. Mannhart, and C. F. Quate. *Phys. Rev. Lett.*, 94:056101, 2005.
30. W. A. Hofer, A. J. Fisher, R. A. Wolkow, and P. Grütter. *Phys. Rev. Lett.*, 87:236104, 2001.
31. J. A. Stroscio, D. T. Pierce, A. Davies, R. J. Celotta, and M. Weinert. *Phys. Rev. Lett.*, 75:2960, 1995.
32. A. Biedermann, R. Tscheliessnig, M. Schmid, and P. Varga. *Appl. Phys. A*, 78:807, 2004.
33. M. Schmid, H. Stadler, and P. Varga. *Phys. Rev. Lett.*, 70:1441, 1993.
34. W.A. Hofer and J. Redinger. *Surf. Sci.*, 447:51, 2000.
35. J. Shen and J. Kirschner. *Surf. Sci.*, 500:300, 2002.
36. J. Stöhr, A. Scholl, T. J. Regan, S. Anders, J. Lining, M. R. Scheinfein, H. A. Padmore, and R. L. White. *Phys. Rev. Lett.*, 83:1862, 1999.
37. S. Heinze, M. Bode, A. Kubetzka, O. Peitzsch, X. Nie, S. Blügel, and R. Wiesendanger. *Science*, 288:1805, 2000.
38. W. A. Hofer and A. J. Fisher. *Surf. Sci. Lett.*, 515:L487, 2002.
39. W. A. Hofer and A. J. Fisher. *Journal of Magnetism and Magnetic Materials*, 267:139, 2003.
40. W. A. Hofer, A. J. Fisher, G. R. Lopinski, and R. A.Wolkow. *Phys. Rev. B*, 63:085314, 2001.
41. R. D. Bringans, R. I. G. Uhrberg, A. Olmstaed, and R. Z. Bachrach. *Phys. Rev. B*, 34:7447, 1986.
42. R. J. Hamers, R. M. Tromp, and J. E. Demuth. *Phys. Rev. B*, 34:5343, 1986.
43. R. A. Wolkow. *Phys. Rev. Lett.*, 68:2636, 1992.
44. P. Badziag, W. S. Verwoed, and M. A. Van Hove. *Phys. Rev. B*, 43:2058, 1991.
45. R. Terborg, M Polcik, J. T. Hoeft, M. Kittel, D. I. Sayago, R. L. Tomes, and D. P. Woodruff. *Phys. Rev. B*, 66:085333, 2002.
46. W. A. Hofer, A. J. Fisher, and R. A. Wolkow. *Surf. Sci.*, 475:83, 2001.
47. F. Wang, D. C. Sorescu, and K. D. Jordan. *J. Phys. Chem.*, 106:1316, 2002.
48. R. A. Wolkow. *Annu. Rev. Phys. Chem.*, 50:413, 1999.
49. G. Held. *J. Phys: Condens. Mat.*, 15:R1501, 2003.
50. W. A. Hofer, A. J. Fisher, T. Bitzer, T. Rada, and N. V. Richardson. *Chem. Phys. Lett.*, 355:347, 2002.

51. A. Bilic, J. R. Reimers, W. A. Hofer, and N. S. Hush. *Chem. Phys. Lett.*, 385:341, 2004.
52. A. L. Linsebigler, G. Lu, and J. T. Yates. *Chem. Rev.*, 95:735, 1995.
53. V. E. Henrich and P. A. Cox. *The Surface Science of Metal Oxides*. University Press, Cambridge, 1996.
54. J. Lausmaa. *J. Elec. Spec. Rel. Phen.*, 81:343, 1996.
55. U. Diebold. *Surf. Sci. Rep.*, 48:53, 2003.
56. U. Diebold, J. F. Anderson, K. O. Ng, and D. Vanderbilt. *Phys. Rev. Lett.*, 77:1322, 1996.
57. G. Charlton, P. B. Howes, C. L. Nicklin, P. Steadman, J. S. G. Taylor, C. A. Muryn, S. P. Harte, J. Mercer, R. McGrath, D. Norman, T. S. Turner, and G. Thornton. *Phys. Rev. Lett.*, 78(3):495, 1997.
58. D. Vogtenhuber, R. Podloucky, A. Neckel, S. G. Steinemann, and A. J. Freeman. *Phys. Rev. B*, 49:2099, 1994.
59. M. Ramamoorthy, R. D. King-Smith, and D. Vanderbilt. *Phys. Rev. B*, 49(11):7709, 1994.
60. M. Ramamoorthy, D. Vanderbilt, and R. D. King-Smith. *Phys. Rev. B*, 49(23):16721, 1994.
61. P. J. D. Lindan, N. M. Harrison, M. J. Gillan, and J. A. White. *Phys. Rev. B*, 55(23):15919, 1997.
62. N. M. Harrison, X. G. Wang, J. Muscat, and M. Scheffler. *Faraday Discussions*, 114:305, 1999.
63. A. T. Paxton and L. Thiên-Nga. *Phys. Rev. B*, 57:1579, 1998.
64. R. Schaub, E. Wahlström, A. Rønnau, E. Lægsgaard, E. Stensgaard, and F. Besenbacher. *Science*, 299:377, 2003.
65. S. P. Bates, G. Kresse, and M. J. Gillan. *Surf. Sci.*, 409:336, 1998.
66. P. J. D. Lindan, N. M. Harrison, and M. J. Gillan. *Phys. Rev. Lett.*, 80:762, 1998.
67. J. Muscat, N. M. Harrison, and G. Thornton. *Phys. Rev. B*, 59:2320, 1999.
68. I. M. Brookes, C. A. Muryn, and G. Thornton. *Phys. Rev. Lett.*, 87:266103, 2001.
69. G. Liu, J. A. Rodriguez, Z. Chang, J. Hrbek, and L. González. *J. Phys. Chem. B*, 106:9883, 2002.
70. E. Wahlström, N. Lopez, R. Schaub, P. Thostrup, A. Rønnau, C. Africh, E. Lægsgaard, J. K. Nørskov, and F. Besenbacher. *Phys. Rev. Lett.*, 90:026101, 2003.
71. K. I. Fukui, H. Onishi, and Y. Iwasawa. *Phys. Rev. Lett.*, 79:4202–4205, 1997.
72. S. H. Ke, T. Uda, and K. Terakura. *Phys. Rev. B*, 65:125417, 2002.
73. A. S. Foster, O. H. Pakarinen, J. M. Airaksinen, J. D. Gale, and R. M. Nieminen. *Phys. Rev. B*, 68:195410, 2003.
74. H. Onishi and Y. Iwasawa. *Chem. Phys. Lett.*, 226:111, 1994.
75. S. A. Chambers, S. Thevuthasan, Y. J. Kim, G. S. Hermann, Z. Wang, E. Tober, R. Ynzunza, and J. Morais. *Chem. Phys. Lett.*, 267:51, 1997.
76. K. I. Fukui, H. Onishi, and Y. Iwasawa. *Chem. Phys. Lett.*, 280:296, 1997.
77. S. P. Bates, G. Kresse, and M. J. Gillan. *Surf. Sci.*, 409:336, 1998.
78. P. Käckell and K. Terakura. *Surf. Sci.*, 461:191, 2000.
79. Q. Guo, I. Cocks, and E. M. Williams. *J. Chem. Phys.*, 106:2924, 1997.
80. A. Gutiérrez-Sosa, P. Martínez-Escolano, H. Raza, R. Lindsay, P. L. Wincott, and G. Thornton. *Surf. Sci.*, 471:163, 2001.

81. A. S. Foster and R. M. Nieminen. *J. Chem. Phys.*, 121:9039, 2004.
82. A. Sasahara, H. Uetsuka, and H. Onishi. *Phys. Rev. B*, 64:121406, 2001.
83. A. Sasahara, H. Uetsuka, and H. Onishi. *Surf. Sci. Lett.*, 481:L437, 2001.
84. A. Sasahara, H. Uetsuka, and H. Onishi. *J. Phys. Chem. B*, 105:1, 2001.
85. H. Onishi, A. Sasahara, H. Uetsuka, and T. Ishibashi. *Appl. Surf. Sci.*, 188:257, 2002.
86. A. Sasahara, H. Uetsuka, T. Ishibashi, and H. Onishi. *Appl. Surf. Sci.*, 188:265, 2002.
87. A. Sasahara, H. Uetsuka, and H. Onishi. *Langmuir*, 19:7474, 2003.
88. A. S. Foster, A. Y. Gal, A. L. Shluger, and R. M. Nieminen. *J. Phys. Chem.*, 109:4554, 2005.
89. R. Bennewitz, A. S. Foster, L. N. Kantorovich, M. Bammerlin, Ch. Loppacher, S. Schär, M. Guggisberg, E. Meyer, and A. L. Shluger. *Phys. Rev. B*, 62:2074, 2000.
90. M. Bammerlin, R. Lüthi, E. Meyer, A. Baratoff, M. Guggisberg, C. Gerber, L. Howald, and H.-J. Güntherodt. *Probe Microscopy*, 1:3, 1997.
91. M. Reichling and C. Barth. Scanning force imaging of atomic size defects on the CaF_2 (111) surface. *Phys. Rev. Lett.*, 83:768, 1999.
92. A. S. Foster, C. Barth, A. L. Shluger, and M. Reichling. Unambiguous interpretation of atomically resolved force microscopy images of an insulator. *Phys. Rev. Lett.*, 86:2373, 2001.
93. C. Barth, A. S. Foster, M. Reichling, and A. L. Shluger. Contrast formation in atomic resolution scanning force microscopy on CaF_2 (111): Experiment and theory. *J. Phys.: Condens. Matter*, 13:2061, 2001.
94. A. S. Foster, C. Barth, A. L. Shluger, R. M. Nieminen, and M. Reichling. *Phys. Rev. B*, 66:235417, 2002.
95. D. J. Binks. PhD Thesis, University of Surrey, 1994.
96. T. S. Bush, J. D. Gale, C. R. A. Catlow, and P. D. Battle. *J. Mater. Chem.*, 4:832, 1994.
97. A. I. Livshits, A. L. Shluger, A. L. Rohl, and A. S. Foster. *Phys. Rev. B*, 59:2436, 1999.

8

Single-Molecule Chemistry

In this chapter we review recent progress made in accounting for the ability of an STM to manipulate and rearrange single molecules at surfaces, which gave rise to a genuinely new research area, the ability to induce and follow chemical reactions in situ and on a genuinely atomic level of resolution. Related to this area is the excitation and observation of single modes of molecular motion described by phonon excitations.

8.1 Introduction

The ability of an STM to induce motion of atomic adsorbates was demonstrated in the early 1990s [1]. Since then, the field has gradually been extended into what one could term a single-molecule laboratory [2]: the ability to detect, to analyze, and to modify adsorbed molecules. This new ability has been exploited in two directions: On the one hand, the arrangement of molecules can be modified so that they start to interact on a surface and form new chemical bonds; in this case the manipulation is tantamount to the induction of chemical reactions in a very controlled environment. On the other hand, the electric current between the molecule at the surface and the STM tip can be regulated in such a way that the energy of transiting electrons is sufficient to excite specific phonon modes of the molecules, in particular stretch modes, torsion modes, and rotational modes. This is similar to the analysis of a surface by infrared spectroscopy experiments, since the energy thresholds are generally very low (in the range of a few hundred meV). However, the unsurpassed local resolution of an STM makes it possible to watch and record these events on the single molecule level. Excellent reviews of these new developments have been written by Lorente et al., Ho, and Ueba [2, 3, 4].

8.2 Manipulation of atoms

Manipulation of single molecules by SFM is in principle possible, but still somewhat restricted by the dynamic motion of the cantilever, which leads to a periodic change of interactions with a surface molecule involving short- and long-range forces alike (see previous sections on SFM theory). At present the most convincing experiment in this field is the change of group conformation of large organic molecules by NC-SFM [5], but the field is under rapid development. In an STM the close vicinity of surface and STM tip leads to a transfer of atoms or molecules from the surface onto the STM tip due to the formation of chemical bonds. In principle, the process is similar to the rupture of a chemical bond, e.g., in an atomic wire, because the attractive forces between a molecule and the surface need to be overcome. It is quite clear from the simulations, e.g., on metal surfaces (see previous chapters), that this process depends on the distance between the molecule and the STM tip, and that it will occur in a range where van der Waals interactions are comparatively small in comparison to interactions due to bond formation. Given the extensive simulations presented in this volume, the distance should be somewhat below 6 Å.

However, this is not the only way to manipulate the position of a molecule. If the attraction between the molecule and the tip is lower that the threshold for atomic transfer, it will remain on the surface. But since the corrugation of the attractive potential is generally very small, in particular on close-packed metal surfaces, the projection of the attractive force onto the surface plane may still be sufficient to move the molecule out of its lateral minimum and to slide it across a maximum of the potential energy surface (PES) into the next minimum. In this case the force between surface and STM is actually used to manipulate the molecule laterally.

Bartels and coworkers have systematically studied the lateral manipulation of single adatoms on metal surfaces [6]. If the STM tip moves across the surface at specific tunneling conditions, i.e., at a defined median distance from the core of surface atoms, then the tip height does not change in a smooth periodic distribution, as it would according to the surface electronic structure and its overlap with the tip wavefunctions. Quite to the contrary, it reveals a distinct maximum whenever the STM tip is above an fcc hollow position, and decreases smoothly to a minimum immediately before the next fcc hollow position (see Figure 8.1).

Depending on the distance between surface and STM tip, two modes of lateral manipulation can be distinguished [6]:

1. A pulling mode, whereby the attraction between tip and adsorbate leads to a discontinuous motion of the adsorbate from one adsorption site to the next.
2. A sliding mode, whereby the adsorbate is trapped underneath the tip apex and follows the motion of the STM tip continuously.

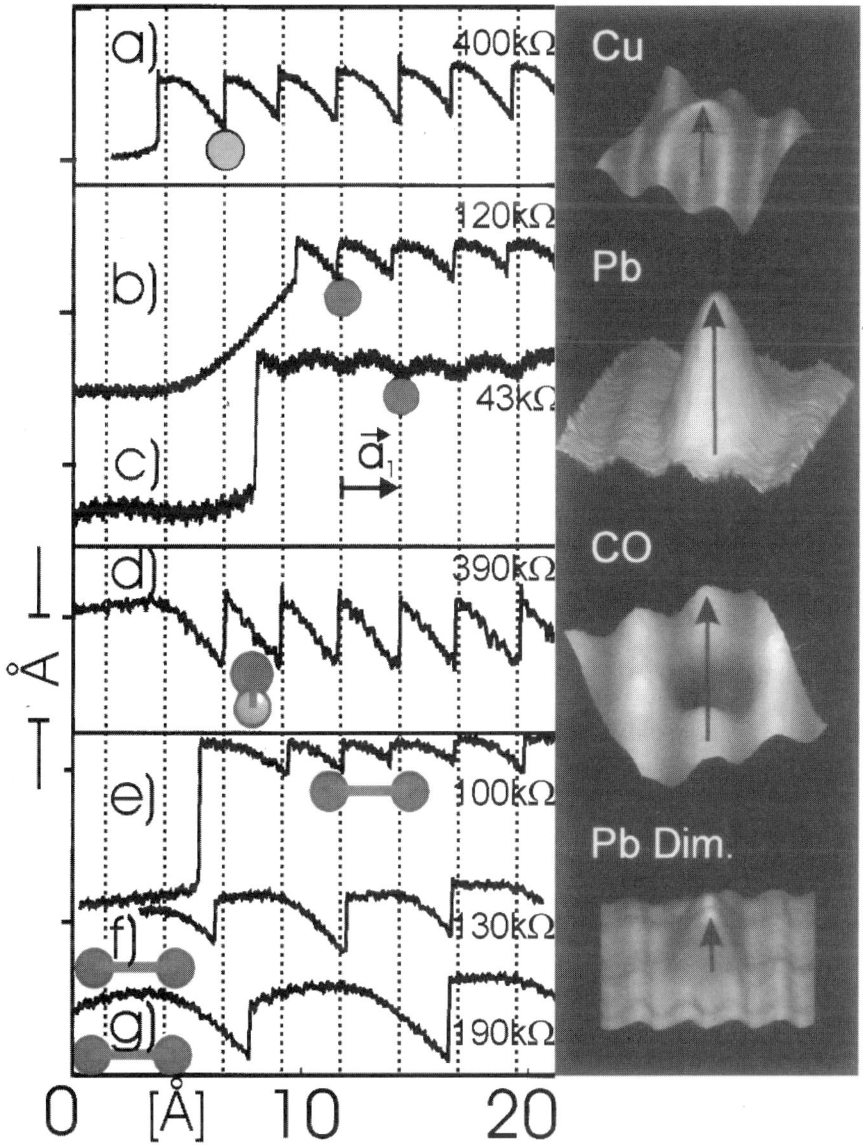

Fig. 8.1. Recorded STM tip position in constant-current mode for a (a) Cu, (b, c) Pb atom, and (d) a CO molecule on Cu(111). The apparent height is a maximum at the fcc hollow site, indicating that at this position of the STM tip the atom (molecule) is directly underneath the tip apex. As the tip moves away from the fcc site, the atom (molecule) remains in the same position, until it moves abruptly to the next fcc site immediately before the STM tip arrives there. The ensuing sawtooth is characteristic of a pulling mode (a, b, d), while the flat line indicates that the atom is trapped underneath the tip apex, in a sliding mode (c). L. Bartels and G. Meyer and K.-H. Rieder, Phys. Rev. Lett. 79, 697 (1997). Copyright (1997) American Physical Society, reprinted with permission.

From a theoretical point of view the surface/adsorbate/tip interface constitutes a rather complicated physical system, which is made difficult to simulate due to the many degrees of freedom introduced by motion and excitations of the adsorbate. If chemical reactions are induced in this way, the difficulty increases even more due to the ability of adsorbed molecules to interact with the surface, with other molecules, or with the STM tip.

8.2.1 Modeling atomic manipulation

Confronted with a complex physical problem, theorists tend to break it down into manageable parts, which can be treated separately, and to resort to approximations for specific applications. A different strategy would be to treat all parts of the problem in the same way; in this case the level of theoretical precision has to be somewhat lower than actual first-principles simulations.
In this spirit, the first simulations of, for instance, the manipulation of Xe atoms on Cu(110) [1] were performed with analytic model potentials for the interaction of the Xe atom with the surface and the STM tip [7]. The potential parameters were fitted to the experimental values for the adsorption energy in the low coverage regime. Even though such a treatment is quite crude, from the viewpoint of sophisticated DFT simulations, Bouju et al. were able to differentiate three distinct distance regimes, which are not too far from a first principles analysis of related problems: (i) if the distance z is greater than 9 Å, the STM tip has no effect on the adsorbed Xe atom; (ii) if 9 Å$>z>$6 Å, the PES of the surface adsorbate system is modified and the Xe atom is able to move out of its ground-state position; (iii) if $z<$6 Å, the Xe atom is transferred to the STM tip. Other simulations with either potentials derived from the embedded atom model or modified extended Hückel potentials have included the motion of Cu adatoms on Cu(111) [8], or Ag adatoms on Si(001) [9]. Since in this case the emphasis was on the interaction energy, the results could not directly be compared to the constant-current signal of STM operation. At present, this seems to be the main obstacle in this field: every position of the STM tip leads to a different interaction energy, and thus a different position of an adatom on the PES. Since every position of the adatom changes the electronic structure of the surface adsorbate system, this means that it also corresponds to a different tunneling current. Connecting the two separate parts of the theoretical models seems to be the main obstacle.
In order to understand the problem and to determine possible ways to overcome it, it seems necessary to look at the situation in more detail. If the physical components of such a system are broken down into its components, one can differentiate between three types of quantities that need to be considered:

Adsorbate surface interactions

This is probably the most extensively modeled part of the problem. Adsorbate surface interactions are routinely calculated in a wide range of dynamic problems, related to the adsorption and dissociation of molecules at surfaces. Sophisticated methods exist today that allow one to construct a full PES from a rather limited number of data points. Since this is done for every problem in dynamics, it seems feasible to adopt the methods for atomic manipulations. In this case one ends up with a three dimensional map of potential energies as a function of adsorbate position.

Adsorbate tip interactions

As shown by model calculations on close-packed metal surfaces and also by the explicit derivation of the interaction energy in terms of the Bardeen matrix element (see Chapter 5, (5.93)),

$$E_{\text{int}} = -4\left(\frac{\hbar^2}{m}\right)^2 \sum_{i,k} \frac{|M_{ki}|^2}{|E_i - E_k|}, \tag{8.1}$$

the interaction energy between a surface configuration and an STM tip can be calculated quite accurately using perturbative methods, based on ground-state DFT. Neglecting second-order terms, i.e., the changes of the electronic structure and their effect on interaction energy, one can thus estimate the shift of the adsorbate from its ground-state position for every position of the STM tip. In principle, this calculation is quite similar to the calculation of the tunneling current with perturbation methods; numerically the effort will be much higher, though, since the matrix elements M_{ik} have to be calculated for a large energy interval.

Adsorbate tip current

If every position of the STM tip leads to a different position of the adsorbate on the surface, then every tip position will lead to a different surface electronic structure. Since every surface electronic structure leads to a different current for the very same STM tip position, the fact that the current is used as the feedback signal in the experiments makes it necessary to introduce a self-consistency cycle at this point.

Assume that we start from an initial guess for the vertical STM position z (the lateral position is usually known, since it can be estimated from the lattice geometry far from the adsorbate). On the basis of this guess we calculate the interaction energy from (8.1), which leads to a specific position of the adsorbate, both in terms of the PES and in terms of the three-dimensional position on the surface. Next we calculate the ground-state electronic structure for this position of the adsorbate using DFT methods. And finally we calculate the tunneling current using first-order scattering approximations (see Chapter 5). The current we obtain is then compared to the experimental current, which leads to a determination of z. This concludes one iteration of the self-consistency cycle, because in the next iteration we vary the vertical position

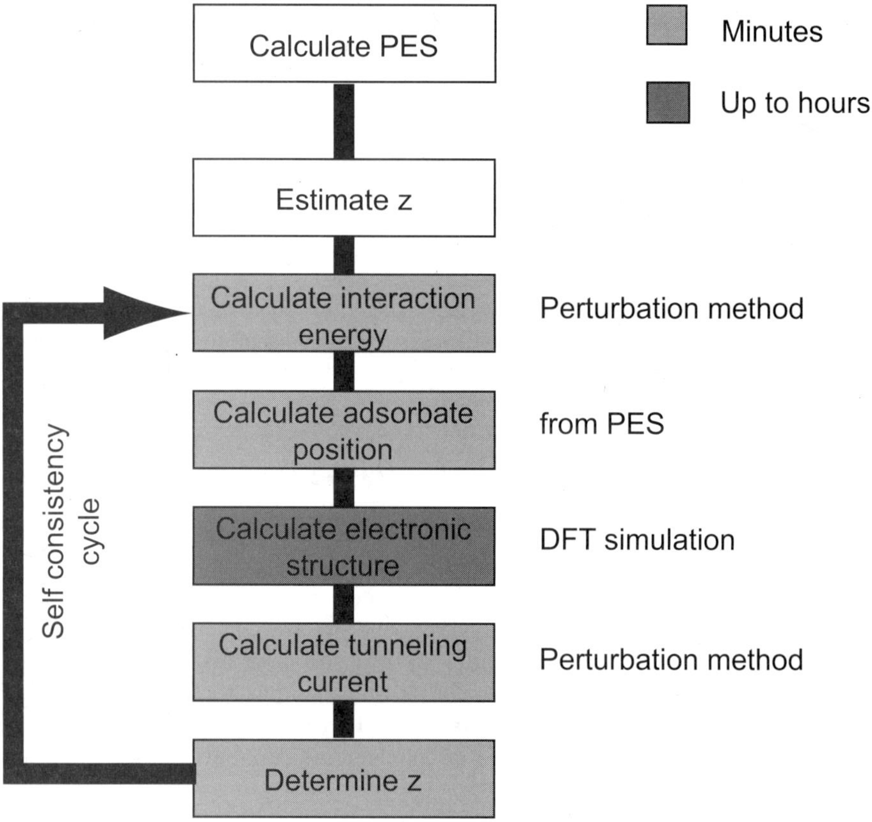

Fig. 8.2. Calculation of tunneling current in a self-consistent cycle including the motion of the adsorbate. If the PES of the adsorbate is calculated by standard DFT methods, an initial guess of the vertical distance z can be gradually refined by first-principles calculations of the interaction energy, the actual adsorbate position, and the tunneling current until self-consistency is achieved. The limiting step in the cycle is the DFT calculation of the electronic structure.

depending on the obtained result z. None of these steps is beyond theoretical means today; in fact, all of them are fairly standard. The scheme is sketched in Figure 8.2. It should be noted that all steps involving the calculation of the interaction energy and the tunneling current can be accomplished within perturbation methods; since they need to be calculated only for a single lateral point, they can be completed in minutes. The limiting step in the calculation is the simulation of the ground-state electronic structure at a shifted adsorbate position. Depending on the DFT method and the required precision, this could lead to a duration of a single cycle of more than one hour. However, even in this case a single scanline should be calculated in a reasonable, less than one week, time scale on fast parallel computers.

8.3 Phonon excitation

While atomic manipulation involves the external degrees of freedom, recent experimental developments also have made it possible to address the internal degrees of freedom, i.e., internal modes of rotation and oscillation. The physical origin of the phenomenon is the existence of phonon modes, in the range of a few to a few hundred meV, which can be targeted by the kinetic energy of tunneling electrons. Once the energy surpasses the excitation energy of a particular mode, the probability that the electron propagates through the lead only after it has excited a phonon is not zero and in the range of a few percent. This change of the tunneling current due to excitations shows up in the tunneling spectrum, in particular as a peak in the second derivative. The reason is that the derivative of the tunneling current will change discontinuously, in the same way as, for example, for a surface state (see the following chapter, on tunneling spectroscopy), if a new pathway for electron propagation is opened up. This happens if the electron energy surpasses the phonon excitation threshold. For a very small bias interval around this level the dI/dV curve due to electron transitions can be assumed to be constant. The additional mode of propagation then shows up as a step function:

$$\frac{dI}{dV} \propto \frac{dI_0}{dV} + \theta(V_{ph} - V) \cdot \frac{\Delta I}{\Delta V}. \tag{8.2}$$

Here we assume that the additional current contribution due to phonon excitation is given by $\Delta I/\Delta V$. Since the derivative of a step function is a delta functional, the second derivative of the current with respect to voltage at the threshold value will be a delta functional:

$$\frac{d^2 I}{dV^2} \propto \delta(V_{ph} - V) \cdot \frac{\Delta I}{\Delta V}. \tag{8.3}$$

In practice, such a peak has been very difficult to detect experimentally, because the noise in an experimental spectrum is quite substantial (see the following chapter). For this reason it was thought, until a proof to the contrary was given by Wilson Ho's group (see [3] and references therein), that it is impossible to detect the signal with suitable precision. In fact, many experimentalists still refrain from entering this field of research, explaining in conversations that it is due to their limited trust in a numerical analysis of experimental data. However, quite a few groups have taken up the challenge in the last few years, and a number of them have succeeded in establishing beyond doubt that measurements of the excitation threshold are both repeatable and correct (see [2] and references therein). A collection of d^2I/dV^2 curves for various molecules, taken from [3], is shown in Figure 8.3. The raw data reveal that finding a phonon mode in the background noise of a spectrum is about as easy as finding the signature of a surface state in the dI/dV spectrum (see next chapter). It should be noted that experiments in this field are done at very low temperatures (about 8 K), and deal exclusively with molecules adsorbed at metal surfaces. Metal surfaces are necessary because the excitation thresholds are very low; on semiconductor surfaces the electron would usually be trapped in the molecule and could not propagate through the surface. In this case it does not lead to a change of the tunneling current and remains therefore undetected.

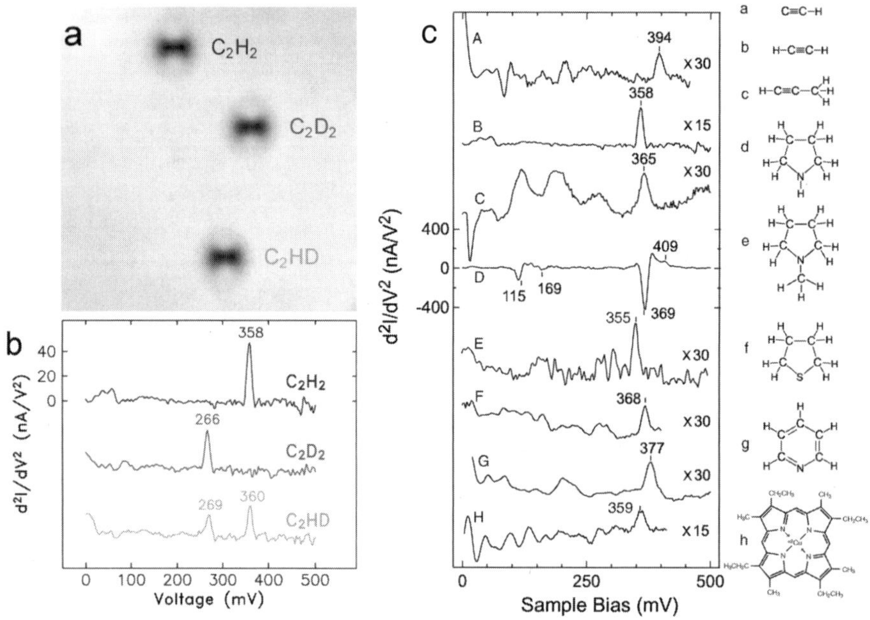

Fig. 8.3. (a) Topographic image of three acetylene molecules on Cu(001). The three molecules, C_2H_2, C_2D_2, and C_2HD are not differentiated in topographical images, since their electronic structures are identical. (b) Single-molecule vibrational spectra on the three molecules: the C-H stretch frequency is at 358 meV, the C-D frequency at 266 meV; the vibrational spectrum for C_2HD reveal both frequencies simultaneously. (c) A collection of vibrational spectra for different hydrocarbon molecules. In all cases the C-H stretch frequency most prominently shows up in the spectrum. W. Ho, J. Chem. Phys. 117, 11033 (2002). Copyright (2002) American Institut of Physics, reprinted with permission.

The vibrational frequency of a particular peak is the signature of a particular phonon excitation. As seen in Figure 8.3, the easiest mode to identify for organic molecules is the C-H stretch frequency. With sophisticated DFT methods, the calculation of phonon modes of a surface or a molecule is theoretical routine today. However, even if fully quantitative predictions of the change of the first and second derivatives of the tunneling current with respect to the bias voltage are possible (see Chapter 5), the theoretical results presented so far have relied on qualitative models. The most sophisticated of these models use many-body theory to determine the change of the surface Green's function at an energy threshold, and estimate the change of the conductance as the bias voltage crosses the threshold of a phonon excitation.

8.3.1 Theoretical procedure

The theoretical framework for calculating the many-body effects due to electron–phonon coupling has been introduced in Section 5.6. In practice, the inelastic contribution to the conductance changes across a vibration threshold is given by (see Chapter 5, (5.104):

$$\Delta\left(\frac{dI}{dV}\right)_{ine} = \frac{\pi}{2}\sum_{k,\lambda} \left| \int d\mathbf{S}\cdot(\delta\psi_\lambda \nabla\chi_k^* - \chi_k \nabla\delta\psi_\lambda^*)\right|^2 \delta(E_F - E_k)\delta(E_F - E_\lambda). \tag{8.4}$$

The χ_k denote, as usual, the tip wavefunctions, while the $\delta\psi_\lambda$ describe the differential changes of the surface wavefunctions due to phonon excitation. The perturbed sample wavefunctions are given in the same chapter, (5.6).
In the quasistatic limit the elastic contribution, which comes from an exchange of electron paths in the limit of infinite time, is formally identical to the inelastic one, the only two differences are that the differential changes of the surface wavefunctions are the imaginary parts of the changes in the elastic case, and that the contribution is negative. Both contributions can be readily evaluated by a modification of existing DFT codes if the overlaps of wavefunctions are explicitly calculated.

8.3.2 Applications

The field is expanding quite rapidly; at present there are several experimental groups in Europe and America that have mastered the intricacies of the experimental method and produce inelastic tunneling spectra on a routine basis. Theoretically, the field is very much dominated by European groups, which first developed a model to include the effect in the Tersoff–Hamann model of tunneling and succeeded in establishing qualitative agreement between the first experiments and simulations based on DFT.

C_2H_2 on Cu(100)

The second derivative of the tunneling current of C_2H_2 on Cu(100) is shown in Figure 8.3. It reveals only a single peak at 358 mV; the change of the conductance is very high and about 10%. This seems quite surprising at first glance, because this change is close to the theoretical limit for the inelastic contribution. The answer to this problem was found by Lorente and Persson [10] by an explicit calculation of vibrational contributions. It turned out that the electrons tunneling via the π^* orbital of the molecule experience roughly equal elastic (change of electron pathways) and inelastic (excitation of phonons) contributions, which renders the total change in conductance close to zero near the Fermi level. The only excitation is an antisymmetric C-H stretch mode of the molecule, where the contributions are purely inelastic and therefore lead to an increase of the conductance close to the theoretical limit. The experimental and theoretical results of this case are shown in Figure 8.4.

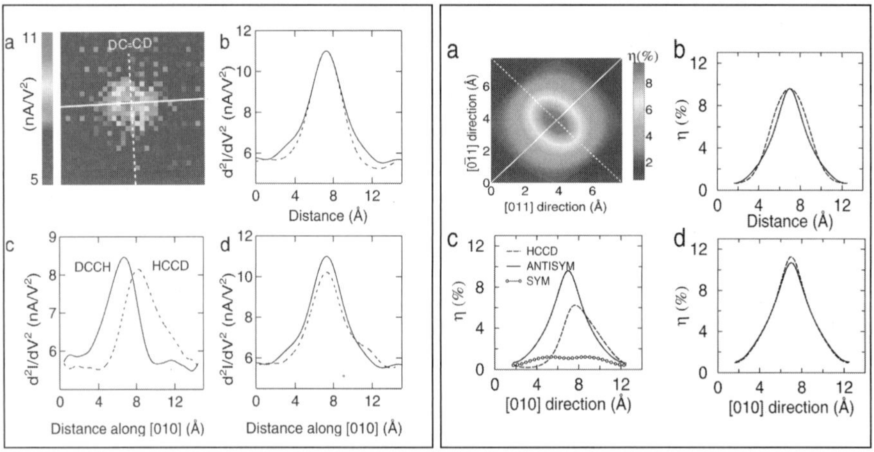

Fig. 8.4. Experimental (left frames) and theoretical (right frames) results of vibrational spectroscopy for acetylene on Cu(100). The vibration signal in the STM spectra shows an elliptic local increase of conductance; the bell-shaped maximum in the second derivative of the tunneling current is well reproduced in the simulations. The increase of conductance is due to an antisymmetric C-H stretch mode. N. Lorente and M. Persson and L. J. Lauhon and W. Ho, Phys. Rev. Lett. 86, 2593 (2000). Copyright (2000) American Physical Society, reprinted with permission.

O_2 on Ag(110

The vibration spectra of acetylene on copper showed a rather featureless, bell-shaped maximum at the position of the molecule as the bias voltage is ramped across the threshold of the C-H stretch mode of the molecule. The electronic distribution in this case is rather featureless and more or less centered at the C-H axis. A different example, and one where an electronic π state is actually seen in the spectrum, was presented a few years ago by Hahn et al. [11]. They measured topographical images and vibration spectra of $^{16}O_2$ and $^{18}O_2$ adsorbed on Ag(110), measurements were again taken at very low temperatures. In Figure 8.5 we show their spectrum for the two molecules.

For reference, the spectrum of the bare Ag(110) surface was also measured. As the difference between the spectra of the adsorbed molecules and the clean surface reveals, the $^{18}O_2$ molecule possesses two distinct phonon modes: an O-O stretch mode at -76.6 meV, and an antisymmetric O_2Ag stretch mode at -35.8 meV. The conductance changes in this case are much lower and only about 2–3%. Correspondingly, the peaks at the energy of the modes are less distinct and become comparable to background noise.

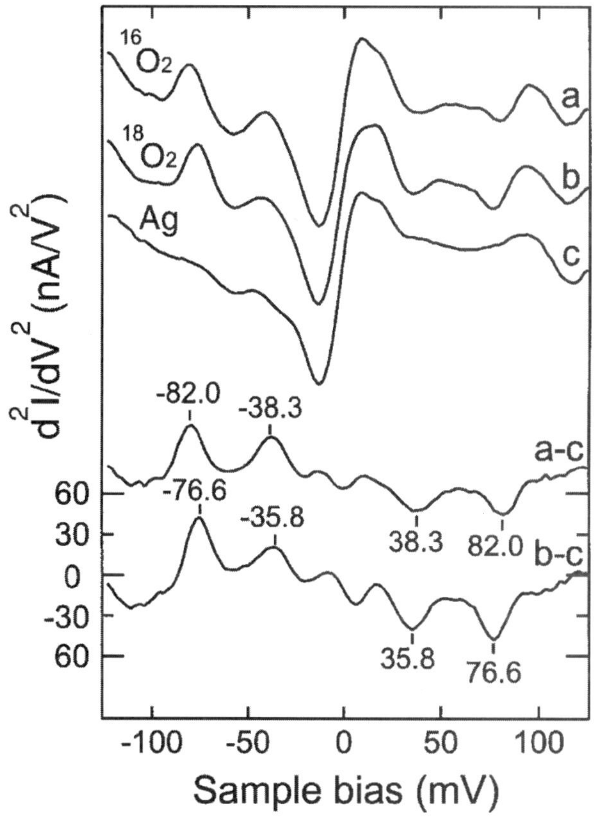

Fig. 8.5. Single-molecule vibration spectrum for $^{16}O_2$ (curve a) and $^{18}O_2$ (curve b), and the clean Ag(110) surface (curve c). The difference spectra a-c and b-c are also shown. The spectra reveal two phonon modes of the molecule: an O-O stretch mode at -76.6 mV (for $^{18}O_2$) and an O_2Ag stretch mode at -35.8 mV (also for $^{18}O_2$). The corresponding values for $^{16}O_2$ are slightly shifted. J.R. Hahn, H.J. Lee and W. Ho, Phys. Rev. Lett. 85, 1914 (2000). Copyright (2000) American Physical Society, reprinted with permission.

Vibration spectra with high local resolution provide additional information about the electronic states. As seen in Figure 8.6, the topographic image is quite different from the vibration spectrum. In this case the orientation of the molecule can be determined (see frames (a) and (b)). A comparison with the vibration spectra (frames (c) to (f)) shows that the phonon excitation concerns only a single molecular orbital, the 1π orbital of the molecule.
Theoretically, the problem of O_2 adsorption on Ag(110) has been analyzed by Olsson et al. [12]. Even though their theoretical treatment did not include an explicit calculation of the vibration spectrum, the conclusion that tunneling is primarily effected via the $1\pi_g^{\perp}$ orbital of the molecule agrees well with the spatial features of the vibrational spectrum in the experiments.
In a further development of chemistry at this level it was shown by Ho's group [3] that oxygen molecules adsorbed on Pt(111) can be dissociated via a voltage pulse of about 0.3 V and 22 nA. The problem had been studied previously [13], where it was found that the anharmonicity of the potential well of oscillating electrons may lead to a rupture of the bond if an electron tunnels into an adsorbate induced resonance that remains populated long enough so that multiple excitations become possible. In this case successive excitations with an ever higher energy level will finally lead to electron energy surpassing the upper limit of the anharmonic potential well, and the electron will ultimately escape, rupturing the chemical bond.

8.4 Summary

In this chapter we reviewed recent work on single molecule manipulation and chemistry. It should be emphasized that bottom-up engineering of new chemical and material structures will ultimately be confronted with this frontier, because the detailed electronic and atomic processes not only need to be understood, but actually to be exploited if genuinely new structures are to be created. In this sense the field covers a substantial part of what one could call atomic-scale engineering. Not surprisingly, it turns out that the SPM is the preferred tool under these conditions.
For theorists, the field provides a substantial challenge. This is mainly due to the interaction between electronic and nuclear degrees of freedom under these conditions. While ground-state DFT provides a clear and precise framework for processes related to electronic behavior, it is generally based on the Born–Oppenheimer approximation: the nuclei are, for all practical purposes, considered unmovable. Molecular dynamics may seem to provide a general solution to these problems, but it is restricted by the short time scales it can simulate, so that the field is, theoretically, still rather in its infancy.
The manipulation of atoms or molecules on a surface, as long as the surface and STM tip atoms are reasonably far apart (about 4–6 Å) should be possible within self-consistency cycles, including groundstate DFT, the calculation of

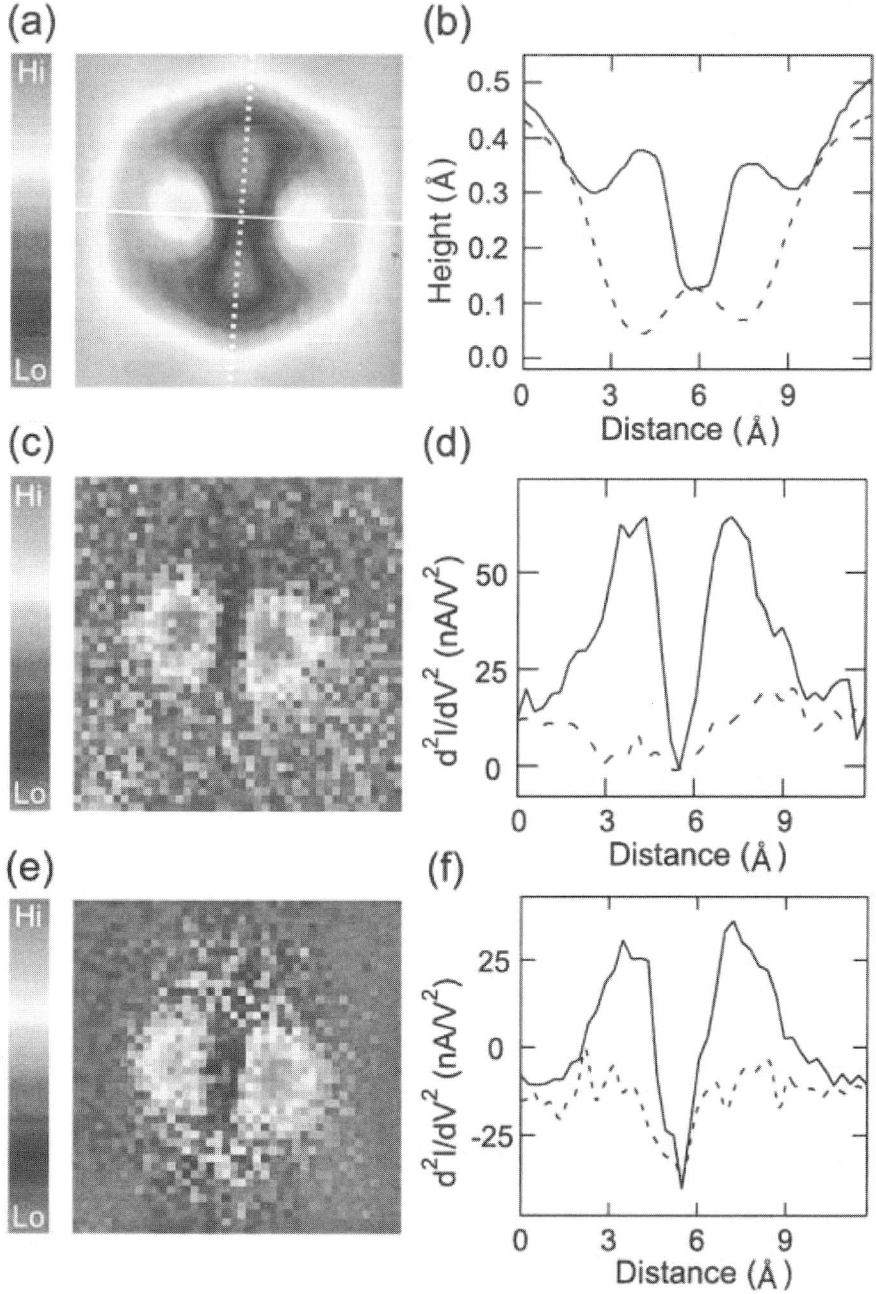

Fig. 8.6. High local resolution topography (a) and vibration spectra ((c) and (e)) of $^{18}O_2$ on Ag(110). The corresponding scanlines are shown in frames (b), (d), and (f). The topographic image shows the molecule as a depression on the silver surface (a); the scanlines (b) allow one to determine its orientation along the [001] direction of the crystal. A vibration spectrum across the threshold of the O-O stretch frequency at -76.6 mV (c) and (d) reveals that the excited electrons originate from the 1π orbital of the molecule. The spectrum (e) was taken at the antisymmetric stretch mode of O_2Ag. [11].

interaction energy, and the determination of tunneling currents. The scheme of such a self-consistency cycle has been developed in this chapter.

Vibrational spectroscopy, with its ability to target specific modes and thus to incite specific reactions [14], has been the subject of intense theoretical efforts in the last years. It is in principle possible to treat the effect on the same theoretical level as other tunneling phenomena. However, a fully first principles approach, including the STM tip, has so far not been developed. The theoretical basis for such an approach can be found in Chapter 5.

References

1. D. M. Eigler and S. Schweizer. *Nature*, 344:524, 1990.
2. N. Lorente, R. Rurali, and H. Tang. *J. Phys: Condens. Mat.*, 2005.
3. W. Ho. *J. Chem. Phys.*, 117:11033, 2002.
4. H. Ueba. *Surf. Rev. Lett.*, 5:771, 2004.
5. C. Loppacher, M. Guggisberg, O. Pfeiffer, E. Meyer, M. Bammerling, R. Luthi, S. Schittler, J. K. Gimzewski, and C. Joachim. *Phys. Rev. Lett.*, 90:066107, 2003.
6. L. Bartels, G. Meyer, and K.-H. Rieder. *Phys. Rev. Lett.*, 79:697, 1997.
7. X. Bouju, C. Joachim, C. Girard, and P. Sautet. *Phys. Rev. B*, 47:7454, 1993.
8. U. Kürpick and T. S. Rahman. *Phys. Rev. Lett.*, 83:2765, 1999.
9. L. Pizzigalli and A. Baratoff. *Phys. Rev. B*, 68:115427, 2003.
10. N. Lorente, M. Persson, L. J. Lauhon, and W. Ho. *Phys. Rev. Lett.*, 86:2593, 2000.
11. J. R. Hahn, H. J. Lee, and W. Ho. *Phys. Rev. Lett.*, 85:1914, 2000.
12. F. E. Olsson, M. Persson, and N. Lorente. *Surf. Sci.*, 522:L27, 2003.
13. B. C. Stipe, M. A. Rezaei, W. Ho, S. Gao, M. Persson, and B. I. Lunqvist. *Phys. Rev. Lett.*, 78:4410, 1997.
14. J. I. Pascual, N. Lorente, Z. Song, H. Conrad, and H. P. Rust. *Nature*, 423:525, 2003.

9

Current and Force Spectroscopy

In this chapter we present theoretical tools for current spectroscopy and look at the present state of the art in force spectroscopy. It will be seen that accounting for current spectra is rather more involved, theoretically, than simulating constant-current contours. It will be explained why the band structure plays such an important role in spectroscopy, and how the energy resolution of a surface band enters the thermal range in spectroscopy, which can then be simulated with any degree of precision required. We shall have a close look at the problem of the tip electronic structure, and develop a differential spectroscopy model that circumvents the problem that surface and tip states are convoluted in a Bardeen approach to spectroscopy. The method will be demonstrated on three noble metal surfaces with surface states, where we find that the present efficiency in electronic structure simulations makes it possible to simulate spectra in the thermal range above 150 K. We shall also look at magnetic surfaces and see that magnetic properties can, in single cases, be resolved at the very small level of single atoms.

For spectroscopies performed by SFM we shall emphasize the experimental conditions and requirements on the instrument, and the advantages it presents in comparisons with theoretical results.

9.1 Current spectroscopy

Experiments in scanning tunneling microscopy (STM) over the last ten years have increasingly moved from the analysis of structure and atomic positions to the analysis of electronic properties and processes. The main reason for this change of focus is the wide availability of low-temperature (about 80–200 K) and even very low temperature (mK to a few K) microscopes. The new level of experimental precision has had some remarkable consequences in research. In particular, the demonstration of standing-wave patterns of surface electrons, e.g., on Ag(111) [1, 2, 3], images still used for the cover design on nanotechnology books, has come to symbolize more than other measurements the instrument's power and precision. In a wider frame of reference tunneling spectroscopy, due to its rich features and resolution power of complex electronic properties, is probably the most expansive field in low-temperature STM experiments today. This evolution clearly indicates that the field has sufficiently matured. High-resolution topographies and numerical simulations are now all but routine, and apart from the so far unsolved problem of assessing the agreement between experimental and simulated STM tip models, the correspondence between experiment and theory is satisfactory if not, in some cases, spectacular [4, 5].

To move theoretically in the same direction as low-temperature experiments, i.e., to go from topography to high-resolution spectroscopy, poses some particular problems once theory tries to progress beyond the description of the surface alone. The surface alone and its characteristics in tunneling spectra can be inferred from the local density of states and its changes with bias voltage. But if a simulation is to include the whole tunneling junction and not just one part of it, it is immediately confronted with the problem of tip representation. It is quite well known that metal surfaces frequently possess a rather complicated band structure near the Fermi level [6]. In an experimental spectrum the features will be determined not only by the surface itself, but also by the STM tip.

Experimentally, a way around this problem is the preparation of specific blunt tips [7], which can be gauged on particular systems, e.g., Cu(111), where the signature of the surface electronic structure is well known. Theoretically, a similar strategy needs to be adopted if one does not know initially the detailed structure of the STM tip used in spectroscopy experiments. In spectroscopy, the parameter space for tip simulations is substantially larger than in topographies, because one crucial property of tips in high-resolution topographic experiments is missing. Spectroscopy tips do not usually yield atomic resolution of a flat metal surface, which suggests that they *cannot* possess a monoatomic apex. This removes many of the constraints used to great profit in topography simulations. Since the current is no longer passing through a single apex, the theoretical problem can no longer be reduced to finding a particular chemical composition, e.g., a contamination on a tungsten base [8], but is also confronted with the exact geometry and termination of a large array of atoms. The parameter space in this case increases beyond the numerically manageable. That is to say, even though one can in principle construct tip models of a few hundred atoms with today's computer codes, it is infeasible to map the ensuing band structure with suitable resolution.

How does the resolution of a spectrum relate to the map of the band structure? Assuming that the energy resolution in today's best experiments is in the range of mV [3], a numerical representation of the experiment has to include, for a specific band, at least one k-point of the surface Brillouin zone every few mV. This makes, on a noble metal surface, for a few thousand k-points of the band structure map. On a known surface one could circumvent the problem by some clever interpolation routine. However, on an unknown or electronically very complicated, e.g., magnetic, surface, this might not work too well and shift the problem mainly to a different arena, without actually reducing the numerical effort. Given this level of precision, the parameter space for accurate tip calculations becomes so large, and the calculations so expensive, that they would actually be infeasible for routine simulations of spectroscopy experiments.

For these reasons we have adopted a two-pronged strategy. On the one hand, we have developed numerical spectroscopy simulations based on the Bardeen method, shown to be very reliable in topography simulations [9, 4]. On the other hand we have also developed a direct way to determine the changes from one bias value to the next. This method allows one to separate the effects of surface and tip electronic structure. The main conclusions of our simulations of noble metal surfaces are that (i) simulated spectra are at present precise enough for detailed comparisons with state-of-the-art experiments in the temperature range above 150 K; and (ii) the electronic structure of the tip models tested is unsuitable for representing the tips in the experiments. This chapter is organized as follows: In section 9.1.1 we shall briefly talk about the computational methods and the main equations of Bardeen and differential spectra. Section 9.1.2 shows our simulated results for the spectra on noble metal (111) surfaces. The last sections present a discussion of results and a summary of the work.

9.1.1 Differential tunneling spectroscopy simulations

The central theoretical formulations of Bardeen's approach to tunneling are well known; their main equation describes the dependency of the tunneling current on the bias voltage V and the position of the STM tip $\mathbf{R}$. It is usually written [9, 4]

$$I(\mathbf{R}, V) = \frac{4\pi e}{\hbar} \sum_{ik} \int_0^{eV} dE \left[f(\mu_S, E - eV) - f(\mu_T, E) \right] \left| -\frac{\hbar^2}{2m} M_{ik} \right|^2 \delta(E_i - E_k). \tag{9.1}$$

Here, $\mu_{S(T)}$ denotes the Fermi function of the surface (tip); the delta functional in this equation is due to energy conservation, which is a central assumption in the derivation of the formula, e.g., from the separate wavefunctions of surface and tip in Chen's modified Bardeen approach [10]. In principle, the relation can be used for any bias voltage within a range of about ± 1 V, where the change of electronic properties of the sample surface due to the potential of the tip is sufficiently small and well below the range of field emission. The set point of a given spectroscopy measurement is determined by the current at a specific bias voltage. Usually, the current feedback loop is then disengaged and the bias voltage changed over a specific range. It is thus only by performing the integration that one actually knows the distance of the STM tip from first principles.

Here, we encounter two fundamental theoretical problems: firstly, the fact that experimental results are usually given in terms of dI/dV curves, which will involve the numerical differentiation of (9.1). The differential value cannot directly be obtained, e.g., from the integral kernel, without some approximations involved, since a change of bias will also change the transition channels by virtue of the delta functional. Such a numerical differentiation, however, will make the calculation inherently unstable due to the Fermi distribution functions. Since the largest slope of the current will be found near the Fermi level, this region will always show up prominently in the computed spectrum. An obvious solution to the problem is to compute the spectra for zero temperature. In this case the Fermi distribution functions are changed to step functions and electrons of both systems will always tunnel strictly from an occupied into an unoccupied state.
The second problem is a problem of control: the numerical differentiation procedure does not allow any distinction between states of the surface and states of the tip. The ensuing $I(V)$ distribution and its numerical derivative are therefore convolutions of surface and tip states. The electronic structure of both sides of the tunneling junction contribute in an equal manner to the result. A solution to this problem would be to revert back to the Tersoff–Hamann model of tunneling [11, 12]. There, the current is assumed proportional to the local density of states of the surface by virtue of an approximation for the tip states, that is, that the only involved tip state is a state of radial symmetry, and that such a tip state exists for every bias voltage. The voltage dependency can be included by a model suggested by Lang [13]:

$$\frac{dI}{dV} \propto e\rho_S(\epsilon_F + eV)T(\epsilon_F + eV, V) + \int_{\epsilon_F}^{\epsilon_F+eV} \frac{d}{dV}\left[\rho_S(\epsilon)T(\epsilon, V)\right] d\epsilon, \quad (9.2)$$

where ρ_S denotes the local density of surface states at the vacuum boundary z_s, and $T(\epsilon, V)$ is the transmission coefficient through an average barrier at a given bias voltage and energy ϵ; it is described by

$$T(\epsilon, V) = \exp\left(-2s\sqrt{2m(\Phi_{av} + \epsilon_F - \epsilon + eV/2)}/\hbar\right), \quad (9.3)$$

with $\Phi_{\text{av}} = (\Phi_S + \Phi_T)/2$ the average workfunction, while s is the vertical distance between the tip apex and the vacuum boundary z_s of the surface. In the limit of zero bias only the first term survives, and the expression reduces to the usual formulation within the Tersoff–Hamann approximation:

$$I(\mathbf{R}, V) \propto \int_0^{eV} dE \sum_i |\psi_i|^2 . \quad (9.4)$$

In this case the integral kernel and thus the dI/dV spectrum can be directly determined since the eigenvalues, or the transition channels, are related only to the bias interval and are incrementally increased for an incremental increase of the bias voltage. However, it has to be clear that the spectrum obtained from the integral kernel is qualitatively different from the spectrum, e.g., obtained by the Bardeen method and a numerical differentiation. While the first could be called a differential spectrum, the second is the derivative of an integral spectrum. For reasons of consistency, the assumption that current is proportional to the local density of states, which is at the base of the Tersoff–Hamann model, finds its expression here in the proportionality of the current derivative and the derivative of the local density of states.

Using the scattering formulation of the tunneling current developed in previous chapters (see Chapter 5), we may include the bias dependency of the spectrum in this model also in a different way. Since the current is proportional to (Chapter 5, (5.77))

$$I_{TH}(\mathbf{R}, V) \propto \int_0^{eV} dE \sum_i \left| \left(-\frac{\hbar^2}{2m} - \frac{eV}{\kappa_i^2 - \kappa_T^2} \right) \psi_i(\mathbf{R}) \right|^2, \tag{9.5}$$

its derivative with respect to the bias voltage contains two separate terms (in atomic units for simplicity):

$$\frac{dI_{TH}(\mathbf{R}, V)}{dV} \propto \sum_i \left| \left(1 + \frac{V}{\kappa_i^2 - \kappa_T^2} \right) \psi_i(\mathbf{R}) \right|^2 + 4 \int_0^V dE \cdot E \sum_i \left| \frac{\psi_i(\mathbf{R})}{\kappa_i^2 - \kappa_T^2} \right|^2. \tag{9.6}$$

The increase of a dI/dV curve with the applied bias voltage should therefore show a quadratic dependency. This is quite different from the relation obtained from the Bardeen integration in the limit of zero bias, where the relation will be linear.

The distinction between differentiation in an $I(V)$ curve and computing the dI/dV curve directly by computing the increment seems an unnecessarily fine distinction, but it is also made in experimental spectra. Early experiments, e.g., by Stroscio and Biedermann [14, 15], determined the dI/dV curve by numerically differentiating a measured $I(V)$ curve. Numerical differentiation of noisy data is generally problematic. It is thus that the method works comparatively well in an ambient environment and for very distinct electronic features of a surface, e.g., a surface state on Fe(001) or Cu(111). It fares far worse in the determination of subtle features, e.g., the onset of a surface state in low temperature experiments or the threshold of a vibrational node in single-molecular spectroscopy. There, the incremental differences are directly determined by oscillating the bias voltage around a median value by about 20 mV [16]. Since the actual changes here occur in a range of only 20 mV, it becomes infeasible to account for experimental results by integral spectroscopy.

A short example might illustrate the issue. It is well known that the surface state of Cu(111) is located at about –450 mV from the Fermi level at the $\bar{\Gamma}$ point of the surface Brillouin zone. It is equally well known that the dispersion of the state leads to a k-vector of about $0.5\pi/a$ at the Fermi level. If the spectrum from –450 mV to 0 mV should possess a resolution better than 10 mV, then the interval must be spanned by at least 100 k-points, due to the requirement of numerical stability in the differentiation. This, in turn, means that the Brillouin zone has to be mapped by more than 100,000 k-points. Not only is it close to impossible to perform such a calculation within density functional theory, given standard-size computer systems, it also seems highly inefficient. The ensuing spectrum, precise though it might be, still does not answer the question, what actually causes a particular feature in the spectrum? The origin still could be both: the copper surface and the STM tip.

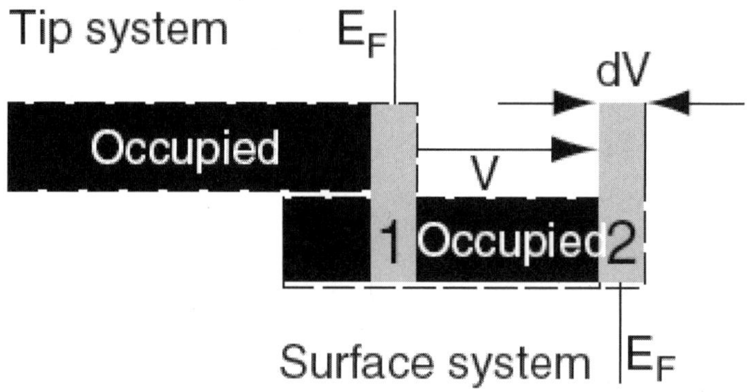

Fig. 9.1. Contributions to the tunneling current at a bias voltage V. The change of the bias by an incremental value dV leads to two distinct additional contributions (gray): one contribution mapping the states at the Fermi level of the tip onto the band structure of the surface (1), and one contribution mapping states at the Fermi level of the surface onto the band structure of the tip (2).

To account for both problems, the problem of high resolution and the problem of identifying a particular feature, we need to develop a method, similar to a determination of the spectrum from the integral kernel alone, but for a calculation including the STM tip. The solution could be called a differential Bardeen spectrum, and it rests on one approximation: a change of the bias by an incremental amount leaves the bulk of the transition channels unchanged. The only changes are the additional transitions due to the change of bias voltage. The incremental current change in this case has two distinct components: one component due to overlaps of states at the Fermi level of the tip and states of the band structure of the surface; and one component mapping states at the Fermi level of the surface to the band structure of the tip (see Figure 9.1). The method deviates to some extent from the method suggested by Hörmandinger a few years ago. There, the derivative of the current with respect to voltage was described by [17]

$$\frac{dI}{dV} = f_T(0) f_S(eV) - \int_0^{eV} dE\, f_T'(E - eV) f_S(E). \tag{9.7}$$

The functions $f_{S(T)}$ of the surface (tip) electronic structure denote the contributions to the tunneling matrix element in the Bardeen approach, and it was assumed that these contributions can be split into products. Here we follow a different approach for two reasons: (i) The expression still contains a derivative with respect to bias of the tip contributions. It will be seen in our analysis of simulated results that such a derivative is inherently numerically unstable and requires a map of the tip band structure beyond today's computational means. (ii) The integral kernel in the Bardeen matrix element contains the expression $(\chi^* \nabla \psi - \psi \nabla \chi^*)$. It seems infeasible to separate the ensuing square of the surface integral in separate surface and tip functions, or only if one assumes a featureless tip that can be described analytically.

Apart from making the separate contributions identifiable, the suggested method for differential spectroscopy has two additional advantages: since it does not involve a numerical differentiation it is also more stable and more precise; and since it involves at each step only a fraction of the actual bias voltage range, it is also considerably faster. The numerical values in such a calculation, however, are still actual values, e.g., in nA/V, and can therefore directly be compared to experimental results. The incremental change in the current due to a change of bias from V to $V + dV$ is then

$$\begin{aligned} dI = & \sum_{i_1 k_1} \left| \left(\frac{\hbar^2}{2m} - \frac{eV}{\kappa_{i_1}^2 - \kappa_{k_1}^2} \right) M(\psi_{i_1}, \chi_{k_1}) \right|^2 \\ & + \sum_{i_2 k_2} \left| \left(\frac{\hbar^2}{2m} - \frac{eV}{\kappa_{i_2}^2 - \kappa_{k_2}^2} \right) M(\psi_{i_2}, \chi_{k_2}) \right|^2 , \end{aligned} \tag{9.8}$$

where the eigenvalues of surface $E_{i_{1(2)}}$ and tip $E_{k_{1(2)}}$ states are within the following intervals:

$$\begin{aligned} E_{i_1} &\in [E_F + eV - edV/2, E_F + eV + edV/2] , \\ E_{k_1} &\in [E_F - edV/2, E_F + edV/2] , \\ E_{i_2} &\in [E_F - edV/2, E_F + edV/2] , \\ E_{k_2} &\in [E_F - eV - edV/2, E_F - eV + edV/2] . \end{aligned} \tag{9.9}$$

Here, E_F denotes the Fermi level of surface and tip system, respectively. Then the total spectrum contains equally two distinct contributions due to the band structure of the surface and the tip system:

$$\frac{dI}{dV} = \frac{1}{V}\sum_{i_1 k_1}\left|\left(\frac{\hbar^2}{2m} - \frac{eV}{\kappa_{i_1}^2 - \kappa_{k_1}^2}\right) M(\psi_{i_1}, \chi_{k_1})\right|^2 + \frac{1}{V}\sum_{i_2 k_2}\left|\left(\frac{\hbar^2}{2m} - \frac{eV}{\kappa_{i_2}^2 - \kappa_{k_2}^2}\right) M(\psi_{i_2}, \chi_{k_2})\right|^2 . \quad (9.10)$$

In this case the simulated values are very stable, since a constant-bias increment has the effect that the surface band structure always maps onto the same STM tip states at the Fermi level. Any changes from one bias value to the next are therefore directly related to the changes of the surface electronic structure. Numerically, the calculation is also very efficient: given a high resolution of the surface and tip Brillouin zone and a small bias increment of about 5–10 mV, the simulation of one point of a spectrum can be performed on a workstation within less than one hour. This, in turn, makes it possible to simulate tunneling spectra also with high local resolution.

Calculating the surface electronic structures

The electronic structures of noble metal surfaces and model tips were calculated with density functional methods. We used the Vienna ab initio Simulation Program (VASP) [18, 19]; the potentials of the ionic cores were modelled with VASP's projector augmented waves (PAW) [20] implementation. The exchange correlation energy was calculated within the PW91 parameterization scheme [21]. Initially, we calculated the electronic structure of 13 layer films, separated by 15 Å of vacuum. We found that in this case the surface states of the film couple through the bulk as well as the vacuum, leading to two surface states on either side with an energy difference of 10 mV. The number of layers was therefore increased to 25 (Ag(111)), or one surface of the film was passivated by hydrogen adsorbed in the threefold hollow sites (for Cu(111) and Au(111)). This measure was suitable either for reducing the splitting of the surface bands to less than 1 meV (Ag(111)), or for quenching the surface electrons on one surface. Similar results were previously found by full potential calculations of the Au(111) and Ag(111) surface states [22]. It should be mentioned that we used the experimental lattice constant in all calculations, since previous experience with a theoretical lattice constant indicates that the surface state of Ag(111), for example, is shifted into the unoccupied range. All surfaces were fully relaxed. The electronic ground state and the charge distribution were calculated with a Monkhorst–Pack grid of 10×10×1 special k-points in the irreducible wedge of the surface Brillouin zone. In the final calculation we used a k-map centered at the $\bar{\Gamma}$ point of 331 special points within the irreducible wedge covering only 25% of the full Brillouin zone. This is sufficient to map the surface bands on all surfaces up to about 1 eV above the Fermi level. The final map amounts to more than two thousand points at the center and a correspondingly high resolution of the metal band structure.

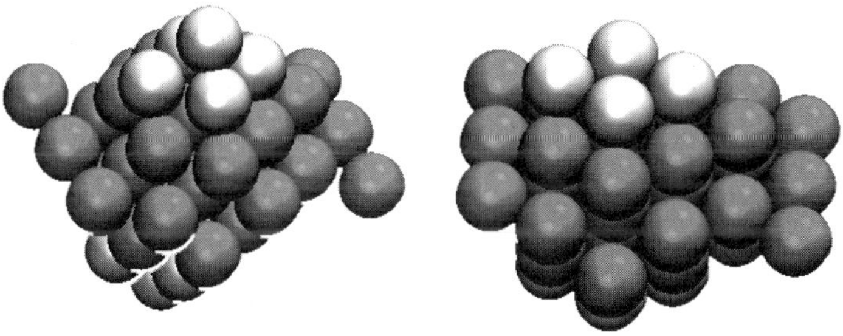

Fig. 9.2. The two STM tip models used in the simulations. The models are based on a three-layer tungsten film in (110) orientation (dark atoms) covered by a tungsten pyramid (left, bright atoms), or a tungsten tetramer (right).

The STM tip models in the simulations were tungsten (110) films terminated either by four tungsten atoms (flat tip) or by four tungsten atoms and a single atom at the apex (sharp tip). Also in this case the systems were fully relaxed. Since the tip unit cells are significantly larger than the surface unit cells, we used only 100 k-points centered at the $\bar{\Gamma}$ point to map the tip electronic structure. The atomic arrangement of the two tip models is shown in Figure 9.2.

Calculating the tunneling spectra

The calculation of tunneling spectra with the method described above is straightforward. The bias voltage is ramped from an initial value (–1 V) to a final value (+1 V) in steps of 20 mV. At every step the incremental change is calculated. We present results only for one STM tip position on the surface: the position on top of surface atoms. The results for the hollow positions differ only slightly by their absolute values (less than 1% in the distance range of evaluation). The set point for the evaluation, which corresponds to the point in the experiments, where the feedback loop is disengaged, was set to a distance of 7.0 Å between the surface and the STM tip. In this range currents are sufficiently small (less than 1 nA at a bias of –1 V) so that interactions between surface and tip can be safely neglected. The dI/dV spectra were finally smoothed with Gaussians of 20 mV, 50 mV, and 80 mV half-width, respectively.

9.1.2 Differential spectra on noble metal surfaces

The (111) surfaces of copper, silver, and gold have been researched extensively due to their importance as model systems for two dimensional electronic structures. The silver (111) surface, where the bottom of the surface state band lies very close to the Fermi level, also allows one to study interference of surface electrons scattered at atomic centers or step edges. The reason, in this case, is the long wavelength of surface electrons at the Fermi level, which is substantially larger than the dimension of the unit cell. In this case the interference pattern dominates STM images, and the decay, e.g., from a step edge, can be used to estimate the lifetime of surface state electrons due to their interactions with electrons and phonons of the crystal. The copper (111) surface is one of the main systems in surface science research for three reasons: it is easy to clean and then provides large flat terraces that can be measured by scanning probe or electron diffraction methods; it does not interact easily with adsorbates and can therefore be used to study the processes occurring during alloying with other metals; and it is one of the easiest systems to measure with STM due to the relatively high corrugation of 10–20 pm. Apart from that, its surface state is easily recognized even under ambient conditions. The gold surface is important in two separate fields: In nanoelectronics it is used as the conducting material, e.g., in circuits comprising organic molecules or single organic films positioned between gold electrodes. Such a setup has potentially important applications in organic transistors. In heterogeneous catalysis it was found recently that small clusters of gold on, e.g., titanium oxide templates provide a much enhanced catalytic performance [23]. There seems to be still some controversy over whether this enhancement is due to catalytic processes at the rims of the cluster, or whether the strain on the gold films changes their electronic properties in such a way that the interaction with physisorbed molecules changes significantly [24]. Copper and gold surface states have been most successfully mapped by angle-resolved photoemission experiments [25], while low-temperature STM measurements have been used to determine the onset and thus the lifetime broadening [3]. In all three cases detailed spectroscopic simulations have so far not been performed; the best theoretical data result from one-dimensional models including many-body effects [26].

Tunneling spectra on Cu(111)

The differential spectra on Cu(111) are shown in Figure 9.3. The top frames represent the surface contributions only. The left frames are the simulated spectra with an atomically sharp tungsten tip; the right frames show the spectra for a flat tip. We simulated the ensuing spectra under different thermal conditions, included in the evaluation by a broadening of spectral features with a Gaussian of 20 mV (gray), 50 mV (full line), and 80 mV (dashed line) half-width. We note that the top frames for the two spectra are initially identical. The only difference is the absolute value, which depends on the number of states at the Fermi level of the STM tip. It can be seen that this has no influence on the onset of the surface state at about −450 mV. The oscillations of the spectrum with a period of about 25 mV denote the energy resolution in the simulation. Decreasing the broadening to less than 20 mV thus will not affect the ensuing spectrum. Considering the overall shape of the spectrum, it can be seen that it is rather similar to a step function. If the thermal broadening is increased to about 50 mV the oscillations due to the discrete k-mesh in the simulation vanish. It may thus be concluded that the spectrum yields an accurate description of experiments at temperatures above 150 K.

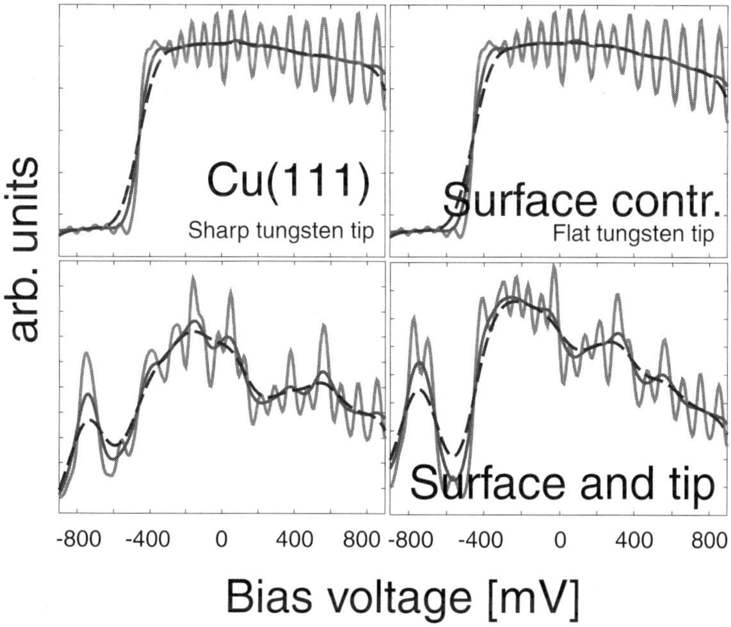

Fig. 9.3. Differential spectrum on Cu(111). The top frames show the surface contribution of the spectrum, simulated with the sharp (left) and flat (right) tip. The bottom frames display the sum of surface and tip contributions. Results are given for three different values of thermal broadening: 20 mV (gray), 50 mV (solid line), and 80 mV (dashed line). It can be clearly seen that the tip contribution for the chosen model tips leads to a peak at about −800 mV, in contrast to experiments.

The bottom frames show the spectra including the band structure of the STM tip. We notice two features that are due to the tip electronic structure alone: (i) a peak at about −800 mV, (ii) a dip at the Fermi level. Both of these features are not confirmed by experimental data [27]. They clearly indicate the limitations of the tip models used in the simulations. Given that the peaks in the full spectra are unequally spaced, the additional features should not be due to the limited k-grid of the tip used in the simulations. The effective area of the k-mesh was limited to about 25% of the tip Brillouin zone. If this were the origin of the additional peaks, then under the condition of equally spaced d-bands the peak would reveal an equal distribution with respect to energy. However, the difference between the first peak at −800 mV and the second or third varies not only in value, but also from one tip to the other. We conclude from this feature that the additional peaks, distorting the spectrum of the surface electronic structure, are due to the limitation of our tip unit cell and the confinement of electrons to comparatively few film layers. The same conclusion could be drawn by a comparison with the spectra on Ag(111) or Au(111), which reveal the same additional peaks, also varying from the sharp to the flat STM tip model. There is, however, one feature of the spectrum including the STM tip that has a profound physical meaning. It can be seen in particular by comparing the full spectrum for the combined flat tip and surface with the spectrum of the surface alone that the former shows a slope towards higher energy values that does not exist in the latter. This effect is genuine: as the surface states map onto tip states with lower energy, and hence the energy is increased, the decay length of tip states becomes shorter and thus the overlap with surface states smaller. The dispersion of the surface state, which leads to a change of the electron momentum in the z direction, and the height of vacuum barrier at a specific energy uniquely characterize the decay length of surface state electrons. Since it remains fairly constant over the bias range, as revealed by the spectrum of the surface state alone, the additional tip contribution leads to a slope toward higher bias values, which is also confirmed by experiments [27].

Tunneling spectra on Ag(111)

On Ag(111) the simulations including only the surface electronic structure and the states at the Fermi level of the STM tip show an onset of the surface state around –80mV to –70mV, indicating the bottom of the surface band. In this case the spectrum also shows a slope toward higher bias values. We conclude from this difference that the one-dimensional picture, which unambiguously assigns a step function behavior to the spectrum of surface states, needs to be modified in a three-dimensional environment. As the decay of the surface state electrons into the vacuum depends on an interplay between the dispersion of the state and the decay characteristics due to the vacuum barrier, it is subject to differences for different metals. In a simplified model we may write for the relation between the vacuum decay constant κ, the energy eigenvalue E, and the reciprocal lattice vector k (in atomic units),

$$\kappa^2 - k^2 = 2E, \qquad E = E_0 + \alpha k^2, \tag{9.11}$$

where α is characteristic of the dispersion of a particular surface state. If we assume that the differential contribution to the tunneling current at a specific bias voltage and distance z_0 depends on the vacuum decay length κ and the area of the Brillouin zone covered in a particular energy interval, the differential change of the current will be

$$dI(k) = \exp\left(-z_0\sqrt{E\frac{1+2\alpha}{\alpha} - \frac{2E_0}{\alpha}}\right) 2k\pi dk. \tag{9.12}$$

The relation between bias voltage and energy is given by

$$E = E_F + V \tag{9.13}$$

Then conductance, or dI/dV will be described by

$$\frac{dI}{dV}(z_0, V) = \frac{\pi}{\alpha} \exp\left(-z_0\sqrt{V\frac{1+2\alpha}{\alpha} + \text{const}}\right). \tag{9.14}$$

The spectrum will show a slope as the bias voltage is increased. The slope depends on the energy of the surface state at the center of the Brillouin zone E_0, and the surface state dispersion α, and will differ for different surface states.

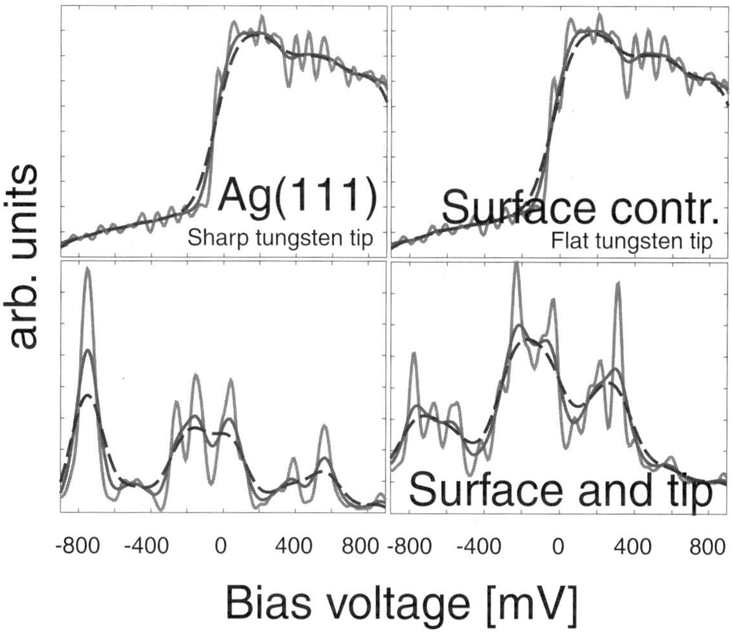

Fig. 9.4. Differential spectrum on Ag(111). The top frames show the surface contribution of the spectrum, simulated with the sharp (left) and flat (right) tips. The bottom frames display the sum of surface and tip contributions. It can be clearly seen that the tip contribution in this case distorts the spectra and makes the onset of the surface state all but unrecognizable.

The results of our simulation on Ag(111) are shown in Figure 9.4. The slope is considerably larger in the positive-bias regime than, e.g., for Cu(111). Also in this case the surface contributions alone are in good agreement with experiments, while including the tip band structure in the differential spectrum leads to artificial peaks at –800 mV, –200 mV, and +500 mV (sharp tip), or –800 mV, –200 mV, and +300 mV (flat tip). But while on Cu(111) the surface state remains recognizable even with the tip contributions, it is distorted beyond recognition on Ag(111). In case of a sharp tip the main peak is at the lower end of the spectrum. In case of a flat tip the energy value of the peak is reproduced, but the onset is substantially extended.

Tunneling spectra on Au(111)

The tunneling spectrum on Au(111) shows the onset of the surface state around –500 mV, in agreement with experiments. Also the steeper slope toward higher-bias voltages as compared to, e.g., Cu(111) is backed by experimental data [26]. In the plots of the surface contributions the energy resolution of the band map with low thermal broadening (Figure 9.5, gray line) can clearly be distinguished. We conclude from the simulations, as in the previous cases, that the band map is precise enough to reproduce experimental spectra in a thermal range above 150 K. Concerning the tip contributions we note the same problem as on the other surfaces: the peaks introduced by the tip electronic structure, in particular its confinement to a few layers of tungsten, have no correspondence in experimental data.

Surface state onset and lifetime broadening

In the past, two methods to determine the lifetime of surface state electrons have been suggested: (i) the slope of the surface state onset in a spectrum, which directly relates to the lifetime broadening due to inelastic electron–electron and electron–phonon interactions [3]; (ii) the decay of standing waves with the distance from a surface terrace [28]. We have calculated the slope of the onset from our spectroscopy simulations and compared the results to experimental low-temperature measurements. In all cases the spectrum was simulated with an energy resolution of 2 mV, and the ensuing spectrum smoothed with a Gaussian of 10 mV half width. The results of this comparison are shown in Figure 9.6. At first glance, the slope differs for Ag(111) and Cu(111) or Au(111) surface states; the results furthermore seem in acceptable agreement with experimental data [26].
However, the theoretical model of our simulations does not include many-body effects. Given that the onset of a surface state should in an ideal case be a step function (see (9.14)), there should be no difference between the onset at the three noble metals. But the picture changes if we assume that the energy resolution in the spectrum, essentially determined by the band structure map, is comparable to the onset we determine in the simulation. Then, the dispersion curve of the surface state will provide the energy difference from one k-point of the band structure map to the next. And if the dispersion curve has a very low curvature at the $\bar{\Gamma}$ point, then the energy resolution will be improved. This makes the simulated onset for Ag(111) only

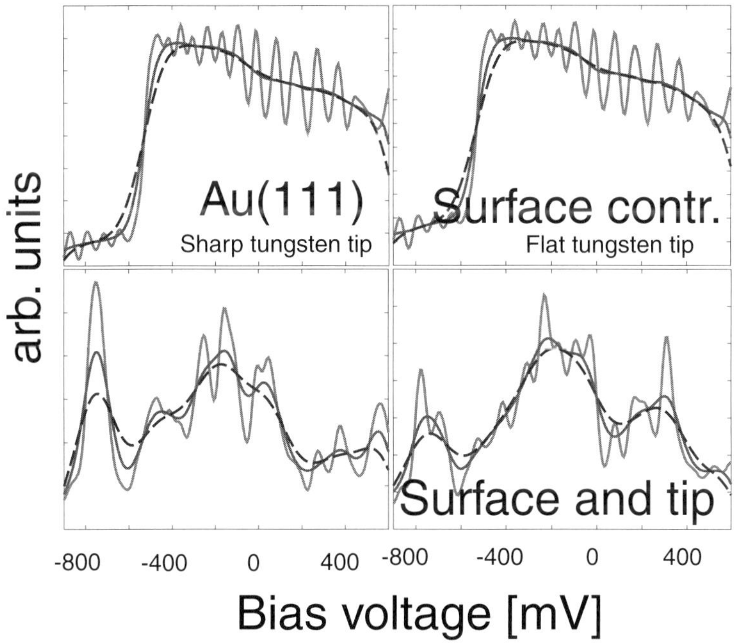

Fig. 9.5. Differential spectrum on Au(111). The top frames show the surface contribution of the spectrum, simulated with the sharp (left) and flat (right) tips. The bottom frames display the sum of surface and tip contributions. It can be clearly seen that the tip contribution also in this case distorts the spectra and makes the onset of the surface state all but unrecognizable.

about half the value obtained for Au(111) and Cu(111). From the viewpoint of comparisons between experiments and simulations with respect to surface state lifetimes, it seems that a fully adequate comparison requires an even higher resolution of the band structure map than used in this work. Including many-body effects in such a simulation seems well beyond computational means today. Considering the problem of simulation, it seems to be safe to conclude that a theoretical treatment of lifetime broadening is best applied to the decay of standing waves, e.g., scattered from a step edge rather than to the slope of the surface state onset. In the latter case it seems unclear, as long as three-dimensional simulations are not viable, how a particular experimental situation will actually influence the achieved onset and thus make a detailed comparison between experiment and theory problematic.

9.1.3 Spectra on magnetic surfaces

In recent years tunneling spectroscopy has gained importance due to two separate developments: (i) the modifications of the instrument itself, making

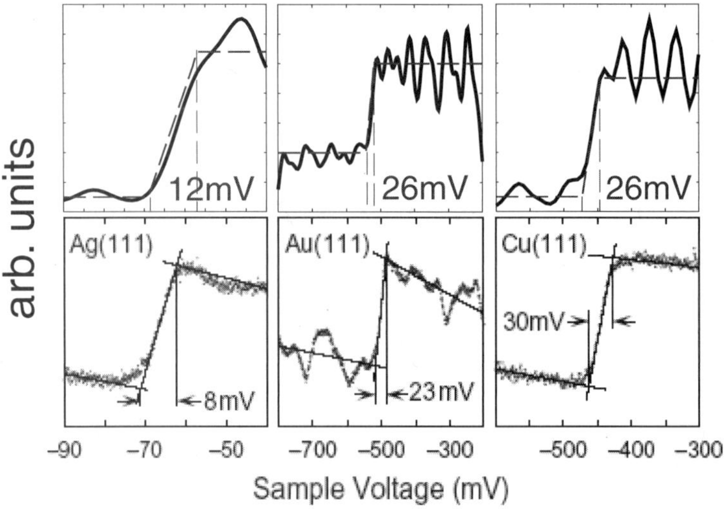

Fig. 9.6. Onset of surface states on noble metal surfaces. The values obtained seem to correspond well with experimental values, even though they relate to the band structure map rather than inelastic effects, as explained in the text. Experimental data (bottom frames): P.M. Echenique et al., Surf. Sci. Rep. 52, 219 (2004). Copyright (2004) Elsevier, reprinted with permission.

it suitable for very low temperatures and bias voltage oscillations around a small value (typically a few to a hundred mV); (ii) functionalizing STM tips with magnetic coating, which allowed the detection of detailed magnetic structures. While the first development has its main applications in surface state analysis and vibration spectra, the second has led to a host of new experiments of locally confined magnetic structures [29].

Fe(001)

The (001) surface of iron possesses a surface state in the spin-down band near the Fermi level at a positive energy of about 150 to 200 meV. The surface is notoriously difficult to simulate, since the change of lattice parameters, the change interlayer spacing, and lattice distortions will all affect the magnetic properties quite dramatically. A first indication of the complexity of this particular surface has been given in the chapter on topographic images. There we found that the distance from the surface can change the shape of a constant current contour quite substantially, since the corrugation is in general very low. Here, we focus on spectroscopy and a detailed comparison with experimental data.

For spectroscopy simulations of complicated metals the precision of the calculation as well as the resolution of the band structure are decisive. If, for

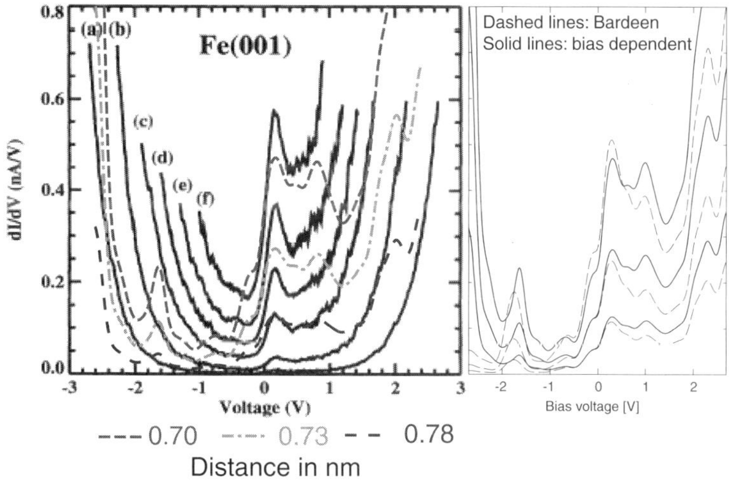

Fig. 9.7. Tunneling spectrum dI/dV of Fe(001) measured with a tungsten tip. (Left) The black graphs show experimental data from [14]. The dashed lines show the simulated results at three different distances: 0.70 nm, 0.73 nm, and 0.78 nm.The agreement between experiment and theory is rather good in the range below 1 V. The spectrum is highly asymmetric, a feature reproduced in the simulations.(Right) Comparison between the Bardeen method, neglecting the bias dependency, and the first-order scattering method, including it. The deviations from the experimental spectrum, due to the neglect of the bias-dependent terms, occur mainly in the negative bias range.

example, the STM is modeled by only a few layers of a metal film, the vertical confinement of the film will introduce a spacing of eigenvalues that is in the range of the bias intervals. In this case the unrealistic setup induces additional variation of the dI/dV curve on top of the variations due to the band structure map of the surface. Here, we note that surface states with a rather flat dispersion curve, such as for transition metal surfaces, actually improve the precision of the simulation, since the same number of k-points in the simulation lead to a higher energy resolution than, e.g., for the surface states of noble metal surfaces. In this case the simulation of the STM tip becomes the important factor in determining the resolution of the spectrum. The spectrum shown in Figure 9.1.3 was obtained with a 13-layer tip model. The mapping of the surface band structure was pushed to its computational limits by using more than 3000 k-points of the Brillouin zone; the tip map in this case was comparatively modest and comprised only 100 k-points at the very center of the tip Brillouin zone.

The result of the simulation is given in Figure 9.1.3. The left graphs show the experimental results obtained by Stroscio et al. [14] under ambient conditions. To account for the thermal environment, the computed dI/dV curves have been smoothed with a Gaussian of 100 mV half width. The agreement between experiments and simulations for three distinct curves is quite good: the onset of the surface state, its change with decreasing distance, and the general trend in the high bias regime are well reproduced. The difference between the absolute values in the high bias regime is probably due to field emission from the STM tip. As the bias range is increased, the shifted eigenvalues of the STM tip approach the vacuum level. For bias voltages comparable to the workfunction of a surface (4–5 eV), field emission is the main component to the tunneling current. However, this effect will already become important at much lower values and could well lead to an increase of the measured current by a substantial amount. At present, we have no accurate method to simulate field emission that can be used in conjunction with DFT structure simulations. Given the problems in the calculation of excited states, this situation is also not expected to change in the near future.

The graphs in the right frame of Figure 9.1.3 show a comparison of simulations with the Bardeen method (the bias voltage shows up only in the energy interval) and the first-order scattering method (the bias voltage is explicitly included). It can be seen that the bias dependency leads to a much better agreement between experiments and simulations, in particular in the negative bias regime.

Spectra on GaMnAs(110) with high local resolution

While we used only a single point of the dI/dV spectrum in the previous examples, the possibility of high local resolution in differential spectroscopy adds a qualitatively new analytic tool. If the spectrum is resolved with a resolution of about 0.2 Å in the lateral direction, we gain insight not only in the energy, but also in the exact local distribution of contributing electron states. A particularly striking example of this ability is provided by a III-V semiconductor surface GaAs(110) if it is doped by a few percent of manganese atoms.

Semiconductors with magnetic properties have increasingly gained importance in research due to their potential importance for applications in magneto electronics [30, 31]. In the past, manipulations of electronic spin in diluted magnetic semiconductors have been focused on the transport of carriers through a magnetic medium, the injection and detection of these carriers, and the coherence length of their spin properties [32, 33]. But since the Curie temperatures of promising systems are usually very low, the crystal magnetic properties do not provide an easy route for technical applications [34, 35, 36, 37]. Recently, however, the properties of an isolated magnetic atom were studied by the use of a scanning tunneling microscope [38]. In that study, Heinrich and coworkers could show that it is possible to detect the energy required to flip the spin of a single manganese atom from "up" to "down.". A number of present and future high-tech applications of semiconductor technology rely on the special properties of the III-V semiconductors like gallium arsenide. Due to the wide

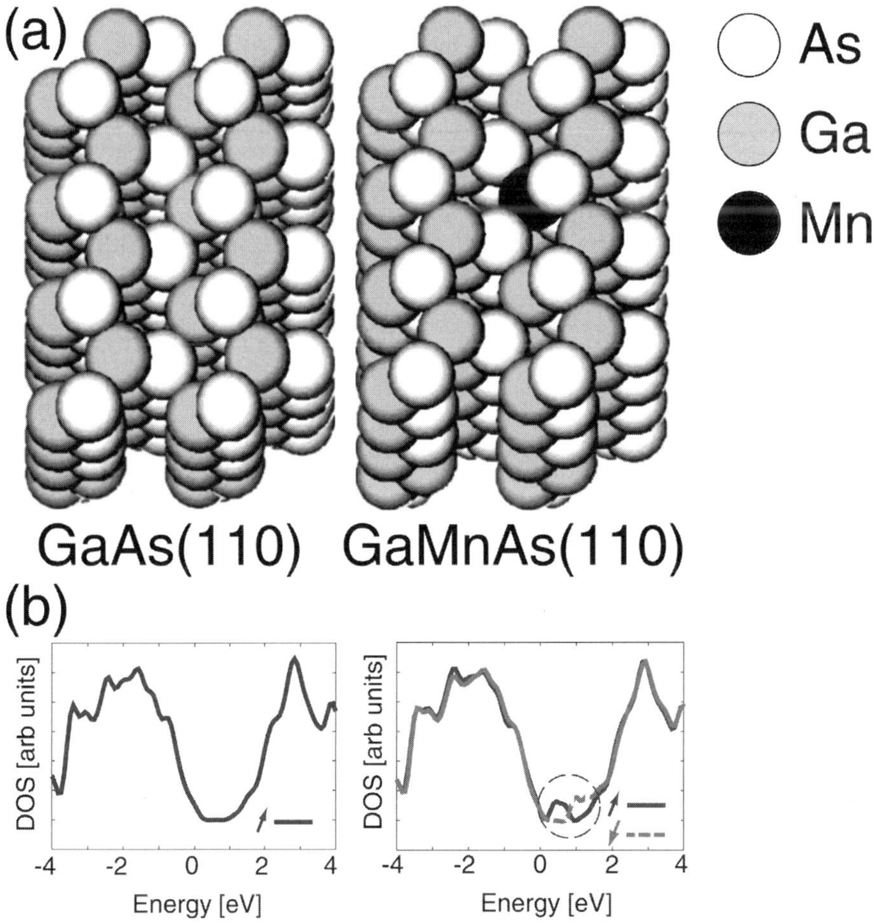

Fig. 9.8. (a) Atomic positions of the calculated GaAs(110) and GaMnAs(110) films. A single Ga atom (gray) at the subsurface is substituted by a Mn impurity (black). (b) Total density of states (DOS) of the films, calculated with 32 k-points in the Brillouin zone of the systems. Energy zero corresponds to the highest occupied state. The band gap of the GaAs lattice is populated by spin polarized (spin-up, solid and spin-down, dashed) Mn related states (see dashed circle). Apart from the bandi gap the electronic structure of the film remains nearly unchanged.

range of III-V materials, and the possibility to easily combine them into more complex compounds and heterostructures, they are among the favorites for nanoscale engineering. This is certainly also the case in magnetic semiconductor technology (or spintronics), where the introduction of magnetic defects has led to some of the first real integrations of magnetic and semiconducting materials. The magnetic defects led to the creation of spin-polarized electrons or holes, which can be propagated through a semiconductor heterostructure [32], and even led to the creation of a new type of ferromagnet with properties that can be controlled by applying small voltages across the structure. Currently, research is focused mainly on manganese [37, 39, 40], which can be produced in a substantial range of small concentrations (0.1–10%[37]). So far, it has remained unclear how the spin-polarized states of the manganese impurity interact with the electronic states of the GaAs lattice. However, the low Curie temperatures of $Ga_{1-x}Mn_x$ where x = 0.01–0.1 , ranging from 40 to 160 K (x = 0.045) [41], indicate that the spin-polarized electrons at different centres of the magnetic structure are only weakly coupled. This suggests that the spin states of GaMnAs must be rather confined locally and the overlap between the wavefunctions at different magnetic centers consequently rather low. While this feature confines the yield of spin-polarized electrons for magneto electronic transport, it raises the possibility of spin detection and spin manipulation on a very restricted local level. Spin-confinement to single atoms has been shown to exist and to be detectable by scanning tunneling experiments in only two cases so far: the anti ferromagnetic ordering of a single manganese layer on tungsten (110) [42] and the detection of the "spin-flip" energy of a single Mn atom on an oxide substrate [38].

Fig. 9.9. Topographies of GaMnAs for negative (left) and positive (right) bias voltages. The insets show images simulated with a tungsten tip and a single k-point. The bias voltage in the constant current contours is −/+ 2 V, the tunneling current 0.12 nA. In the negative-bias range the As atoms appear as protrusions, in the positive bias range the Ga atoms. Due to the low k-space resolution of the tip, the experimental details cannot be fully resolved.

The surface structure of the Mn-doped GaAs(110) film is shown in Figure 9.8(a). The magnetic ground state of the GaMnAs system was calculated for both positive and negative magnetic moments. In this case we find that the energy of both states is equal to within the resolution limits of the density functional calculation. The GaAs(110) possesses a band gap at the Fermi level of about 1.0 eV, as seen in Figure 9.8(b). In the GaMnAs case, it can clearly be seen that the bandgap is filled. The magnetic state of the GaMnAs surface is degenerate with respect to positive and negative magnetic moments. We obtain a ground state energy that is the same for a magnetic moment of +2.94 μB and −2.94 μB at the Mn atom. To estimate the energy barrier for transitions from one magnetic state to the other, one may, as the simplest estimate, simulate the nonmagnetic state. The energy barrier in this case is 160 meV. Physically, this barrier describes the gradual reversal of magnetic momentum due to the replacement with electrons of the opposite spin. In tunneling topographies the As positions will show up as protrusions in STM images; for positive bias the protrusions correspond to Ga. On GaMnAs we note a different picture (see Figure 9.9). Here, the Mn position is revealed unambiguously as a protrusion in a filled state (−2 V) and empty state (+2 V) images. The reason can be only that electronic states exist in the valence as well as the conductance band at the position of Mn atoms. The results of the differential spectroscopy simulations with a tungsten tip for GaAs and GaMnAs are shown in Figures 3a and 3b, respectively. Individual curves describe the spectrum for the position of As, Ga, and Mn. As a comparison with experimental spectra (dashed graph) shows, the simulation agrees very well with experimental data. Analyzing the individual spectra we note that the values at the position of As are higher in the negative bias range, with a distinct peak around −1 V. In the positive bias range the contributions from the position of the Ga atoms dominate. The spectra also give a first indication of the local confinement of states at the Mn atoms.

Tunneling spectra with high local resolution and, initially, with a nonmagnetic tip, lead to a more detailed understanding of the electronic effects visible in the dI/dV graphs. Tunneling spectra were simulated by increasing the bias voltage from −1.5 V to +1.5 V in steps of 20 mV at every point as the tip scans across the surface. Since the surface scans were performed with an increment of about 0.2 Å in both the x and y directions, the unique feature of the method is to yield high energy resolution as well as high local resolution of the spectral features. In that it surpasses the accuracy of most experimental methods available today. The ensuing differential spectra were directly visualized, which allows one to refer a particular feature in an experimental spectrum to the location and the spatial distribution of the contributing electronic states. The results of these simulations for selected bias values are shown in Figure 9.10. It can be seen that the states at a negative bias range of about −1.0 V to −0.5 V do not reveal a substantial maximum at the positions of the Mn atoms. Instead, we note protrusions at the position of the As atoms. States in this energy range are essentially states of the semiconductor lattice. In the range between –0.5 and +0.6 V we note distinct protrusions at the position of the Mn atoms. This feature indicates a localization of states in the band gap of the semiconductor and states associated essentially with a single atomic impurity, manganese. In the positive range above +0.6 V the protrusions are at the position of Ga atoms, since the states in this energy range are mainly associated with Ga. It should be clear that we obtain identical results for the spectra simulated with a tungsten tip whether we simulate the GaMnAs surface with a positive or a negative magnetic moment.

This result changes if we simulate spectra using a magnetic tip. In this case the densities of spin-up and spin-down states at the Fermi level of the tip differ significantly, and this difference will translate into a change of the signal at the position of the manganese atom, depending on its spin state. The simulations were performed with an iron tip [43]. We assumed ferromagnetic ordering in the tunneling junction, i.e., a transfer of spin-up electrons of the GaMnAs surface into spin-up states of the tip, and identically for spin-down electrons. Results of our simulations for positive and negative magnetic momenta are shown in Figure 9.11. The most striking difference from the tungsten case is the difference of the tunneling signal at a bias voltage of about –0.2 V. The reason for this difference is that the density of state of an iron tip possesses a distinct maximum in the minority band close to the Fermi level, while the density of majority states in this range is comparatively small [43]. Thus the surface with a positive magnetic moment yields a close to negligible contribution, while a negative magnetic moment leads to a signal well surpassing even the signal of the semiconductor surface itself.

9.1.4 Present limitations in current spectroscopy

Concerning the theoretical scheme developed for differential tunneling spectroscopy, it seems accurate enough to describe the detailed features of a spectrum with a resolution of 20–50 mV. For most spectra, this level of accuracy seems sufficient. However, in very detailed low-temperature experiments, experimental accuracy is above this level. In principle, it is possible to increase the number of k-points to about 10,000 in the Brillouin zone, and to map the band structure with sufficient accuracy. This has not been done as yet. Given the theoretical results presented previously, it seems fair to consider differential spectroscopy, even if it is based on an approximation for the bulk of electron transitions–that they will not be influenced substantially by the change of transition channels due to the offset of bias–as a quite accurate tool in simulating tunneling spectra. The additional advantages, i.e., the detailed analysis of surface and tip contributions, the comparatively high efficiency, and the numerical stability, more than balance the approximation involved.

Concerning the tip electronic structure, the simulations reveal a lack of detailed understanding. It was shown in previous simulations that tunneling topographies can be accurately simulated with STM tip models of a rather limited size. This, however, is not possible in tunnelling spectroscopy or only

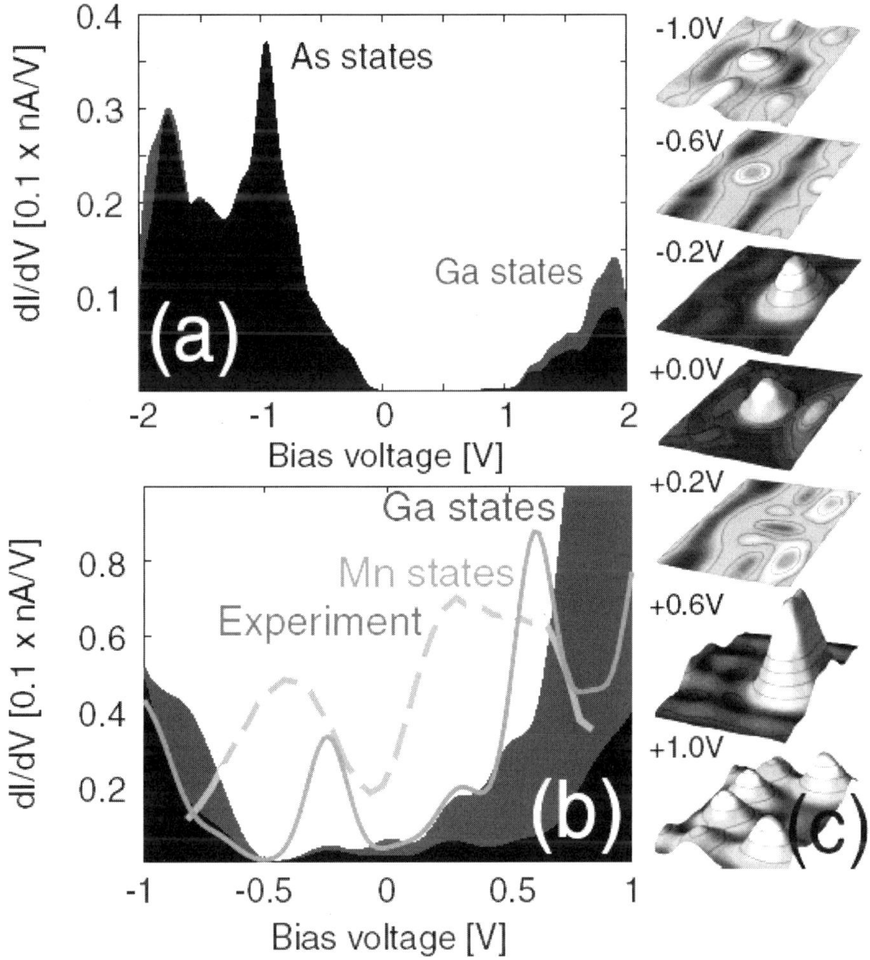

Fig. 9.10. Differential spectra with a non-magnetic tungsten tip. (a) Single point spectra on GaAs. The main contributions in the negative bias range are located at the position of the As atom; in the positive bias range we observe a small surplus of states at the Ga atoms. (b) Spectra on GaMnAs. While the spectra at the positions of As and Ga remain unchanged, an additional peak arises from contributions at the Mn atom. The energy of Mn states is in the bandgap of the GaAs semiconductor crystal. Experimental spectra measured by E. Lundgren are given for comparison (dashed graph) (c) Tunneling spectra with high lateral resolution from the GaMnAs(110) surface at selected voltages.

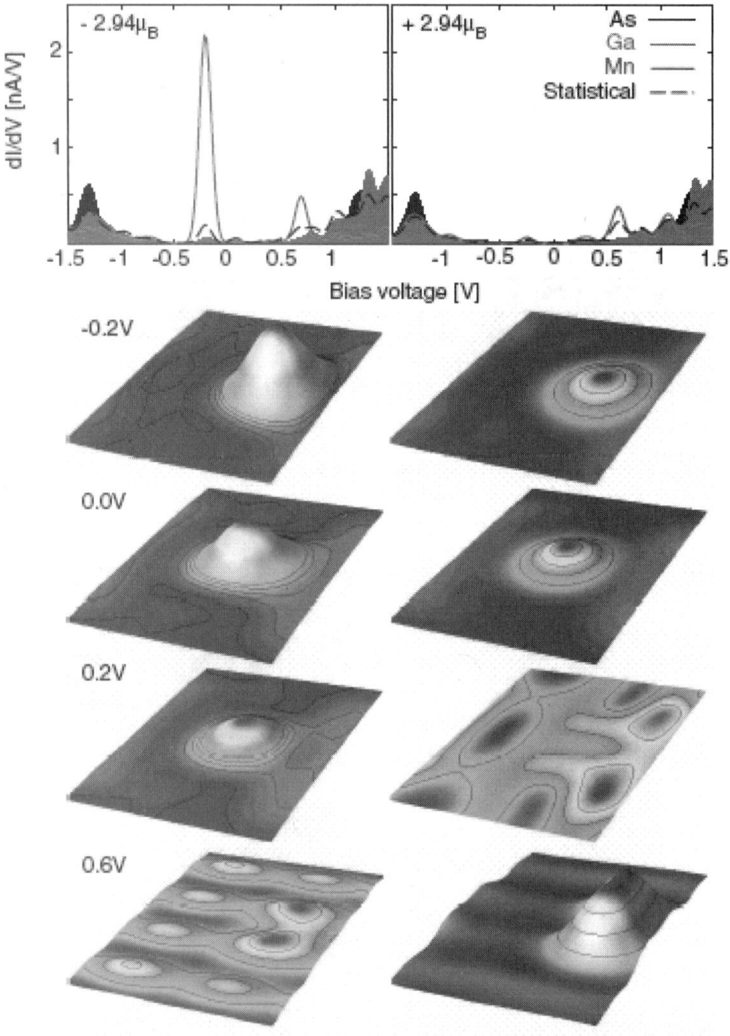

Fig. 9.11. Differential spectra on GaMnAs simulated with an iron tip, for negative (left) and positive (right) magnetic moment of the Mn atom. The spectra for the position of As, Ga, and Mn as well as a statistical evaluation of all points are shown in the top figures. The bottom figures display the locally resolved spectra at four distinct bias voltages (–0.2, 0.0, 0.2, and 0.6 V) near the Fermi level. The signal from the Mn impurity with a negative magnetic moment is substantially higher than the signal from the impurity with positive moment.

under specific conditions. If, for example, the density of states at the Fermi level of the surface system is very low, the second term in (9.8) becomes very small compared to the first term. Since this second term describes the contributions of the tip band structure to the differential spectrum, it will become negligible compared with the overlap of the surface electronic structure and tip states near the Fermi level. In this case the simulated tunneling spectrum will be essentially the spectrum of the surface alone. This applies in particular to metal surfaces with a surface state far from the Fermi level and semiconductor surfaces. Simulations of GaMnAs(110), as one striking example, were presented in the previous section.

If the problem of suitable tip electronic structures analyzed with this method were to remain unsolved, the method of Lang [13], using only the surface electronic structure, would be the endpoint of any analysis. But the advantage of the suggested scheme is that it allows quantifying a large class of materials where the tip electronic structure, even with Bardeen's method, is not too relevant: these are all materials like semiconductors or certain magnets, where the surface density of states at the Fermi level is very low by comparison. It also tells us, via the analysis of surface states and their numerical representation, that the band structure of a surface is correctly represented only, if the calculation involves at least 23 layers. And finally, it allows a qualification of tip electronic structures that will faithfully represent even complicated surfaces. This is always the case if the tip electronic structure has a sharp peak at the Fermi level. In this case the second term in (9.8) is equally negligible. That such tip structures may exist can be inferred from the reduction of the valence band if atoms have a decreased number of nearest neighbors, such as, for example, at surfaces. Reducing this number further, e.g. for a tip apex of very few atoms, and increasing the number of layers as well as the size of the tip unit cell should eventually produce such a structure in simulations. It thus provides a clear research program: how we can include the tip electronic structure and what these tips, geometrically and electronically, will look like. Improving the electronic structure of the STM tip relies on an accurate map of a semi-infinite tip crystal with a distinct apex structure of a few atoms. Given the numerical limitations, such a tip cannot, at present, be calculated within standard DFT methods, relying, e.g., on a three-dimensional repeat unit. Given the fact that the surface of, e.g., Ag(111) has to be simulated by a film of at least 23 atomic layers, and considering that a single layer of a model tip includes at least eight atoms, an exact map with a high number of k-points is infeasible with current simulation methods. A potential way to improve this situation is to compute the tip electronic structure with nonstandard methods, e.g., the Korringa–Kohn–Rostoker method, which can routinely be applied to semi-infinite systems. However, this method is usually limited to only a few atoms per layer and a comparatively low number of k-points. Whether it can be applied to the specific geometry and composition of the STM tip has to be determined in future calculations.

9.2 Force spectroscopy

It is evident from the previous chapters of this book that SFM is very sensitive to short-range chemical interactions, and in principle should be able to measure directly over specific sites on the surface. In Chap. 6 we demonstrated that it is possible to directly extract the force from the experimentally measured frequency change, but as examples, we considered only measurements of the long-range macroscopic forces. Measurement of the force over specific atoms/sites in the surface, or *force spectroscopy*, (so far) requires experiments at low temperature and has been possible only in a few cases, e.g., [44, 45, 46, 47].
Low temperature SFM was introduced as a general method for improving the sensitivity and controllability of experiments [48]. The SFM is placed in contact with a cryostat providing operating temperatures down to about 10 K. This provides significant improvements in several aspects of the measurement (see also Allers *et al.* in [49]):

- *Thermal drift.* It is very difficult to maintain a precise constant temperature in a real experiment due to the movement of atoms. This can be characterized by the thermal expansion coefficient, typically $10\text{–}20\times10^{-6}$ K^{-1} in bulk solids. Thermal drift in a standard room temperature setup is on the order of 1–10 mK/min. This means that it is impossible to keep the tip over a specific site long enough to measure the force interaction directly.
  Temperature changes also change the spring constant of the cantilever due to thermal expansion. For the drift discussed previously, this results in changes in the resonance frequency of the cantilever of about 10–100 mHz/min, which approaches measured contrast in many cases.
  Performing experiments at low temperature reduces the thermal drift to about 50 μK/min, greatly reducing these effects and allowing a full site-specific force curve to be measured.
- *Noise.* There are three major sources of noise in experiment: mechanical noise, which is not strongly dependent on temperature; amplitude noise, which is on the order of only 10 fm even at room temperature (using models from [50]); and frequency noise. Generally, instrumental frequency noise does not depend on temperature, but thermal frequency noise scales with the square root of T.
- *Instabilities.* As emphasized throughout this book, atomic instabilities on the tip and surface are one of the major obstacles maintaining stable high-resolution imaging. This problem is exagerrated when one attempts to approach closely to a specific site with the tip in order to measure an atomic force curve. At low temperatures most thermally activated processes are frozen, and the probability of atomic instabilities is significantly reduced.

Practically, site-specific measurements are achieved by first obtaining an atomically resolved image, then retracting the tip to a preset distance and finally positioning the tip over the site according to the lattice seen in the image. The frequency change is then measured as the tip approaches the surface. A usual procedure after measuring a series of curves over different sites is to again image the surface in atomic resolution to check that no significant tip changes have occurred.

In order to understand the pontential of force spectroscopy in improving understanding of the tip–surface interaction, and surface processes in general, here we consider three different systems to which it has been successfully applied. In all three cases the combination of experimental site-specific force curves and theoretical modelling was able to provide direct interpretation of experimental images.

9.2.1 Silicon 7 × 7 (111) surface

In Chapter 2 we already briefly introduced the low-temperature images of the silicon 7×7 (111) surface, but here we will discuss in more detail how they have provided a much better understanding of the tip-surface interaction and also confirmed the theoretical interpretation.

Initial theoretical progress on the silicon (111) surface was made by simulating the interaction of a hydrogen-terminated 10-atom silicon tip with the silicon (111) 5×5 surface [51] using the DFT theory. The simulations predicted that the force, and contrast in images, would be dominated by the onset of covalent bonding between the dangling bond at the tip apex and dangling bonds in the surface. Since the surface dangling bonds, by definition, appear above the silicon atomic sites and decay very rapidly as one moves from the surface, then the different relative heights of atoms in the surface should provide the atomic contrast. This qualitative interpretation was consistent with the experimental images, where protruding adatoms appear brighter than rest atoms (assuming that the tip does not enter the repulsive interaction region) [52].

This interpretation was made even stronger by a detailed low-temperature experimental study of the surface [44, 53]. The greater sensitivity of the low-temperature techniques allowed force curves to be taken over specific surface sites, which could then be compared to theoretical curves [54] over the same sites. Figure 9.12(a) gives an example of the high quality of images achieved at low temperature on silicon, and Figure 9.12(b) shows the magnitude of the measured force over the adatom site labeled 1 in Figure 9.12(a). For comparison with the theoretical curve, the long-range part of the experimental curve has been fitted to a simple model for the van der Waals force (see Chapter 6) and removed from the total force to leave the short-range force (see Figure 9.12(b)). This method is not perfect and can lead to errors on the order of 40% in the absolute force magnitude. However, the inset in Figure 9.12 shows that theoretical and experimental force curves agree well, both predicting a maximum force of about 2 nN, and a short decay length. The

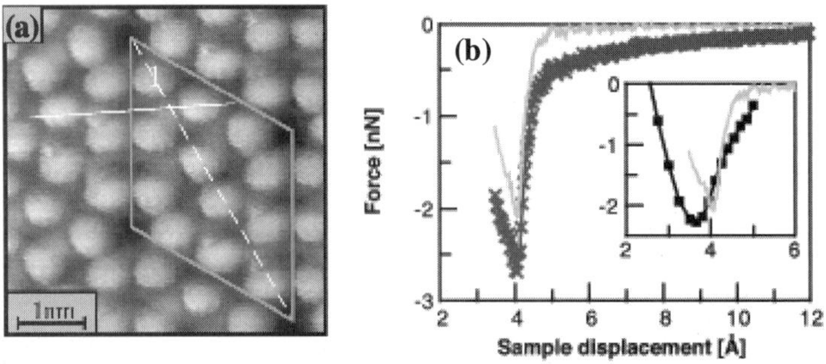

Fig. 9.12. Low-temperature SFM on silicon (111) 7 × 7. (a) Atomically resolved image of the surface with the surface unit cell shown by the white diamond [44] (Δf = −31 Hz, f_0 = 155 kHz, Q = 370,000, k = 28.6 N/m), and (b) force curves over the adatom site labeled 1 in (a). The dark curve is the total force and the light curve is the estimated short-range force alone. The inset shows a comparison between the experimental and theoretical predictions for the force over the adatom. Reprinted with permission. Copyright 2001 AAAS [53].

experimental curve demonstrates a narrower minimum than the theoretical result, but this is likely due to the simplistic nature of relaxations allowed in such a small tip model.

The agreement between experiment and theory provides very strong evidence that bright spots in images are really adatoms, and that the tip consists of some form of silicon nanocluster presenting a dangling bond toward the surface. This a major development in both understanding of SFM experiments on silicon, but also, more importantly, offers the potential to identify the tip structure on the atomic scale.

Further confirmation of the strength of site-specific force curves in SFM interpretation was found in another series of experiments on the silicon surface using a silicon oxide tip [55]. The unprecedented control offered by low-temperature operation, means that experimentalists can have much greater confidence that the tip is not contaminated. Since an untreated tip is almost certainly oxidized silicon, it will remain that way during scanning if it does not touch the surface, and should produce very different results from those of a clean silicon tip. Specifically, an oxide tip should be unable to form covalent bonds with the surface, and the interactions should be dominated by much weaker electrostatic forces. Site-specific measurements with an untreated silicon oxide tip [55] demonstrated an order of magnitude smaller forces than for the silicon tip, demonstrating that it is possible to control the tip structure at low temperature.

9.2.2 Calcium Difluoride (111) surface

Following the successful interpetation of atomic contrast in images of the CaF_2 (111) surface (see Chapter 7), an obvious next step in developing complete understanding was to attempt force spectroscopy [47]. This would provide both a confirmation of the qualitative interpretation of contrast and a quantitative challenge to theoretical modeling of the site-specific force curves.
Experiments were performed with a home-built low-temperature DFM described in detail in [56]. The tips used were sharp unsputtered Si tips covered by native oxide. In scanning, the utmost care was taken to avoid any tip change during the force spectroscopy measurements, and the tip was replaced if any obvious contact was observed. Although this reduced the possibility of contamination, it cannot be completely excluded. Many of the obtained topographic images yielded two basic types of contrast pattern: circular and triangular. Careful analysis of these images and corresponding scanlines demonstrated that they are very similar to those observed in the earlier room-temperature measurements discussed in Chapter 7. This agreement confirmed the theoretical model, and allowed the chemical identity of the image features to be taken directly from analysis of the measured image pattern and scanlines. It further demonstrated that whatever the complexity of the atomic structure of the tip apex, it is likely to produce either a positive or negative local electrostatic potential, and hence either a triangular or disklike contrast pattern respectively.

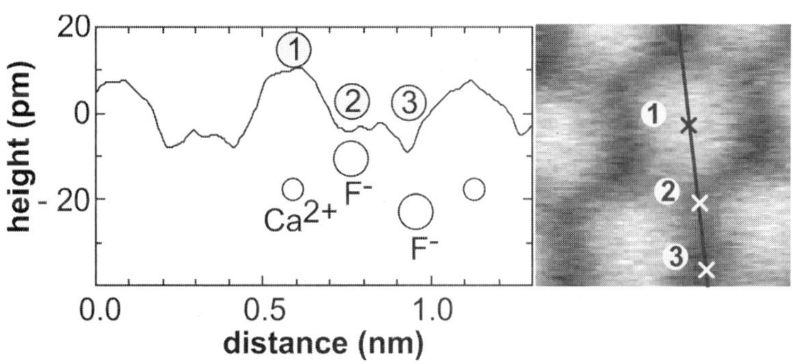

Fig. 9.13. Spectroscopy positions above the calcium ion (1), high-fluorine ion (2) and low-fluorine ion (3) are defined in the cross section along the [221] direction and a high-resolution image ($A_0 = 6.1$ nm, $\Delta f = -25$ Hz).

Force spectroscopy was then performed on an atomically resolved image demonstrating disklike contrast, where the brightest spot in each feature is associated with the position of a surface Ca^{2+} ion (see Figure 9.13). Spectroscopic measurements were performed exactly above the center of the Ca^{2+} site (position 1) and at two inequivalent high-symmetry sites equidistant from the

Ca^{2+} ions (positions 2 and 3). The measurement was started with a careful approach above site 2 and was stopped when the maximum attractive force to avoid tip instability upon further approach. The measurement was repeated above site 1, where the tip approached 0.06 nm closer to the surface. Note that high- and low-fluorine positions differ only by 0.08 nm in height from each other, but are discernible in the image. Residual thermal drift prevented, however, the subtle differences between the two force curves over sites 2 and 3 from being unambiguously resolved. Systematic distortion of force curves due to the residual thermal drift turned out to be the limiting factor determining the precision of these experiments.

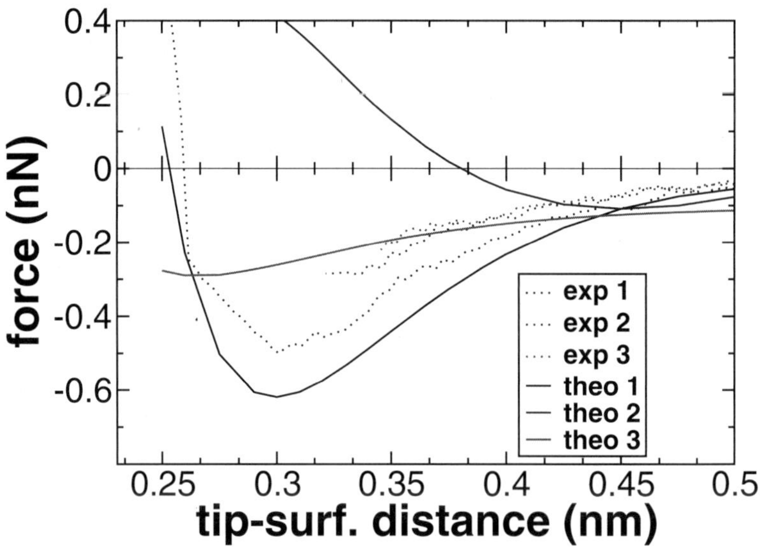

Fig. 9.14. Experimental results of force–distance curves at positions and compared to theoretical predictions for the respective locations. The large deviation between experiment and theory for position 3 is explained in the text.

The measured forces are shown in Figure 9.14. Although the chemical nature of the atomic sites is known, the physical meaning of the force curves remains undefined unless the structure of the tip apex responsible for this short-range interaction is clarified. In order to find a good model for the tip, several different tip clusters were considered, specifically an MgO cluster (as used for CaF_2 in Chapter 7), a silicon cluster (see Chapter 7) , a silicon cluster terminated by oxygen (see Figure 9.15(a)), and an hydrogen saturated silicon dioxide cluster (see Figure 9.15(b)). The forces produced by these models over sites 1, 2, and 3 were then compared quantitatively with the experimental curves. The oxygen-terminated MgO cluster could reproduce the observed contrast pattern, but was excluded due to an overestimation of the magnitude of forces by a factor of two to three due to its ionicity. The silicon cluster model could not reproduce the observed disklike contrast, since it always produced strongest interaction with cations in the surface. A clean silicon tip was also an unlikely possibility since the tip was exposed to air and therefore covered by a native oxide. The oxygen-terminated silicon tip showed a much better quantitative agreement with the experimental forces. Figure 9.14 shows that, in fair agreement with experiment, the maximum force over the Ca^{2+} ion at site 1 is about 0.6 nN, and weaker interaction is seen over site 3. However, there is an obvious deviation for the fluorine 2 position. The calculated force curves predict repulsion in a region where the experiment yields attraction.

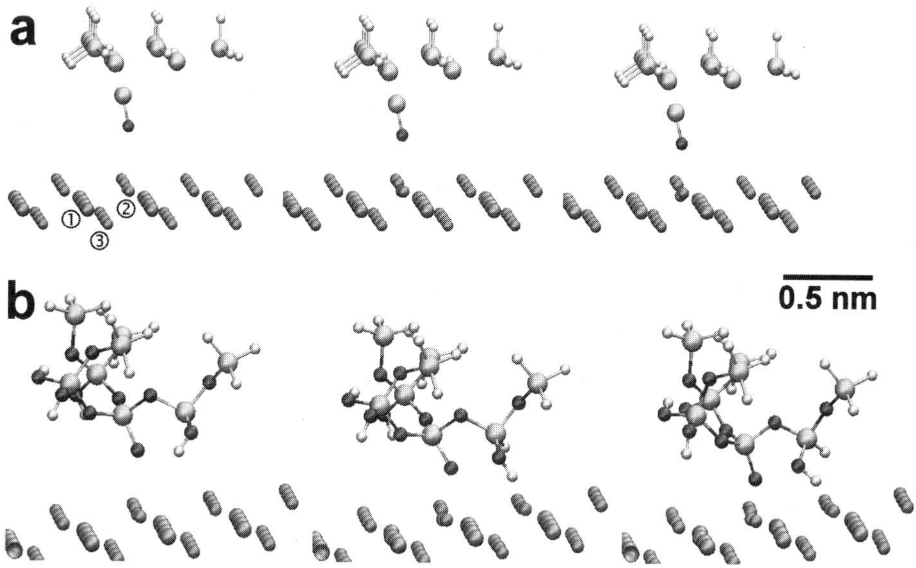

Fig. 9.15. Simulations demonstrating atomic relaxation of tip and surface atoms during approach of the tip to the CaF_2 (111) surface above the high fluorine ion (position 2). Model tips are a hydrogen -saturated Si tip with a single oxygen atom at the end (a) and a SiO_2 tip with hydrogen saturated dangling bonds (b). Images shown are snapshots taken from a series of calculations for varying tip–surface distance (a: 0.375, 0.340, 0.300 nm and b: 0.375, 0.300, 0.240 nm). Series (a) demonstrates strongly repulsive interaction between the tip-terminating oxygen and the high fluorine ion, resulting in an inward relaxation of the ion. In series (b), there is an additional attractive interaction between the foremost hydrogen atom that is strongly pulled toward the nearest-neighbor high fluorine ion exhibiting outward relaxation.

In order to try to explain why no net repulsion was observed in experiments, a more complex silicon-dioxide-based tip model was considered. This model was based on a model generated via simulated annealing of a larger SiO_2 cluster, with all its dangling bonds terminated by hydrogen [57]. Due to the more complex structure of this cluster, the tip apex is much less well defined at the atomic scale and instead of one atom clearly protruding and dominating the interaction with the surface, other atoms are now contributing significantly to the chemical interaction. This becomes immediately evident from the comparison of results for both tip models as shown in the two series of Figures 9.15(a, b). These sketches of the configurations of tip and surface ions calculated for tips positioned at various distances above position 2 reveal significant atomic relaxation. For the simple oxygen-terminated silicon tip, the oxygen mainly experiences repulsion from the high fluorine, causing significant inward relaxation of the latter ion. For the silica tip cluster, there is an additional attractive interaction of the most-protruding hydrogen atom with the neighboring high fluorine, and the overall interaction between tip cluster and surface is attractive, although the interaction between the tip terminating oxygen and the adjacent high fluorine is clearly repulsive. Calculating a force curve for this silica tip yields, in fact, a shallow minimum in a region where we find repulsion for the oxygen terminated silicon tip and thus a possible explanation for our experimental finding of similar force curves above low and high fluorines. However, due to the strong interaction of two atomic probes from this tip, the overall magnitude of the force is much larger than for the oxygen-terminated tip, and therefore the real tip lies between these two extremes.

The application of force spectroscopy on the CaF_2 (111) surface provides possibly the best analysis of the real tip structure in SFM on insulators. However, this was facilitated by a huge body of experimental and theoretical work laying the foundations of understanding, and still the final model cannot explain fully the experimental force curves.

9.2.3 Potassium bromide (100) surface

As briefly mentioned in the discussion of cubic crystals in Chapter 2, interpretation of images of such regular sublattices is very difficult. In fact, the only interpretation of atomically resolved images of a cubic insulating surface was performed using force spectroscopy on the KBr (100) surface [46].

Figure 9.16 shows atomically resolved images of the KBr surface, and force curves taken at three positions: 1, a contrast minimum; 2, contrast maximum; and 3, 4, bridge positions between maxima. Due to the symmetry of the surface (see, for example, Figure 6.8) and lack of knowledge of the tip polarity, it is impossible to identify from the image whether the contrast maxima are due to K^+ or Br^- ions, or even interstitial positions. Atomistic modeling of the tip–surface interaction using K- and Br-terminated KBr tips showed that, as expected, there is no discernible difference in contrast patterns with different polarities. The calculated forces themselves (see Figure 9.17) did show significant differences for each termination, suggesting the possibility that

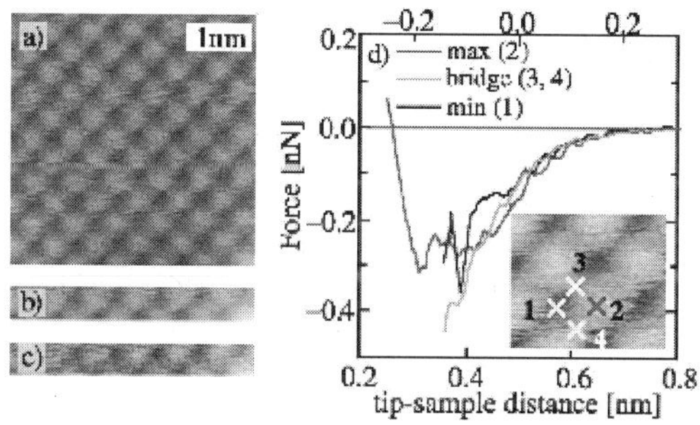

Fig. 9.16. Topographic images recorded at a constant frequency shift (a) before, (b) between, and (c) after all frequency versus distance measurements ($A_0 = 6.1$ nm, $\Delta f = -19.5$ Hz, $f_0 = 160$ kHz, $k = 40$ N/m). Note that (b) and (c) cover the bottom part of (a) and show that the contrast of (a) is reproduced precisely. (d) Force versus distance data obtained from measurements above the maximum, above the minimum, and above the bridge positions indicated in the inset. The horizontal scale is given by the sample displacement apart from an unknown offset. The origin of the top scale marks the mean location of the image in the inset. Reproduced with permission. Copyright 2004 by the American Physical Society [46].

force spectroscopy of the surface may provide immediate interpretation. However, the measured experimental curves, as for CaF_2 in the previous section, demonstrated weaker interaction than predicted by theory and quantitatively matched neither model tip.

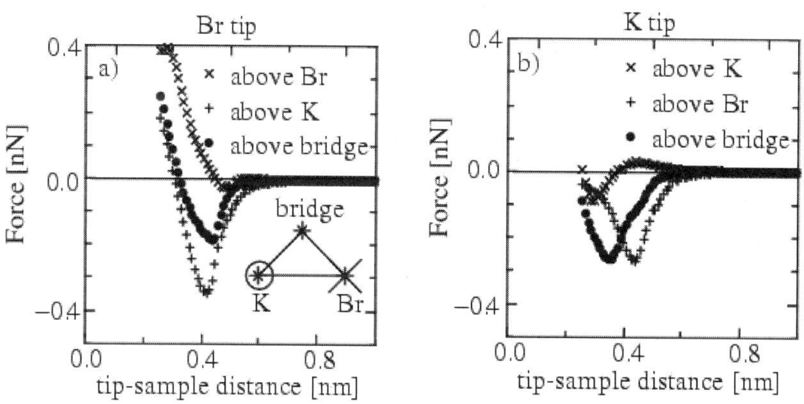

Fig. 9.17. Short-range force versus distance calculated for a (a) Br- and (b) K-terminated tip above a K^+ ion, above a Br^- ion, and above the bridge position. Reproduced with permission. Copyright 2004 by the American Physical Society [46].

Since the assumption that the tip in experiments was covered by KBr due to contact seemed reasonable, qualitatively the calculated force curves were expected to contain relevant information. However, two obvious factors limited any quantitative agreement: (i) the tip is unlikely to be terminated by the ideal cubic cluster used in the simulations; and (ii) the macroscopic van der Waals force removed from the experimental force curves to produce the microscopic forces is an estimate, and may deviate significantly from reality at small distances. The first problem cannot be eliminated, but by taking the differences in force curves over different sites the second factor can be removed, since the macroscopic forces do not change across the sites and will cancel. Figure 9.18(a,b) show that much better agreement between experiment and theory was achieved for the difference curves, but that either tip termination produced similar results. However, Figure 9.18(c) shows that only the K-terminated can match the experimental force difference over sites 2 and 3, and provides strong evidence that the experiments were imaging Br^- ions in the surface as bright with a positively terminated tip.

This study demonstrated the power of force spectroscopy in combination with simulations in providing interpretation even for a very symmetric system, and further encourages the application of this method to other cubic (and noncubic) crystals. However, one should note that, as discussed in Chapter 6, more complex tip models can significantly change the tip–surface interaction, and perhaps weaken any interpretation made via force difference curves alone.

9.3 Summary

In the first part of this chapter we presented a method to calculate the tunneling spectra on metal and semiconductor surfaces. The advantage of the method is that it includes the full electronic structure of the surface and the STM tip. It makes it possible, however, to distinguish between surface and tip contributions in the ensuing dI/dV map. The method was applied to differential spectra on (111) noble metal surfaces, on magnetic Fe(001) and Cr(001) surfaces, and a magnetic semiconductor, GaMnAs(110). It was shown that we obtain good agreement between theory and simulations. We also showed that the usual STM tip models, successful, for example, for tunneling topographies, are unsuitable for detailed comparisons between experiment and theory in this case. The reason, identified by a careful analysis of the ensuing spectra, is the local limitation of STM tip structures, leading to confinement of tip electrons and artifacts in the electronic tip structures. A potential improvement of the method is to base the STM tip structure on simulations of semi-infinite systems.

The final part of the chapter was dedicated to force spectroscopy in SFM, and successful experiments on the Si(111), $CaF_2(111)$, and KBr(001) surfaces. As well as reducing noise and increasing the general quality of images, the increased detail provided by force curves over specific atomic sites offers the

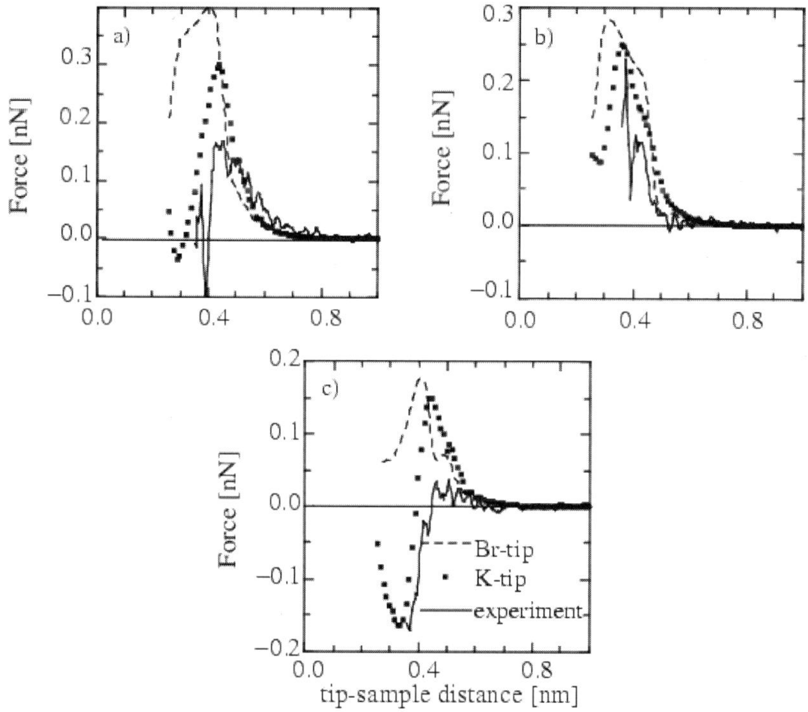

Fig. 9.18. Comparison of the calculated and the measured force differences for the force above (a) sites 1 and 2, (b) sites 1 and 3, (c) sites 3 and 2 identified in Figure 9.16(d). Reproduced with permission. Copyright 2004 by the American Physical Society [46].

opportunity for a much more quantitative comparison with simulations. In principle, this should greatly reduce uncertainity in tip structure and imaged features, and for the examples presented here it is clear that force spectroscopy has added significantly to our understanding. However, the large error involved in extracting microscopic forces from the background interactions means that real quantitative information, such as specific tip structures and absolute distances, remains difficult to justify. In the future, as more laboratories gain access to low temperature instruments and force spectroscopy, it is likely that we will see rapid development of this powerful technique and its integration into standard SFM operation.

References

1. D. M. Eigler and E. K. Schweitzer. *Nature*, 344:524, 1990.

2. K. F. Braun and K. H. Rieder. *Phys. Rev. Lett.*, 88:096801, 2002.
3. J. Kliewer, R. Berndt, E. V. Chulkov, V. M. Silkin, P. M. Echenique, and S. Crampin. *Science*, 288:1399, 2000.
4. W. A. Hofer and A. J. Fisher. *Phys. Rev. Lett.*, 91:036803, 2003.
5. F. Calleja, A. Arnau, J. J. Hinarejos, A. L. Vazquez de Parga, W. A. Hofer, P. M. Echenique, and R. Miranda. *Phys. Rev. Lett.*, 92:206101, 2004.
6. V. L. Moruzzi, J. F. Janak, and A. R. Williams. *Calculated Electronic Properties of Metals*. 1978.
7. R. F. Feenstra. *Surf. Sci.*, 299/300:965, 1994.
8. W.A. Hofer and J. Redinger. *Surf. Sci.*, 447:51, 2000.
9. J. Bardeen. *Phys. Rev. Lett.*, 6:57, 1961.
10. C. J. Chen. *Introduction to Scanning Tunneling Microscopy*. Oxford University Press, Oxford, 1993.
11. J. Tersoff and D. R. Hamann. *Phys. Rev. Lett.*, 50:1998, 1985.
12. J. Tersoff and D. R. Hamann. *Phys. Rev. B*, 31:805, 1985.
13. N. D. Lang. *Phys. Rev. B*, 34:5947, 1986.
14. J. A. Stroscio, D. T. Pierce, A. Davies, R. J. Celotta, and M. Weinert. *Phys. Rev. Lett.*, 75:2960, 1995.
15. A. Biedermann, O. Genser, W. Hebenstreit, M. Schmid, R. Podloucky J. Redinger, and P. Varga. *Phys. Rev. Lett.*, 76:4179, 1996.
16. J. Li, W.-D. Schneider, R. Berndt, and B. Delley. *Phys. Rev. Lett.*, 80:2893, 1998.
17. G. Hörmandinger. *Phys. Rev. B*, 49:13897, 1994.
18. G. Kresse and J. Hafner. *Phys. Rev. B*, 47:558, 1993.
19. G. Kresse and J. Furthmüller. *Phys. Rev. B*, 54:11169, 1996.
20. P. Blöchl. *Phys. Rev. B*, 50:17953, 1994.
21. J. P. Perdew, S. H. Vosko, K. A. Jackson, D. J. Singh, and C. Fiolhais. *Phys. Rev. B*, 46:6671, 1992.
22. G. Nicolay, F. Reinert, S. Hüfner, and P. Blaha. *Phys. Rev. B*, 65:033407, 2001.
23. L. M. Molina, M. D. Rasmussen, and B. Hammer. *J. Chem. Phys.*, 120:7673, 2004.
24. D. C. Meier and D. W. Goodman. *J. Am. Chem. Soc.*, 126:1892, 2004.
25. R. Reinert. *J. Phys: Condens. Mat.*, 15:S693, 2003.
26. P. M. Echenique, R. Berndt, E. V. Chulkov, T. Fauster, A. Goldmann, and U. Höfer. *Surf. Sci. Rep.*, 52:219, 2004.
27. A. L. Vazquez de Parga, O. S. Hernan, R. Miranda, A. Levy Yegati, N. Mingo, A. Martyn-Rodero, and F. Flores. *Phys. Rev. Lett.*, 80:357, 1998.
28. L. Burgi, H. Brune, O. Jeandupeux, and K. Kern. *J. Electr. Spec. Rel. Phen.*, 109:33, 2000.
29. M. Bode. *Rep. PRogr. Phys.*, 66:523, 2003.
30. D. D. Awschalom and J. M. Kikkawa. *Physics Today*, 52:33, 1999.
31. G. A. Prinz. *Science*, 282:1660, 1998.
32. Y. Ohno. *Nature*, 402:790, 1999.
33. B. Grandidier. *Appl. Phys. Lett.*, 77:4001, 2000.
34. B. Beschoten, P. A. Crowell, I. Malajowich, and D. D. Awschalom. *Phys. Rev. Lett.*, 83:3073, 1999.
35. Van Esch. *Phys. Rev. B*, 56:13103, 1997.
36. H. Ohno. *J. Magn. Magn. Mater.*, 200:110, 1999.
37. A. Mikkelsen. *Phys. Rev. B*, 70:085411, 2004.

38. A. J. Heinrich, J. A. Gupta, C. P. Lutz, and D. M. Eigler. *Science*, 2004.
39. F. Tuomisto, K. Pennanen, K. Saarinen, and J. Sadowski. *Phys. Rev. Lett.*, 93:055505, 2004.
40. F. Glas, G. Patriarche, L. Largeau, and A. Lemaitre. *Phys. Rev. Lett.*, 93:086107, 2004.
41. F. Matsukara, H. Ohno, A. Shen, and Y. Sugawara. *Phys. Rev. B*, 57:2037, 1999.
42. S. Heinze, M. Bode, A. Kubetzka, O. Peitzsch, X. Nie, S. Blügel, and R. Wiesendanger. *Science*, 288:1805, 2000.
43. W. A. Hofer and A. J. Fisher. *J. Magn. Magn. Mater.*, 267:139, 2003.
44. M. A. Lantz, H. J. Hug, P. J. A. van Schendel, R. Hoffmann, S. Martin, A. Baratoff, A. Abdurixit, H. J. Güntherodt, and Ch. Gerber. *Phys. Rev. Lett.*, 84:2642, 2000.
45. S. M. Langkat, H. Hölscher, A. Schwarz, and R. Wiesendanger. *Surf. Sci.*, 527:12, 2003.
46. R. Hoffmann, L. N. Kantorovich, A. Baratoff, H. J. Hug, and H. J. Güntherodt. *Phys. Rev. Lett*, 92:146103, 2004.
47. R. Hoffmann, C. Barth, A. S. Foster, A. L. Shluger, H. J. Hug, H. J. Güntherodt, R. M. Nieminen, and M. Reichling. *J. Am. Chem. Soc.*, 127:17863, 2005.
48. W. Allers, A. Schwarz, U. D. Schwarz, and R. Wiesendanger. *Rev. Sci. Instrum.*, 69:221, 1998.
49. S. Morita, R. Wiesendanger, and E. Meyer, editors. *Noncontact Atomic Force Microscopy*. Springer, Berlin, 2002.
50. F. J. Giessibl, H. Bielefeldt, S. Hembacher, and J. Mannhart. *Appl. Surf. Sci.*, 140:352, 1999.
51. R. Pérez, M. C. Payne, I. Stich, and K. Terakura. *Phys. Rev. Lett.*, 78:678, 1997.
52. F. J. Giessibl. *Science*, 267:68, 1995.
53. M. A. Lantz, H. J. Hug, R. Hoffman, P. J. A. van Schendel, P. Kappenberger, S. Martin, A. Baratoff, and H. J. Güntherodt. *Science*, 291:2580, 2001.
54. R. Pérez, I. Stich, M. C. Payne, and K. Terakura. *Phys. Rev. B*, 58:10835, 1998.
55. M. A. Lantz, H. J. Hug, R. Hoffmann, S. Martin, A. Baratoff, and H. J. Güntherodt. *Phys. Rev. B*, 68:035324, 2003.
56. H. J. Hug, B. Stiefel, P. J. A. van Schendel, A. Moser, S. Martin, and H. J. Güntherodt. *Rev. Sci. Instrum.*, 70:3625, 1999.
57. Y. Ma, A. S. Foster, and R. M. Nieminen. *J. Chem. Phys.*, 122:144709, 2005.

10

Outlook

The development of the field of scanning probe microscopy can be best seen in the increasing detail and variety of physical information extracted from experiments. In the last decade SPM has progressed from poorly resolved images of a few simple surfaces to tunneling and force spectroscopy on specific atoms; high resolution of adsorbed molecules and clusters; atomic resolution across step edges and kinks; observation of atomic and molecular diffusion; observation and manipulation of chemical reactions; and imaging of molecular orbitals. This flexibility has established SPM as the preeminent tool for nanoscale studies, and it remains a driving force in the development of many applications in nanoscience. However, if SPM is to become a common tool in surface science and maintain its lead in nanoscience then we must see development both in the experimental techniques themselves and in the information we can reliably extract from images. Here we list the challenges for SPM that we feel will significantly advance the field, and its general use in nanoscale science.

10.1 Challenges

- *Tip preparation.* The heart of many difficulties in interpreting high-resolution SPM images, and also in the applicability of SPM in general, remains the lack of information about the tip apex at the atomic scale. In both STM and SFM, the results presented in this book demonstrate clearly the dependency of high-resolution images on the detailed chemical and atomic structure of the tip. If SPM wishes to be a source of reliable chemical information at the atomic scale, then developments must be made in identifying and controlling the tip apex.
  If we consider SPM as a whole, then the requirements of a good tip are that the nature of the tip apex be known; that the tip apex can be controlled during scanning, either the tip-surface interaction is weak enough so that tip contamination can be easily avoided, or following inadvertant

contact with the surface or ambient contaminants the tip can be easily recovered via some cleaning process; that the tip–surface interaction be strong enough that the force contrast in SFM is significant; that the tip be conducting to provide contrast in STM. Clearly a single candidate is unlikely to fulfill all these characteristics, and compromises must be made. A universal approach would be to develop a conducting probe that could serve as a tip in both STM and SFM. Metallic tips are routine in STM, but they have been little studied in SFM; the possibility of cleaning them via voltage pulses makes them attractive as recoverable and reusable probes. However, the inertness of metal tips may make SFM imaging difficult. Similar problems are likely if we consider some conducting organic coating, since C-H bonds are one of the most inert in nature. As discussed in Chapter 2, some experiments have been performed with *clean* silicon tips, but the successful results in SFM have been achieved on silicon surfaces, and efforts on more relevant insulating surfaces found only very poor contrast. The main barrier here is that the best silicon tip for simplistic interpretation would be a reactive one, but this is intrinsically difficult to keep clean. Promising progress is being made in bonding nanotubes to standard SPM tips, and they have demonstrated success in low-resolution imaging. For high-resolution imaging, the structure of the nanotube apex becomes the essential unknown, but more general efforts in nanotube functionalization may provide some helpful suggestions for controlling the tube end.

If we separate STM and SFM (most definitely a very short term approach), then we can consider the tip candidates that have provided success previously. For STM, a metal tip seems obvious, and the only issues are to establish reliable techniques for cleaning the apex. In SFM, the lack of tip cleaning and repeated contact with the surface strongly suggests that the tip be either oxidized silicon or coated by surface material. Assuming that the main interest in SFM is insulating surfaces, then controllably forming an insulating probe would be a good choice. Two possible approaches can be attempted: firstly, heating a standard tip to over 1000 K should desorb most of the associated and dissociated water on the surface, and provide a relatively clean silicon oxide tip; secondly, attaching insulating nanoclusters, such as MgO, to the tip should provide a more controllable route to an insulating nanoprobe than *kissing* the surface.

- *High resolution in extreme conditions.* Although UHV has offered many scientific insights that would otherwise be inaccessible, many recent and predicted applications require studies in more extreme conditions, without sacrificing high resolution. This is especially true in the field of biology, where liquid environments are almost essential if we are to study application-relevant properties. If SPM is to make advances in nanotechnology as well as nanoscience, then we must break the enviromental barrier; we need high-resolution imaging in air and liquid, and at variable temperatures and pressures. Flexibility in imaging conditions would also allow more systematic studies of, for example, diffusion, adsorption, and reactivity. Several examples of *extreme* SPM exist, but they are far from routine and the techniques remain available to an exclusive few.

- *Dissipation.* In Chapter 6 we outlined the progress made in understanding the mechanism of dissipation in SFM, and also discussed the development of a general model for dissipation contrast. As yet this has not been applied successfully, but in principle dissipation imaging offers several important new avenues of study: the increased sensitivity of dissipated energy to atomic displacements, which are especially pronounced at low-coordinated sites, seems to offer great promise in improving resolution of nanostructures. For example, understanding the atomic structure of metal nanoclusters adsorbed onto oxide surfaces is a crucial topic in catalyst research, and dissipation imaging may offer improved imaging of these low-coordinated systems; the dependence of the dissipated energy on the mass of atoms at the tip apex and surface offer a possible route to chemical identification; the fact that energy dissipates as phonons in the surface gives SFM access to another source of physical information about local atomic-scale processes in the surface.
- *Systematic defect studies.* Although many experiments have demonstrated atomically resolved images of defect species, systematic studies of their properties remain scarce. Most analysis consists in using some very simplistic analysis based on contrast; for example, a dark spot on a row of bright contrast is a vacancy. In the best case, simulations can provide a set of possible defects that would match the experimental results, but without further information a conclusive identity cannot be established. A systematic study would require that the identity and density of surface defects be well controlled; either via system preparation or due to the intrinsic properties of the sample. This then greatly reduces the possible defects that could match the experimental results, and provides strong justification for building defect models for simulations. Although unambiguous interpretation is always difficult, a systematic approach to SPM defect measurements will greatly narrow the range of candidate defects, and hence make it easy to suggest a follow-up experiment or technique likely to complete interpretation.
- *Manipulation.* In STM, manipulation of absorbed molecules on the surface has become almost routine, and challenges lie mainly in developing applications. In high-resolution SFM, manipulation remains a significant challenge, and only on semiconducting surfaces, where STM would likely prove easier, has any significant progress been made. Efforts are in many ways hindered by the lack of understanding in images of defects and adsorbed molecules themselves, so that it is not clear what you are actually manipulating. A successful study will require a systematic identification of an adsorbed species, followed by a manipulation mechanism that is clearly related to the interaction of the tip.

- *Chemical identification.* A general methodological goal of SPM is the establishment of the chemical identity of imaged species without resorting to extensive simulations, i.e., identification would be possible during the experiment itself. During the course of the book, and even in this final chapter, many possibilities for achieving this have been suggested: controlling the tip apex on the atomic scale during preparation and scanning; force difference curves measured via low-temperature force spectroscopy; referencing the tip first by imaging on a sample or defect/molecule where the contrast pattern directly identifies the character of the tip apex; and dissipation images. In STM, the general assumption that chemical identity in images can be readily established from a simple density of states is now being regularly challenged, and can almost be held as an equivalent approximation to assuming that SFM images the physical topography of the system. The failure of simple approximations in understanding experimental results is really just another indicator of the increased complexity of the systems being imaged and the interactions being probed. However, without systematic developments in identification and image interpretation on a general level, the establishment of SPM as a *standard* surface science tool will be severely hampered.
- *Exchange contrast.* As discussed in Chapter 2, for the past five years several leading SFM groups have focused on measuring the so-called exchange contrast on the antiferromagnetic NiO surface, i.e., the difference in spin-polarized tip-magnetic surface interaction over Ni atoms with opposite spins. The most spectacular way to demonstrate experimentally the effect with SFM would be to measure the difference in contrast along parallel rows of Ni ions with antiparallel spins. However, since the surface sublattice seen as bright in images is unknown *a priori*, a more likely approach is to measure force vs. distance curves above different sublattice sites via low-temperature force spectroscopy. The short range of the exchange interaction means that to reliably measure the difference one may need to advance the tip very close to the surface. This could lead to instability in SFM operation, caused by a too-strong tip–surface interaction and by large displacements of the surface/tip atoms and their adhesion to the tip/surface. Preliminary simulations [1] indicated a maximum range of 0.375 nm for detecting the difference, which is very close to the point at which tip-induced atomic displacements become significant. However, the lack of experimental success [2, 3, 4, 5], and the relative simplicity of the theoretical approaches means that the topic remains an open challenge in SFM. Experimentally, suggestions for progress include replacing the tip material (usually an iron-coated silicon tip) with a more controllable option, or by using a magnetic surface that provides a stronger *exchange* interaction. A more likely route to success must involve a systematic combination of theory and experiment, with theory testing different tip/surface combinations and predicting an ideal experimental setup.

10.2 The future

Achieving any of the challenges presented in the previous section will provide breakthroughs both in SPM methodology and in science at the nanoscale, but the future of SPM can also be seen at a more general level. In fact, many of the problems faced by experimentalists and theoreticians in SPM can be overcome by thinking beyond SPM for a solution, or at least by really thinking about *SPM.* Here we are intimating SPM's role as a component in the standard surface science lab. At the most basic level, carrying out combined STM/SFM studies on conducting/semiconducting samples greatly increases the physical information and interpretational possibilities compared to using a single technique. A comprehensive approach to SPM studies would include other standard surface science techniques (XPS, Auger, vibrational and mass spectroscopy, etc.) as part of experimental routine. Imagine an SPM study of surface defects where preparation and spectroscopy have narrowed the defect chemical identity to a matter of specific bonding configuration; at this point the local information provided by SPM, combined with simulation, can pinpoint the exact species. The future of SPM must lie in first establishing confidence in its reliability, and then reinvigorating surface science itself by using SPM as the flagship technique in a comprehensive surface science lab.

References

1. A. S. Foster and A. L. Shluger. *Surf. Sci.*, 490:211, 2001.
2. H. Hosoi, K. Sueoka, K. Hayakawa, and K. Mukasa. *Appl. Surf. Sci.*, 157, 2000.
3. W. Allers, S. Langkat, and R. Wiesendanger. *Appl. Phys. A*, 72:S27, 2001.
4. R. Hoffmann, M. A. Lantz, H. J. Hug, P. J. A van Schendel, P. Kappenberger, S. Martin, A. Baratoff, and H. J. Güntherodt. *Phys. Rev. B*, 67:085402, 2003.
5. S. M. Langkat, H. Hölscher, A. Schwarz, and R. Wiesendanger. *Surf. Sci.*, 527:12, 2003.

Appendix

In this appendix we describe in detail the unified approach for scattering and perturbation, solving all the integrals and taking the limit to zero for the infinitesimal complex components of the single-particle wavefunctions. The notes are added for students and experts alike who want to gain a clear impression of how the Keldysh formalism, or the Landauer–Büttiker equation, is evaluated in practice.

A.1 Green's functions in the interface

A.1.1 Green's function and spectral function

We start with an eigenvector expansion of the two surface Green's functions:

$$\begin{aligned} G_S^{\pm}(\mathbf{r}_1, \mathbf{r}_2; E) &= \lim_{\eta \to +0} \sum_i \frac{\psi_i(\mathbf{r}_1)\psi_i^*(\mathbf{r}_2)}{E - E_i \pm i\eta} \\ &= \sum_i \psi_i(\mathbf{r}_1)\psi_i^*(\mathbf{r}_2) \left[\mathbb{P}\left(\frac{1}{E - E_i}\right) \mp i\pi\delta(E - E_i) \right], \qquad \text{(A.1)} \\ G_T^{\pm}(\mathbf{r}_1, \mathbf{r}_2; E) &= \lim_{\epsilon \to +0} \sum_j \frac{\chi_j(\mathbf{r}_1)\chi_j^*(\mathbf{r}_2)}{E - E_j \pm i\epsilon} \\ &= \sum_j \chi_j(\mathbf{r}_1)\chi_j^*(\mathbf{r}_2) \left[\mathbb{P}\left(\frac{1}{E - E_j}\right) \mp i\pi\delta(E - E_j) \right], \qquad \text{(A.2)} \end{aligned}$$

where S and T denote the surface and tip, respectively. The spectral functions are defined as

$$A_S(\mathbf{r}_1,\mathbf{r}_2;E) = i\left[G_S^+(\mathbf{r}_1,\mathbf{r}_2,E) - G_S^-(\mathbf{r}_1,\mathbf{r}_2,E)\right]$$
$$= \lim_{\eta\to+0} 2\eta \sum_i \frac{\psi_i(\mathbf{r}_1)\psi_i^*(\mathbf{r}_2)}{(E-E_i)^2+\eta^2} = 2\pi \sum_i \psi_i(\mathbf{r}_1)\psi_i^*(\mathbf{r}_2)\delta(E-E_i), \tag{A.3}$$

$$A_T(\mathbf{r}_1,\mathbf{r}_2;E) = i\left[G_T^+(\mathbf{r}_1,\mathbf{r}_2,E) - G_T^-(\mathbf{r}_1,\mathbf{r}_2,E)\right]$$
$$= \lim_{\epsilon\to+0} 2\epsilon \sum_j \frac{\chi_j(\mathbf{r}_1)\chi_j^*(\mathbf{r}_2)}{(E-E_j)^2+\epsilon^2} = 2\pi \sum_j \chi_j(\mathbf{r}_1)\chi_j^*(\mathbf{r}_2)\delta(E-E_j). \tag{A.4}$$

A.1.2 Contacts

Let us now investigate the transmission probability through the vacuum barrier in our system,

$$Tr\left[\Gamma_T(E)G^+(E)\Gamma_S(E)G^-(E)\right], \tag{A.5}$$

for the zero-order Green's function where the contacts (Γ) are defined with the help of the spectral functions as

$$A_S(E) = G_S^+(E)\Gamma_S(E)G_S^-(E) = G_S^-(E)\Gamma_S(E)G_S^+(E), \tag{A.6}$$
$$A_T(E) = G_T^+(E)\Gamma_T(E)G_T^-(E) = G_T^-(E)\Gamma_T(E)G_T^+(E), \tag{A.7}$$

or can be obtained using the fact that

$$Tr\left[\Gamma_S(E)G_S^+(E)\Gamma_S(E)G_S^-(E)\right] = Tr\left[\Gamma_S(E)A_S(E)\right] = 1. \tag{A.8}$$

A.1.3 Electron density

The electron density can be written as

$$n(\mathbf{r}) = n_S(\mathbf{r}) + n_T(\mathbf{r}), \tag{A.9}$$

which is true in the whole space. The definitions of the electron densities with help of the Green's functions are as follows

$$n(\mathbf{r}) = \mp\frac{1}{\pi}\int_{-\infty}^{\infty} dE\,\Im\left[G^{\pm}(\mathbf{r},\mathbf{r};E)\right], \tag{A.10}$$

$$n_S(\mathbf{r}) = \mp\frac{1}{\pi}\int_{-\infty}^{\infty} dE\,\Im\left[G_S^{\pm}(\mathbf{r},\mathbf{r};E)\right], \tag{A.11}$$

$$n_T(\mathbf{r}) = \mp\frac{1}{\pi}\int_{-\infty}^{\infty} dE\,\Im\left[G_T^{\pm}(\mathbf{r},\mathbf{r};E)\right]. \tag{A.12}$$

A.1.4 Zero-order Green's function

This suggests that the zero-order approximation for Green's function of the system $G^{\pm}$ can be written as the sum of the surface and tip Green's functions:

$$G^{\pm}_{(0)}(\mathbf{r}_1, \mathbf{r}_2; E) = G^{\pm}_S(\mathbf{r}_1, \mathbf{r}_2; E) + G^{\pm}_T(\mathbf{r}_1, \mathbf{r}_2; E), \tag{A.13}$$

which is justified on the one hand by the following derivation,

$$\left(-\frac{\hbar^2}{2m}\nabla^2 + V(\mathbf{r}_1)\right) G(\mathbf{r}_1, \mathbf{r}_2) = -\delta(\mathbf{r}_1 - \mathbf{r}_2), \tag{A.14}$$

where

$$V(\mathbf{r}) = V_S(\mathbf{r}) + V_T(\mathbf{r}), \tag{A.15}$$

and the standard condition of vanishing overlap between surface and tip potentials is used:

$$V_S(\mathbf{r})V_T(\mathbf{r}) = 0. \tag{A.16}$$

If $\mathbf{r}_1 = \mathbf{r}_2$ then the zero-order Green's function describes just the fact that the composite charge in the interface is the sum of surface and tip contributions. If $\mathbf{r}_1 \neq \mathbf{r}_2$, then the following applies:

$$\begin{aligned}
&\left(-\frac{\hbar^2}{2m}\nabla^2 + V_S(\mathbf{r}_1) + V_T(\mathbf{r}_1)\right)(G_S(\mathbf{r}_1, \mathbf{r}_2) + G_T(\mathbf{r}_1, \mathbf{r}_2)) = 0, \\
&\underbrace{\left(-\frac{\hbar^2}{2m}\nabla^2 + V_S(\mathbf{r}_1)\right) G_S(\mathbf{r}_1, \mathbf{r}_2)}_{=0} + \underbrace{\left(-\frac{\hbar^2}{2m}\nabla^2 + V_T(\mathbf{r}_1)\right) G_T(\mathbf{r}_1, \mathbf{r}_2)}_{=0} \\
&+ V_S(\mathbf{r}_1)G_T(\mathbf{r}_1, \mathbf{r}_2) + V_T(\mathbf{r}_1)G_S(\mathbf{r}_1, \mathbf{r}_2) = 0.
\end{aligned} \tag{A.17}$$

A.1.5 Consistency check: Schrödinger equation

From the above, the following has to be satisfied,

$$V_S(\mathbf{r}_1)G_T(\mathbf{r}_1, \mathbf{r}_2) + V_T(\mathbf{r}_1)G_S(\mathbf{r}_1, \mathbf{r}_2) = 0, \tag{A.18}$$

which, in turn, can be rewritten by multiplying this equation by V_S or V_T and using (A.16), and the following is obtained:

$$V_S^2(\mathbf{r}_1)G_T(\mathbf{r}_1, \mathbf{r}_2) = 0, \tag{A.19}$$

$$V_T^2(\mathbf{r}_1)G_S(\mathbf{r}_1, \mathbf{r}_2) = 0. \tag{A.20}$$

The last two equations have to be satisfied in the zero order approximation of the Green's function of the system, see (A.13). Let us check these conditions in the three regions of the system.

In the surface region, $V_S^2 G_T \approx 0$ because $G_T \sim \frac{e^{-\kappa_T z}}{z} \approx 0$; $V_T^2 G_S = 0$ because V_T in this region is zero. In the vacuum region, $V_S = 0$ and $V_T = 0$, thus (A.19) and (A.20) hold. In the tip region, $V_S^2 G_T = 0$ because $V_S = 0$ and $V_T^2 G_S \approx 0$ because $G_S \sim \frac{e^{-\kappa_S z}}{z} \approx 0$.

A.1.6 Consistency check: definition of Green's functions

From a different angle, we check whether the zero-order approximation of the Green's function obeys the well-known property

$$\frac{\partial G(z)}{\partial z} = -G^2(z). \tag{A.21}$$

Substituting (A.13) into this equation, we obtain

$$\begin{aligned}\frac{\partial G(z)}{\partial z} &= \frac{\partial G_S(z)}{\partial z} + \frac{\partial G_T(z)}{\partial z} = -G_S^2(z) - G_T^2(z) \\ &= -G^2(z) = -\left[G_S(z) + G_T(z)\right]^2 = -G_S^2(z) - G_T^2(z) - 2G_S(z)G_T(z),\end{aligned} \tag{A.22}$$

from which the following condition is obtained:

$$G_s(z)G_T(z) = 0. \tag{A.23}$$

It can be shown that although this expression is not exactly zero, it is very small in the whole system: in more detail, in the surface region $G_T \sim \frac{e^{-\kappa_T z}}{z} \approx 0$, in the tip region $G_S \sim \frac{e^{-\kappa_S z}}{z} \approx 0$, and in the vacuum region $G_S G_T \sim \frac{e^{-\kappa_S(z-z')}}{|z-z'|}\frac{e^{-\kappa_T(z-z')}}{|z-z'|} \approx 0$. The choice of the zero order approximation for the barrier Green's function is therefore satisfied by two independent calculations.

A.2 Transmission probability

A.2.1 Contacts

In order to avoid the singularity, we calculate the transmission probability with a given imaginary part of the energy, and the limit of zero will be taken at the very end. Using the definition of the contact as given in (A.6), the ansatz $\Gamma_S(\mathbf{r}_1, \mathbf{r}_2; E) = C_S(E) \sum_j \psi_j(\mathbf{r}_1)\psi_j^*(\mathbf{r}_2)$, the following can be carried out:

$$\begin{aligned}&A_S(\mathbf{r}_1, \mathbf{r}_2; E) \\ &= \iint d^3r_3 d^3r_4 G_S^{\pm}(\mathbf{r}_1, \mathbf{r}_3; E)\Gamma_S(\mathbf{r}_3, \mathbf{r}_4; E)G_S^{\mp}(\mathbf{r}_4, \mathbf{r}_2; E) \\ &= \int\int d^3r_3 d^3r_4 \sum_i \frac{\psi_i(\mathbf{r}_1)\psi_i^*(\mathbf{r}_3)}{E - E_i \pm i\eta} C_S(E) \sum_j \psi_j(\mathbf{r}_3)\psi_j^*(\mathbf{r}_4) \sum_k \frac{\psi_k(\mathbf{r}_4)\psi_k^*(\mathbf{r}_2)}{E - E_k \mp i\eta} \\ &= C_S(E) \sum_k \frac{\psi_k(\mathbf{r}_1)\psi_k^*(\mathbf{r}_2)}{(E - E_i)^2 + \eta^2},\end{aligned} \tag{A.24}$$

which compared to (A.3) results in

$$\varGamma_S(\mathbf{r}_1, \mathbf{r}_2; E) = 2\eta \sum_k \psi_k(\mathbf{r}_1)\psi_k^*(\mathbf{r}_2), \tag{A.25}$$

and similarly,

$$\varGamma_T(\mathbf{r}_1, \mathbf{r}_2; E) = 2\epsilon \sum_i \chi_i(\mathbf{r}_1)\chi_i^*(\mathbf{r}_2). \tag{A.26}$$

A.2.2 Tunneling current of zero order

The transmission probability is then given by

$$\begin{aligned}
&Tr\left[\varGamma_T(E)G^+(E)\varGamma_S(E)G^-(E)\right] = \iiiint d^3r_1 d^3r_2 d^3r_3 d^3r_4 \\
&\quad\times \varGamma_T(\mathbf{r}_1, \mathbf{r}_2; E)\left[G_S^+(\mathbf{r}_2, \mathbf{r}_3; E) + G_T^+(\mathbf{r}_2, \mathbf{r}_3; E)\right]\varGamma_S(\mathbf{r}_3, \mathbf{r}_4; E)\times \\
&\quad\left[G_S^-(\mathbf{r}_4, \mathbf{r}_1; E) + G_T^-(\mathbf{r}_4, \mathbf{r}_1; E)\right] \\
&\quad= 4\eta\epsilon \sum_{i,k} |A_{ik}|^2 \left[\frac{1}{(E-E_k)^2+\eta^2} + \frac{1}{(E-E_k+i\eta)(E-E_i-i\epsilon)}\right. \\
&\quad\left. + \frac{1}{(E-E_k-i\eta)(E-E_i+i\epsilon)} + \frac{1}{(E-E_i)^2+\epsilon^2}\right].
\end{aligned} \tag{A.27}$$

Here, we have to take the limit $\eta \to +0$ and $\epsilon \to +0$. The sum in brackets can be written as

$$\begin{aligned}
&4\eta\epsilon\left[\frac{1}{(E-E_k)^2+\eta^2} + \frac{1}{(E-E_k+i\eta)(E-E_i-i\epsilon)}\right. \\
&+ \left.\frac{1}{(E-E_k-i\eta)(E-E_i+i\epsilon)} + \frac{1}{(E-E_i)^2+\epsilon^2}\right] \\
&= 4\eta\epsilon \frac{(E-E_k+E-E_i)^2+(\eta+\epsilon)^2}{[(E-E_k)^2+\eta^2][(E-E_i)^2+\epsilon^2]} \\
&= \left[\underbrace{\frac{4\eta\epsilon(E-E_k+E-E_i)^2}{[(E-E_k)^2+\eta^2][(E-E_i)^2+\epsilon^2]}}_{\to 4\pi^2\delta(E-E_k)\delta(E-E_i)(E-E_k+E-E_i)^2}\right. \\
&+ \left.\underbrace{\frac{4\eta^3\epsilon+4\eta\epsilon^3}{[(E-E_k)^2+\eta^2][(E-E_i)^2+\epsilon^2]}}_{\to 0} + \underbrace{\frac{8\eta^2\epsilon^2}{[(E-E_k)^2+\eta^2][(E-E_i)^2+\epsilon^2]}}_{\to 0 \text{ if } E\neq E_k\neq E_i\ ;\ \to 8 \text{ if } E=E_k=E_i}\right]
\end{aligned} \tag{A.28}$$

Here, only the first line will contribute to the energy integral, since the terms in the second line are either zero, or constant for an energy interval of width zero. Writing the zero-order current yields

$$I_{(0)} = \frac{2e}{h} \int_{-\infty}^{\infty} dE[f(\mu_S, E) - f(\mu_T, E)] \times \tag{A.29}$$

$$\times \sum_{i,k} \left| \frac{M_{ik}}{\kappa_i^2 - \kappa_k^2} \right|^2 4\pi^2 \delta(E - E_k)\delta(E - E_i)(E - E_k + E - E_i)^2$$

$$= \underbrace{\frac{8\pi^2 e}{h}}_{=\frac{4\pi e}{\hbar}} \sum_{i,k} [f(\mu_S, E_k) - f(\mu_T, E_i)] \left| \frac{(E_k - E_i)M_{ik}}{\kappa_i^2 - \kappa_k^2} \right|^2 \delta(E_i - E_k).$$

Taking into account the applied bias voltage V, we have to include the energy shift between an eigenstate with a certain decay and the energy value in the tunneling current by

$$-\frac{\hbar^2 \kappa_i^2}{2m} = E_i - \frac{eV}{2}, \tag{A.30}$$

$$-\frac{\hbar^2 \kappa_k^2}{2m} = E_k + \frac{eV}{2}, \tag{A.31}$$

and the zero-order current expression can be rewritten as

$$I_{(0)} = \frac{4\pi e}{\hbar} \sum_{i,k} [f(\mu_S, E_k) - f(\mu_T, E_i)] \left| \left(-\frac{\hbar^2}{2m} - \frac{eV}{\kappa_i^2 - \kappa_k^2} \right) M_{ik} \right|^2 \delta(E_i - E_k).$$

For zero-bias voltage ($V = 0$), $E_i = -\frac{\hbar^2 \kappa_i^2}{2m}$ and $E_k = -\frac{\hbar^2 \kappa_k^2}{2m}$, and we recover the standard Bardeen result

$$I_{(0)} = \frac{4\pi e}{\hbar} \sum_{i,k} [f(\mu_S, E_k) - f(\mu_T, E_i)] \left| -\frac{\hbar^2}{2m} M_{ik} \right|^2 \delta(E_i - E_k). \tag{A.32}$$

A.3 First-order Green's function

The first-order Green's function is given by the Dyson expansion:

$$\begin{aligned} G_{(1)} &= G_{(0)} + G_S V_T G_S + G_T V_S G_T + G_S V_S G_S + G_T V_T G_T \\ &\quad + G_S V_T G_T + G_S V_S G_T + + G_T V_S G_S + G_T V_T G_S \end{aligned} \tag{A.33}$$

We do not treat the terms in the first line, since they lead to terms that are not related to tunneling transitions. The terms in the second line are

$$
\begin{aligned}
&G_S V_T G_T + G_S V_S G_T + G_T V_S G_S + G_T V_T G_S \qquad \text{(A.34)}\\
&= \sum_{i,k} \frac{\psi_i(\mathbf{r}_1)\chi_k^*(\mathbf{r}_2)}{(E-E_i+i\eta)(E-E_k+i\epsilon)} \int_{\Omega_S+\Omega_T} d^3r \psi_i^*(\mathbf{r}) V_T(\mathbf{r}) \chi_k(\mathbf{r})\\
&+ \sum_{i,k} \frac{\psi_i(\mathbf{r}_1)\chi_k^*(\mathbf{r}_2)}{(E-E_i+i\eta)(E-E_k+i\epsilon)} \int_{\Omega_S+\Omega_T} d^3r \psi_i^*(\mathbf{r}) V_S(\mathbf{r}) \chi_k(\mathbf{r})\\
&+ \sum_{i,k} \frac{\chi_k(\mathbf{r}_1)\psi_i^*(\mathbf{r}_2)}{(E-E_i+i\eta)(E-E_k+i\epsilon)} \int_{\Omega_S+\Omega_T} d^3r \chi_k^*(\mathbf{r}) V_S(\mathbf{r}) \psi_i(\mathbf{r})\\
&+ \sum_{i,k} \frac{\chi_k(\mathbf{r}_1)\psi_i^*(\mathbf{r}_2)}{(E-E_i+i\eta)(E-E_k+i\epsilon)} \int_{\Omega_S+\Omega_T} d^3r \chi_k^*(\mathbf{r}) V_T(\mathbf{r}) \psi_i(\mathbf{r}).
\end{aligned}
$$

A.4 Recovering the Bardeen matrix elements

In general, we can employ the condition used in the standard derivation of the Bardeen approach, e.g., by Chen, that the potentials V_S and V_T exist only in one half-space of the system, while they are zero in the other half-space:

$$V_S(\mathbf{r}) = 0 \text{ if } \mathbf{r} \in \Omega_T, \qquad V_T(\mathbf{r}) = 0 \text{ if } \mathbf{r} \in \Omega_S. \qquad \text{(A.35)}$$

The integration then has to be performed only for the half-space, in which the potential is not zero. The integrals can be rewritten as surface integrals with the help of the Schrödinger equation:

$$\left(-\frac{\hbar^2}{2m}\nabla^2 + V_S\right)\psi_i = E_i\psi_i, \qquad \left(-\frac{\hbar^2}{2m}\nabla^2 + V_T\right)\chi_i = E_i\chi_i. \qquad \text{(A.36)}$$

Calculating the first term explicitly, we obtain for the integral

$$
\begin{aligned}
&\int_{\Omega_T} d^3r \quad \psi_i^*(\mathbf{r}) V_T(\mathbf{r}) \chi_k(\mathbf{r})\\
&= \int_{\Omega_T} d^3r \left[\psi_i^*(\mathbf{r}) E_k \chi_k(\mathbf{r}) + \psi_i^*(\mathbf{r}) \frac{\hbar^2}{2m}\nabla^2\chi_k(\mathbf{r})\right]\\
&= \int_{\Omega_T} d^3r \left[\chi_k(\mathbf{r}) E_k \psi_i^*(\mathbf{r}) + \psi_i^*(\mathbf{r}) \frac{\hbar^2}{2m}\nabla^2\chi_k(\mathbf{r})\right]\\
&\underbrace{=}_{E_k-E_i} \int_{\Omega_T} d^3r \left[\chi_k(\mathbf{r}) E_i \psi_i^*(\mathbf{r}) + \psi_i^*(\mathbf{r}) \frac{\hbar^2}{2m}\nabla^2\chi_k(\mathbf{r})\right]\\
&\underbrace{=}_{V_S=0} \int_{\Omega_T} d^3r \left[-\chi_k(\mathbf{r}) \frac{\hbar^2}{2m}\nabla^2\psi_i^*(\mathbf{r}) + \psi_i^*(\mathbf{r}) \frac{\hbar^2}{2m}\nabla^2\chi_k(\mathbf{r})\right]\\
&= -\frac{\hbar^2}{2m}\int_S dS \left[\chi_k(\mathbf{r})\nabla\psi_i^*(\mathbf{r}) - \psi_i^*(\mathbf{r})\nabla\chi_k(\mathbf{r})\right] = -\frac{\hbar^2}{2m} M_{ki}^*,
\end{aligned}
$$

Similarly, for the second term,

$$\begin{aligned}
\int_{\Omega_S} & d^3r \quad \chi_k(\mathbf{r}) V_S(\mathbf{r}) \psi_i^*(\mathbf{r}) \\
&= \int_{\Omega_S} d^3r \left[\chi_k(\mathbf{r}) E_i \psi_i^*(\mathbf{r}) + \chi_k(\mathbf{r}) \frac{\hbar^2}{2m} \nabla^2 \psi_i^*(\mathbf{r}) \right] \\
&= \int_{\Omega_S} d^3r \left[\psi_i^*(\mathbf{r}) E_i \chi_k(\mathbf{r}) + \chi_k(\mathbf{r}) \frac{\hbar^2}{2m} \nabla^2 \psi_i^*(\mathbf{r}) \right] \\
&\underbrace{=}_{E_i = E_k} \int_{\Omega_S} d^3r \left[\psi_i^*(\mathbf{r}) E_k \chi_k(\mathbf{r}) + \chi_k(\mathbf{r}) \frac{\hbar^2}{2m} \nabla^2 \psi_i^*(\mathbf{r}) \right] \\
&\underbrace{=}_{V_T = 0} \int_{\Omega_S} d^3r \left[-\psi_i^*(\mathbf{r}) \frac{\hbar^2}{2m} \nabla^2 \chi_k(\mathbf{r}) + \chi_k(\mathbf{r}) \frac{\hbar^2}{2m} \nabla^2 \psi_i^*(\mathbf{r}) \right] \\
&= -\frac{\hbar^2}{2m} \int_S dS \left[\psi_i^*(\mathbf{r}) \nabla \chi_k(\mathbf{r}) - \chi_k(\mathbf{r}) \nabla \psi_i^*(\mathbf{r}) \right] = -\frac{\hbar^2}{2m} M_{ki}^*.
\end{aligned}$$

In the last line we took into account that the surfaces of the two integrations point in opposite directions. The first-order Green's function of the interface is consequently

$$G_{(1)}^{\mathrm{R(A)}} = G_{(0)}^{\mathrm{R(A)}} - \frac{\hbar^2}{m} \sum_{i,k} \frac{\psi_i(\mathbf{r}_1) M_{ki}^* \chi_k^*(\mathbf{r}_2) + \chi_k(\mathbf{r}_1) M_{ki} \psi_i^*(\mathbf{r}_2)}{(E - E_i \pm i\eta)(E - E_k \pm i\epsilon)}. \quad \text{(A.37)}$$

A.5 Interaction energy

The density matrix is calculated from the first-order term of the Green's function, since zero order terms will only lead to a shift of eigenvalues in the presence of the opposite lead. The explicit form of the density matrix ($f_{ik}^{\pm} = (E - E_i \pm i\eta)(E - E_k \pm i\epsilon)$) is as follows:

$$\hat{n}(\mathbf{r}_1, \mathbf{r}_2, E) = -\frac{i}{2\pi} \sum_{i,k} \left(\frac{1}{f_{ik}^-} - \frac{1}{f_{ik}^+} \right) \left[\psi_i(\mathbf{r}_1) M_{ki}^* \chi_k^*(\mathbf{r}_2) + \chi_k(\mathbf{r}_1) M_{ki} \psi_i^*(\mathbf{r}_2) \right]. \quad \text{(A.38)}$$

The trace $Tr[\hat{n}V]$ leads to four terms:

$$Tr[\hat{n}V] = -\frac{i\hbar^2}{2\pi m}\sum_{i,k} M_{ki}^* \left[\underbrace{\int d^3 r \psi_i(\mathbf{r}) V_S \chi_k^*(\mathbf{r})}_{=-\hbar^2/2mM_{ki}} + \underbrace{\int d^3 r \psi_i(\mathbf{r}) V_T \chi_k^*(\mathbf{r})}_{=-\hbar^2/2mM_{ki}} \right]$$
$$-\frac{i\hbar^2}{2\pi m}\sum_{i,k} M_{ki} \left[\underbrace{\int d^3 r \psi_i^*(\mathbf{r}) V_S \chi_k(\mathbf{r})}_{=-\hbar^2/2mM_{ki}^*} + \underbrace{\int d^3 r \psi_i^*(\mathbf{r}) V_T \chi_k(\mathbf{r})}_{=-\hbar^2/2mM_{ki}^*} \right]$$
$$= \frac{i}{\pi}\left(\frac{\hbar^2}{m}\right)^2 |M_{ki}|^2. \tag{A.39}$$

The energy terms $f_{ik}^{\pm}$ lead to the following result:

$$\frac{1}{f_{ik}^-} - \frac{1}{f_{ik}^+} = \frac{(E - E_i + i\eta)(E - E_k + i\epsilon) - (E - E_i - i\eta)(E - E_k - i\epsilon)}{[(E - E_i)^2 + \eta^2][(E - E_k)^2 + \epsilon^2]}$$
$$= 2i \frac{\eta}{[(E - E_i)^2 + \eta^2]} \frac{E - E_k}{[(E - E_k)^2 + \epsilon^2]}$$
$$+2i \frac{\epsilon}{[(E - E_k)^2 + \epsilon^2]} \frac{E - E_i}{[(E - E_i)^2 + \eta^2]}. \tag{A.40}$$

In the limit $\eta, \epsilon \to 0+$ this gives

$$\lim_{\eta,\epsilon\to 0+} = 2i \frac{\delta(E - E_i)}{E - E_k} + 2i \frac{\delta(E - E_k)}{E - E_i}. \tag{A.41}$$

The energy integration now has to be performed over the infinite energy interval. The only terms to consider are

$$2i \int_{-\infty}^{+\infty} dE \left[\frac{\delta(E - E_i)}{E - E_k} + \frac{\delta(E - E_k)}{E - E_i} \right]. \tag{A.42}$$

Here we suppose that, physically speaking, all transitions across the barrier will lead to an increase of bonding and thus interaction energy. We therefore count every component separately:

$$2i \int_{-\infty}^{+\infty} dE \left[\frac{\delta(E - E_i)}{E - E_k} + \frac{\delta(E - E_k)}{E - E_i} \right] = \frac{2i}{|E_i - E_k|} + \frac{2i}{|E_k - E_i|} = \frac{4i}{|E_i - E_k|}. \tag{A.43}$$

The final result for the interaction energy to first order is therefore

$$E_{\text{int}} = -\frac{4}{\pi}\left(\frac{\hbar^2}{m}\right)^2 \sum_{i,k} \frac{|M_{ki}|^2}{|E_i - E_k|}. \tag{A.44}$$

A.6 Trace to first order

The trace in the conductance equation consequently has three new terms:

$$Tr = Tr[\Gamma_T g_1^R \Gamma_S g_0^A] + Tr[\Gamma_T g_0^R \Gamma_S g_1^A] + Tr[\Gamma_T g_1^R \Gamma_S g_1^A] := A + B + C. \quad \text{(A.45)}$$

The three terms are the following integrals:

$$\begin{aligned} A = Tr[\Gamma_T g_1^R \Gamma_S g_0^A] &= -\frac{\hbar^2}{m} 4\epsilon\eta \sum_{i,k,l,m,p} \iiiint dr_1 dr_2 dr_3 dr_4 \\ &\times \chi_i(\mathbf{r}_1)\chi_i^*(\mathbf{r}_2) \left[\frac{\psi_k(\mathbf{r}_2) M_{lk}^* \chi_l^*(\mathbf{r}_3)}{f_{kl}^+} + \frac{\chi_l(\mathbf{r}_2) M_{lk} \psi_k^*(\mathbf{r}_3)}{f_{kl}^+} \right] \\ &\times \psi_m(\mathbf{r}_3)\psi_m^*(\mathbf{r}_4) \left[\frac{\psi_p(\mathbf{r}_4)\psi_p^*(\mathbf{r}_1)}{(E - E_p - i\eta)} + \frac{\chi_p(\mathbf{r}_4)\chi_p^*(\mathbf{r}_1)}{(E - E_p - i\epsilon)} \right], \end{aligned} \quad \text{(A.46)}$$

$$\begin{aligned} B = Tr[\Gamma_T g_0^R \Gamma_S g_1^A] &= -\frac{\hbar^2}{m} 4\epsilon\eta \sum_{i,k,l,m,p} \iiiint dr_1 dr_2 dr_3 dr_4 \\ &\times \chi_i(\mathbf{r}_1)\chi_i^*(\mathbf{r}_2) \left[\frac{\psi_k(\mathbf{r}_2)\psi_k^*(\mathbf{r}_3)}{(E - E_k + i\eta)} + \frac{\chi_k(\mathbf{r}_2)\chi_k^*(\mathbf{r}_3)}{(E - E_k + i\epsilon)} \right] \\ &\times \psi_l(\mathbf{r}_3)\psi_l^*(\mathbf{r}_4) \left[\frac{\psi_m(\mathbf{r}_4) M_{pm}^* \chi_p^*(\mathbf{r}_1)}{f_{mp}^-} + \frac{\chi_p(\mathbf{r}_4) M_{pm} \psi_m^*(\mathbf{r}_1)}{f_{mp}^-} \right], \end{aligned} \quad \text{(A.47)}$$

$$\begin{aligned} C = Tr[\Gamma_T g_1^R \Gamma_S g_1^A] &= +\left(\frac{\hbar^2}{m}\right)^2 4\epsilon\eta \sum_{i,k,l,m,p,s} \iiiint dr_1 dr_2 dr_3 dr_4 \\ &\times \chi_i(\mathbf{r}_1)\chi_i^*(\mathbf{r}_2) \left[\frac{\psi_k(\mathbf{r}_2) M_{lk}^* \chi_l^*(\mathbf{r}_3)}{f_{kl}^+} + \frac{\chi_l(\mathbf{r}_2) M_{lk} \psi_k^*(\mathbf{r}_3)}{f_{kl}^+} \right] \\ &\times \psi_m(\mathbf{r}_3)\psi_m^*(\mathbf{r}_4) \left[\frac{\psi_p(\mathbf{r}_4) M_{sp}^* \chi_s^*(\mathbf{r}_1)}{f_{ps}^-} + \frac{\chi_s(\mathbf{r}_4) M_{sp} \psi_p^*(\mathbf{r}_1)}{f_{ps}^-} \right]. \end{aligned} \quad \text{(A.48)}$$

The three terms each yield three components, which have to be evaluated separately. We start with the integration of the three terms over space, solving the fourfold integrals.

A.6.1 Term A

The innermost integral over r_4 yields

$$\begin{aligned} \int dr_4 &= \frac{\psi_p^*(\mathbf{r}_1)}{(E - E_p - i\eta)} \int dr_4 \psi_m^*(\mathbf{r}_4)\psi_p(\mathbf{r}_4) \\ &+ \frac{\chi_p^*(\mathbf{r}_1)}{(E - E_p - i\epsilon)} \int dr_4 \psi_m^*(\mathbf{r}_4)\chi_p(\mathbf{r}_4) \\ &= \frac{\psi_p^*(\mathbf{r}_1)\delta_{mp}}{(E - E_p - i\eta)} + \frac{\chi_p^*(\mathbf{r}_1) A_{pm}^*}{(E - E_p - i\epsilon)} =: K(\mathbf{r}_1). \end{aligned} \quad \text{(A.49)}$$

The integral over r_3 is

$$\begin{aligned}\int dr_3 &= \frac{\psi_k(\mathbf{r}_2)M_{lk}^*}{f_{kl}^+}\int dr_3\chi_l^*(\mathbf{r}_3)\psi_m(\mathbf{r}_3) + \frac{\chi_l(\mathbf{r}_2)M_{lk}}{f_{kl}^+}\int dr_3\psi_k^*(\mathbf{r}_3)\psi_m(\mathbf{r}_3)\\ &= \frac{\psi_k(\mathbf{r}_2)M_{lk}^*A_{lm}}{f_{kl}^+} + \frac{\chi_l(\mathbf{r}_2)M_{lk}\delta_{km}}{f_{kl}^+}. \qquad (A.50)\end{aligned}$$

The integral over r_2 is

$$\begin{aligned}\int dr_2 &= \frac{M_{lk}^*A_{lm}}{f_{kl}^+}\int dr_2\chi_i^*(\mathbf{r}_2)\psi_k(\mathbf{r}_2) + \frac{M_{lk}\delta_{km}}{f_{kl}^+}\int dr_2\chi_i^*(\mathbf{r}_2)\chi_l(\mathbf{r}_2)\\ &= \frac{M_{lk}^*A_{lm}A_{ik}}{f_{kl}^+} + \frac{M_{lk}\delta_{km}\delta_{il}}{f_{kl}^+}. \qquad (A.51)\end{aligned}$$

The last integral over r_1 is

$$\begin{aligned}\int dr_1 &= \int dr_1 K(\mathbf{r}_1)\chi_i(\mathbf{r}_1) \qquad (A.52)\\ &= \frac{\delta_{mp}}{(E-E_p-i\eta)}\int dr_1\chi_i(\mathbf{r}_1)\psi_p^*(\mathbf{r}_1)\\ &\quad + \frac{A_{pm}^*}{(E-E_p-i\epsilon)}\int dr_1\chi_i(\mathbf{r}_1)\chi_p^*(\mathbf{r}_1)\\ &= \left[\frac{A_{ip}^*\delta_{mp}}{(E-E_P-i\eta)} + \frac{A_{pm}^*\delta_{ip}}{(E-E_p-i\epsilon)}\right]\left[\frac{M_{lk}^*A_{lm}A_{ik}}{f_{kl}^+} + \frac{M_{lk}\delta_{km}\delta_{il}}{f_{kl}^+}\right]\end{aligned}$$

Term A consequently has four components; two of them are products of only two transition matrices; the other two are products of four transition matrices. We simplify the indices using the Kronecker deltas

$$\begin{aligned}\frac{M_{lk}A_{ip}^*}{(E-E_p-i\eta)f_{kl}^+}\delta_{mp}\delta_{km}\delta_{il} &= \frac{A_{ik}^*M_{ik}}{(E-E_k-i\eta)f_{ki}^+},\\ \frac{M_{lk}A_{pm}^*}{(E-E_p-i\eta)f_{kl}^+}\delta_{ip}\delta_{km}\delta_{il} &= \frac{A_{ik}^*M_{ik}}{(E-E_i-i\epsilon)f_{ki}^+},\\ \frac{M_{lk}^*A_{lm}A_{ik}A_{ip}^*}{(E-E_p-i\eta)f_{kl}^+}\delta_{mp} &= \frac{A_{ik}M_{lk}^*A_{lm}A_{im}^*}{(E-E_m-i\eta)f_{kl}^+},\\ \frac{M_{lk}^*A_{lm}A_{ik}A_{pm}^*}{(E-E_p-i\epsilon)f_{kl}^+}\delta_{ip} &= \frac{A_{ik}M_{lk}^*A_{lm}A_{im}^*}{(E-E_i-i\epsilon)f_{kl}^+}. \qquad (A.53)\end{aligned}$$

The trace over the product, or term A, is then described by

$$\begin{aligned}A = &-\frac{\hbar^2}{m}\sum_{ik}\left[\frac{4\eta\epsilon}{(E-E_k-i\eta)f_{ki}^+} + \frac{4\eta\epsilon}{(E-E_i-i\epsilon)f_{ki}^+}\right]A_{ik}^*M_{ik} \qquad (A.54)\\ &-\frac{\hbar^2}{m}\sum_{i,k,l,m}\left[\frac{4\eta\epsilon}{(E-E_m-i\eta)f_{kl}^+} + \frac{4\eta\epsilon}{(E-E_i-i\epsilon)f_{kl}^+}\right]A_{ik}M_{lk}^*A_{lm}A_{im}^*.\end{aligned}$$

A.6.2 Term B

The innermost integral over r_4 yields

$$\int dr_4 = \frac{\chi_p^*(\mathbf{r}_1)M_{pm}^*}{f_{mp}^-}\int dr_4\psi_l^*(\mathbf{r}_4)\psi_m(\mathbf{r}_4) + \frac{M_{pm}\psi_m^*(\mathbf{r}_1)}{f_{mp}^-}\int dr_4\psi_l^*(\mathbf{r}_4)\chi_p(\mathbf{r}_4)$$
$$= \frac{\chi_p^*(\mathbf{r}_1)M_{pm}^*\delta_{lm}}{f_{mp}^-} + \frac{\psi_m^*(\mathbf{r}_1)M_{pm}A_{pl}^*}{f_{mp}^-} =: K(\mathbf{r}_1). \tag{A.55}$$

The integral over r_3 is

$$\int dr_3 = \frac{\psi_k(\mathbf{r}_2)}{(E-E_k+i\eta)}\int dr_3\psi_k^*(\mathbf{r}_3)\psi_l(\mathbf{r}_3)$$
$$+\frac{\chi_k(\mathbf{r}_2)}{(E-E_k+i\epsilon)}\int dr_3\chi_k^*(\mathbf{r}_3)\psi_l(\mathbf{r}_3)$$
$$= \frac{\psi_k(\mathbf{r}_2)\delta_{kl}}{(E-E_k+i\eta)} + \frac{\chi_k(\mathbf{r}_2)A_{kl}}{(E-E_k+i\epsilon)}. \tag{A.56}$$

The integral over r_2 is

$$\int dr_2 = \frac{\delta_{kl}}{(E-E_k+i\eta)}\int dr_2\chi_i^*(\mathbf{r}_2)\psi_k(\mathbf{r}_2)$$
$$+\frac{A_{kl}}{(E-E_k+i\epsilon)}\int dr_2\chi_i^*(\mathbf{r}_2)\chi_k(\mathbf{r}_2)$$
$$= \frac{A_{ik}\delta_{kl}}{(E-E_k+i\eta)} + \frac{A_{kl}\delta_{ik}}{(E-E_k+i\epsilon)}. \tag{A.57}$$

The last integral over r_1 is

$$\int dr_1 = \int dr_1 K(\mathbf{r}_1)\chi_i(\mathbf{r}_1) = \frac{M_{pm}^*\delta_{lm}}{f_{mp}^-}\int dr_1\chi_i(\mathbf{r}_1)\chi_p^*(\mathbf{r}_1)$$
$$+\frac{M_{pm}A_{pl}^*}{f_{mp}^-}\int dr_1\chi_i(\mathbf{r}_1)\psi_m^*(\mathbf{r}_1)$$
$$= \left[\frac{M_{pm}^*\delta_{lm}\delta_{ip}}{f_{mp}^-} + \frac{M_{pm}A_{pl}^*A_{im}^*}{f_{mp}^-}\right]$$
$$\times\left[\frac{A_{ik}\delta_{kl}}{(E-E_k+i\eta)} + \frac{A_{kl}\delta_{ik}}{(E-E_k+i\epsilon)}\right]. \tag{A.58}$$

Term B consequently has four components; two of them are products of only two transition matrices; the other two are products of four transition matrices. We simplify the indices using the Kronecker deltas

$$
\begin{aligned}
\frac{M^*_{pm} A_{ik}}{(E - E_k + i\eta) f^-_{mp}} \delta_{lm} \delta_{ip} \delta_{kl} &= \frac{M^*_{ik} A_{ik}}{(E - E_k + i\eta) f^-_{ki}}, \\
\frac{M^*_{pm} A_{kl}}{(E - E_k + i\epsilon) f^-_{mp}} \delta_{ik} \delta_{lm} \delta_{ip} &= \frac{M^*_{ik} A_{ik}}{(E - E_i + i\epsilon) f^-_{ki}}, \\
\frac{M_{pm} A_{ik} A^*_{pl} A^*_{im}}{(E - E_k + i\eta) f^-_{mp}} \delta_{kl} &= \frac{M_{lk} A^*_{ik} A_{im} A^*_{lm}}{(E - E_m + i\eta) f^-_{kl}}, \\
\frac{M_{pm} A_{kl} A^*_{pl} A^*_{im}}{(E - E_k + i\epsilon) f^-_{mp}} \delta_{ik} &= \frac{M_{lk} A^*_{ik} A_{im} A^*_{lm}}{(E - E_i + i\epsilon) f^-_{kl}}.
\end{aligned} \tag{A.59}
$$

The trace over the product, or term B, is then described by

$$
\begin{aligned}
B = &- \frac{\hbar^2}{m} \sum_{ik} \left[\frac{4\eta\epsilon}{(E - E_k + i\eta) f^-_{ki}} + \frac{4\eta\epsilon}{(E - E_i + i\epsilon) f^-_{ki}} \right] M^*_{ik} A_{ik} \qquad \text{(A.60)} \\
&- \frac{\hbar^2}{m} \sum_{i,k,l,m} \left[\frac{4\eta\epsilon}{(E - E_m + i\eta) f^-_{kl}} + \frac{4\eta\epsilon}{(E - E_i + i\epsilon) f^-_{kl}} \right] A_{im} A^*_{lm} M_{lk} A^*_{ik}.
\end{aligned}
$$

A.6.3 Term C

The innermost integral over r_4 yields

$$
\begin{aligned}
\int dr_4 &= \frac{\chi^*_s(\mathbf{r}_1) M^*_{sp}}{f^-_{ps}} \int dr_4 \psi^*_m(\mathbf{r}_4) \psi_p(\mathbf{r}_4) + \frac{\psi^*_p(\mathbf{r}_1) M_{sp}}{f^-_{ps}} \int dr_4 \psi^*_m(\mathbf{r}_4) \chi_s(\mathbf{r}_4) \\
&= \frac{\chi^*_s(\mathbf{r}_1) M^*_{sp} \delta_{mp}}{f^-_{ps}} + \frac{\psi^*_p(\mathbf{r}_1) M_{sp} A^*_{sm}}{f^-_{ps}} =: K(\mathbf{r}_1).
\end{aligned} \tag{A.61}
$$

The integral over r_3 is

$$
\begin{aligned}
\int dr_3 &= \frac{\psi_k(\mathbf{r}_2) M^*_{lk}}{f^+_{kl}} \int dr_3 \chi^*_l(\mathbf{r}_3) \psi_m(\mathbf{r}_3) + \frac{\chi_l(\mathbf{r}_2) M_{lk}}{f^+_{kl}} \int dr_3 \psi_m(\mathbf{r}_3) \psi^*_k(\mathbf{r}_3) \\
&= \frac{\psi_k(\mathbf{r}_2) M^*_{lk} A_{lm}}{f^+_{kl}} + \frac{\chi_l(\mathbf{r}_2) M_{lk} \delta_{km}}{f^+_{kl}}.
\end{aligned} \tag{A.62}
$$

The integral over r_2 is

$$
\begin{aligned}
\int dr_2 &= \frac{M^*_{lk} A_{lm}}{f^+_{kl}} \int dr_2 \chi^*_i(\mathbf{r}_2) \psi_k(\mathbf{r}_2) + \frac{M_{lk} \delta_{km}}{f^+_{kl}} \int dr_2 \chi^*_i(\mathbf{r}_2) \chi_l(\mathbf{r}_2) \\
&= \frac{M^*_{lk} A_{lm} A_{ik}}{f^+_{kl}} + \frac{M_{lk} \delta_{km} \delta_{il}}{f^+_{kl}}.
\end{aligned} \tag{A.63}
$$

The last integral over r_1 is

$$\begin{aligned}\int dr_1 &= \int dr_1 K(\mathbf{r}_1)\chi_i(\mathbf{r}_1) \\ &= \frac{M^*_{sp}\delta_{mp}}{f^-_{ps}}\int dr_1\chi_i(\mathbf{r}_1)\chi^*_s(\mathbf{r}_1) + \frac{M_{sp}A^*_{sm}}{f^-_{ps}}\int dr_1\chi_i(\mathbf{r}_1)\psi^*_p(\mathbf{r}_1) \\ &\Rightarrow \left[\frac{M^*_{lk}A_{lm}A_{ik}}{f^+_{kl}} + \frac{M_{lk}\delta_{km}\delta_{il}}{f^+_{kl}}\right]\left[\frac{M^*_{sp}\delta_{mp}\delta_{is}}{f^-_{ps}} + \frac{M_{sp}A^*_{sm}A^*_{ip}}{f^-_{ps}}\right].\end{aligned} \tag{A.64}$$

Term C consequently has four components; one of them is the product of only two transition matrices, two are products of four transition matrices, and one is the product of six transition matrices. We simplify the indices using the Kronecker deltas

$$\begin{aligned}\frac{M_{lk}M^*_{sp}}{f^+_{kl}f^-_{ps}}\delta_{km}\delta_{il}\delta_{mp}\delta_{is} &= \frac{M^*_{ik}M_{ik}}{f^+_{ki}f^-_{ki}}, \\ \frac{M_{lk}M_{sp}A^*_{sm}A^*_{ip}}{f^+_{kl}f^-_{ps}}\delta_{km}\delta_{il} &= \frac{M_{lk}M_{sp}A^*_{sk}A^*_{lp}}{f^+_{kl}f^-_{ps}}, \\ \frac{M^*_{lk}A_{lm}A_{ik}M^*_{sp}}{f^+_{kl}f^-_{ps}}\delta_{mp}\delta_{is} &= \frac{A_{ik}M^*_{lk}A_{lm}M^*_{im}}{f^+_{kl}f^-_{mi}}, \\ \frac{A_{ik}M^*_{lk}A_{lm}A^*_{sm}M_{sp}A^*_{ip}}{f^+_{kl}f^-_{ps}} &= \frac{A_{ik}M^*_{lk}A_{lm}A^*_{sm}M_{sp}A^*_{ip}}{f^+_{kl}f^-_{ps}}.\end{aligned} \tag{A.65}$$

The trace over the product, or term C, is then described by

$$\begin{aligned}C = &+\left(\frac{\hbar^2}{m}\right)^2\sum_{i,k}\left[\frac{4\eta\epsilon}{f^+_{ki}f^-_{ki}}\right]|M_{ik}|^2 \\ &+\left(\frac{\hbar^2}{m}\right)^2\sum_{i,k,l,m}\left[\frac{4\eta\epsilon}{f^+_{kl}f^-_{mi}}\right]A^*_{ik}M_{lk}A^*_{lm}M_{im} \\ &+\left(\frac{\hbar^2}{m}\right)^2\sum_{i,k,l,m}\left[\frac{4\eta\epsilon}{f^+_{kl}f^-_{mi}}\right]A_{ik}M^*_{lk}A_{lm}M^*_{im} \\ &+\left(\frac{\hbar^2}{m}\right)^2\sum_{i,k,l,m,p,s}\left[\frac{4\eta\epsilon}{f^+_{kl}f^-_{ps}}\right]A_{ik}M^*_{lk}A_{lm}A^*_{sm}M_{sp}A^*_{ip}.\end{aligned} \tag{A.66}$$

A.6.4 Taking the decay into account

The expressions can be simplified, if we consider that A_{ik} and M_{ik} are related by

$$A_{ik} = \frac{1}{\kappa_i^2 - \kappa_k^2}M_{ik} \tag{A.67}$$

The result of the derivation will therefore be a product of two, four, and six transition matrices, multiplied by energy terms, which have to be integrated in the limit of $\eta, \epsilon \to 0$.

Index